The Fate of Liberty

The Fate of

Liberty

Abraham Lincoln and Civil Liberties

MARK E. NEELY, JR.

New York   Oxford
OXFORD UNIVERSITY PRESS
1991

Oxford University Press

Oxford   New York   Toronto
Delhi   Bombay   Calcutta   Madras   Karachi
Petaling Jaya   Singapore   Hong Kong   Tokyo
Nairobi   Dar es Salaam   Cape Town
Melbourne   Auckland

and associated companies in
Berlin   Ibadan

Copyright © 1991 by Mark E. Neely, Jr.

Published by Oxford University Press, Inc.,
200 Madison Avenue, New York, New York 10016

Oxford is a registered trademark of Oxford University Press

Library of Congress Cataloging-in-Publication Data
Neely, Mark E.   The fate of liberty:
Abraham Lincoln and civil liberties
Mark E. Neely, Jr.
p.   cm.   Includes bibliographical references.
ISBN 0-19-506496-8
1. Lincoln, Abraham, 1809–1865—Views on civil rights.
2. Habeas corpus—United States—History—19th century.
3. Civil rights—United States—History—19th century.
4. United States—History—Civil War, 1861–1865—Law and legislation.
I. Title.   E457.2.N46   1991   973.7'092—dc20   90-31907

2 4 6 8 9 7 5 3 1

Printed in the United States of America
on acid-free paper

For Sylvia

Acknowledgments

Michael Musick, of the National Archives, made this book possible by locating the arrest records on which it is based. William Gienapp, now of Harvard University, gave me my first opportunity to discuss the subject in a public forum back in 1981. Eight years later, he gave the manuscript a careful and astute reading. John Y. Simon, editor of *The Papers of Ulysses S. Grant,* also read the manuscript with care and offered useful suggestions. Sarah McNair Vosmeier and Matthew N. Vosmeier, now graduate students at Indiana University, between them, read one thousand of the cases on which the book is based. We have had numerous conversations on the subject that benefited me a great deal. Sarah also read the final manuscript. Frank J. Williams, president of the Abraham Lincoln Association, read the manuscript closely, and my frequent co-author, Harold Holzer, Executive Vice President for Public Affairs of the New York State Urban Development Corporation, offered his usual good advice. David H. Donald granted me permission to use the J. G. Randall Papers. Robert J. Chandler, of the Wells Fargo Bank's Historical Department; Richard N. Current, now of South Natick, Massachusetts; and Don E. Fehrenbacher, professor emeritus at Stanford University, offered encouragement and read parts of the manuscript.*

Sylvia E. Neely read it all, over and over again, offering ten years of unstinting encouragement, clever advice, and genuine understanding.

Fort Wayne, Indiana   M.E.N., Jr.
February 1990

---

*I have rendered quotations as they appeared in the original sources with the occasional exceptions of terminal punctuation, initial capitalization, and the variant texts of Gideon Welles's diary, as edited by Howard K. Beale. Changes are for the sake of readability alone.

Contents

Introduction

"The Union with him in sentiment, rose to the sublimity of a religious mysticism." With those words, Alexander H. Stephens, the former Confederate vice president, dismissed the constitutional ideas of Abraham Lincoln. To Stephens, Lincoln's almost superstitious love of the Union caused him to misunderstand its Constitution and to destroy the country's liberties.[1] Though Stephens's words are often quoted, Lincoln's constitutional views were neither mystical nor mysterious, and their nature should no longer be in doubt. The tedious historical debate over whether or not President Lincoln's policies were constitutional is a legacy of the brittle party platforms of a bygone era and the constitutional moralizing of sore losers like Stephens.

Rather than continue the fruitless debate over the constitutionality of Lincoln's acts, this book will examine instead the practical impact on civil liberties of the policies Lincoln developed to save the Union. The numerous arrests of civilians by Northern military authorities during the Civil War sparked controversy then and continue to do so. Except for a handful of celebrated cases, however, no one knows exactly who was arrested or how or when. No one knows whether the controversial suspension of the writ of habeas corpus had the results Lincoln intended.[2] Without ignoring what Lincoln and other major political protagonists of the day said, this book will focus on what they in fact did.

This might well be called writing constitutional history from the bottom up. The phrase is hardly original, but its application to constitutional history is somewhat novel. Naturally preoccupied with politicians, judges, and lawyers and what they have said, constitutional historians have rarely looked at what happened after those august persons spoke. In the case of the Civil War, no one has ever systematically examined records of civilians arrested by military authority after Abraham Lincoln suspended the writ of habeas corpus.

The Historical Problem

At the time, the Democratic party maintained that it was the principal victim of Lincoln's policies. If the accusation could be substantiated, then the arrests perpetrated an unconscionable injustice, for the Democratic party in the North formed a loyal opposition, a "respectable minority." The shrewd and pain-staking work of historian Frank L. Klement over the last thirty years has proved, beyond any reasonable doubt, that no systematic, organized disloyal opposition to the war existed in the North.[3]

Lincoln scholars have been slow to grasp the profound change this new view must bring to historical interpretation of the Republican administration during the Civil War and its treatment of the opposition party. For it is well known that President Lincoln suspended the writ of habeas corpus early in the war and thereafter managed the home front, in part, by means of military arrests of civilians—thousands and thousands of them.

As long as historians believed, as the Republicans themselves did in the nineteenth century, that large and dangerous groups of disloyal citizens lurked in the North during the war, Lincoln's tough policies on civil liberties seemed fully justified. But when those supposedly disloyal groups more or less evaporated under the scrutiny of modern scholars, an awkward question arose: if not the disloyal, then who was being arrested? If not organized and treacherous Copperheads, who?

Even in wartime, rounding up and imprisoning political opponents of the party in power is unacceptable in America, a country that enjoys the oldest two-party system in the world. Still, the political parties were relatively young in Lincoln's day, and many Republicans at the time believed the Democrats, or substantial numbers of them, to be at heart disloyal. Did President Lincoln succumb to the temptation to eliminate the political opposition by the use of military arrests? Did subordinates attempt it without his knowledge?

The testimony of the politicians, judges, and lawyers of the day is conflicting. Lincoln ran successfully for reelection in the midst of the war, and his Dem-ocratic opponent, George B. McClellan, never cried foul over the result. "The people have decided with their eyes wide open," the general said two days after his defeat.[4] On the other hand, the two best-known cases of civilians arrested by military authority during the Civil War, those of Ohio's Clement L. Vallandigham in 1863 and Indiana's Lambdin P. Milligan in the presidential election year 1864, involved Democratic politicians, and one of them was arrested for what he said in a speech before a partisan crowd. These justifiably famous examples raise suspicions that Republicans attempted to combat the opposing political party by means other than the ballot. Democrats in Maryland and Kentucky maintained that such cases were the tip of an iceberg of Re-publican interference at the polls that froze the political opposition.

Ultimately, the answer lies in the arrest records and only there. Were the masses of citizens arrested active Democrats? Even if the victims of military arrest were mainly Democratic partisans, why were some arrested and not others? Lincoln's defenders have often pointed to the shrill criticism of the president found in the Northern Democratic press throughout the war as proof that dissent was never really stifled. Who did get arrested? Surely there were never enough disguised Confederate agents, spies, saboteurs, and individual traitors to equal the thousands of civilians known to have occupied dank cells in notorious old garrisons like Fort Lafayette in New York harbor or makeshift military prisons like the Old Capitol in Washington and Myrtle Street prison in St. Louis.

This unsettling problem of large numbers of civilian prisoners in the midst of a substantially loyal population threatens to undermine Lincoln's reputation. It would be a shame to allow it to sink by mere conjecture, though, especially when a substantial body of evidence, the arrest records themselves, has gone without systematic statistical investigation. This book will use those documents to reveal what civilians were arrested by Northern military authority during the Civil War; on the answer stands or falls, to a great degree, the reputation of Abraham Lincoln as a legitimate political leader.

The investigation will take the reader far from the somewhat narrow confines of Lincolniana and involve much more than abstract statistics. It will lead to ships running the blockade on the fog-bound coast of North Carolina and to the raw guerrilla strife in Missouri. The geographical boundaries of inquiry must reach from the Canadian border to the Union picket outposts deep in occupied Confederate territory. It will require an examination of international law, military conscription, and long-forgotten judicial opinions. A better understanding of the administration of government under Lincoln will emerge— not of the president's office alone but also of those who conducted the war effort from day to day, from the War and State departments to the soldiers on campaign. Ultimately, the old constitutional questions will come alive in flesh-and-blood incidents of courage and cowardice, of political martyrdom and drunken folly. This book will reevaluate figures like William H. Seward and Ulysses S. Grant, and it will introduce a myriad of obscure individuals, whose brief moment on the stage of history is preserved for historians only in the papers stemming from their arrests.

Habeas Corpus Before the War

All of these incidents, individuals, and issues were somehow tied together by the fateful suspension of the writ of habeas corpus. The suspension was unprecedented, like the Civil War itself, and therefore Lincoln's actions are not much illuminated by surveying earlier American history. The development of Lincoln's own constitutional ideas before the presidency will be discussed in detail later, but in order to understand the sensibilities which Americans brought to the habeas-corpus question in the Civil War, a brief look at the prewar past is helpful. In the decade of the 1850s, the formation of the Republican party intensified partisan debate about freedom. The Republicans' antislavery impulse reinforced the vein of traditional libertarianism in their party ideology. Leaders like Abraham Lincoln looked back to the revolutionary doctrines of the country's founders, drawing particular inspiration from the Declaration of Independence.

Habeas corpus had been important to the founders of the country and owned a hallowed place in American law and myth long before there was an active antislavery movement. It was a part of the American birthright. Historian Milton Cantor points out that though it was not regarded as a natural right, the writ of habeas corpus was the only common-law process mentioned in the United States Constitution, surely an index of its importance in the eyes of the country's founders. The delegates to the constitutional convention at first voted unanimously that "the privilege of the writ of Habeas Corpus shall not be suspended." Later, Gouverneur Morris introduced a qualifying clause adapted from the Massachusetts constitution: "unless in cases of rebellion or invasion the public safety may require it." In qualified form, then, the provision passed by a vote of seven states to three and became part of Article I, Section 9.[5]

By the 1850s, Republicans thought of the writ, if they considered it at all, primarily as a device to free fugitive slaves, a liberating tool for the antislavery crusade. Salmon P. Chase, who would serve as secretary of the treasury from 1861 until mid-1864, probably knew more about the writ of habeas corpus than Lincoln or any other member of Lincoln's cabinet. As a lawyer in Ohio, he had used it and other legal gambits so often in fugitive slave cases that he came to be known as the "Attorney General for Runaway Negroes." No such nickname ever attached itself to Abraham Lincoln, who had once represented the master in a habeas-corpus hearing in Illinois brought by two local abolitionists trying to free a fugitive slave. Like most lawyers, Lincoln took clients as they came to him, be they masters or slaves; but as a politician, he held consistent antislavery views. These were sharpened and intensified by the slavery controversies of the 1850s, and Lincoln's ideas came more and more to emphasize natural rights and freedom.[6]

The "great writ" was certainly biased toward freedom. It was meant to help people wrongfully detained. An apprentice, for example, could employ it to seek freedom from a master, but a master could not use it to recover an apprentice. Nevertheless, the writ of habeas corpus never became a central issue in the slavery controversy in the nineteenth century. Great as it was, the writ of habeas corpus coexisted comfortably with slavery in the United States. White Southerners embraced it as dearly as Northerners did. Slaves, of course, did not enjoy the privilege of the writ, which was one among many rights and privileges denied them by the peculiar institution. Southern courts ruled that a jury trial was necessary to deprive a master of his slave; a judge could not do so by issuing a simple writ. Thus, a Florida court in 1859 took the view widely held in the South that the "writ of habeas corpus is not the proper method of trying the right of a negro to Freedom. The doctrine of the *court is,* that the person claiming him cannot be deprived of his *property* without jury trial."[7]

Though biased toward freedom, the writ of habeas corpus in the Northern states at first served as a neutral legal instrument available to blacks and slave-catchers alike. In the natural tension that developed between the desire to recover property escaping from slave states and the free states' obligation to protect their free black citizens from kidnapping, the writ of habeas corpus became a two-edged sword. Antislavery forces employed writs to free black people claimed by Southerners in free states, but a New York law of 1828, for example, also allowed claimants to request the issuance of writs that would cause the sheriff to bring in an alleged runaway. The lawmakers drew on an analogy with child custody cases, in which writs of habeas corpus were similarly employed.[8]

Resistance stiffened a bit in Northern states in the 1840s, and many Northerners sought laws that would lead to jury trials—before generally anti-Southern juries, of course—to determine the status of an alleged slave. Though less cooperative, few Northern laws aimed deliberately at nullifying the obligation placed on the free states by the U.S. Constitution (in Article IV, Section 3) to permit recovery of any "Person held to Service or Labour in one State,... escaping into another." On the other hand, state constitutions, like the national one, also guaranteed the writ of habeas corpus, and conflicts continued.[9]

The great writ retained its liberating power for antislavery Northerners until the political Compromise of 1850. The revision of the fugitive-slave law in that sectional compromise package gave to special federal commissioners the power to issue certificates to persons seeking their runaway slaves. And these certificates, the law held, "shall prevent all... molestation of said... persons by any process issued by any court, judge, magistrate, or other person whomsoever."[10] In abolitionists' eyes, the Slave Power had scored another victory over traditional American liberties.

In a lithographed political cartoon of the period, "Black Dan" Webster, who had surprised many of his Massachusetts constituents by supporting the Compromise of 1850, leads slaveholders in pursuit of a black family. One of the owners stumbles over a book of "Massachusetts Laws" and a "Writ." To antislavery zealots the great writ had been trampled upon by the proslavery forces. Yet the writ lay in an obscure corner of the cluttered cartoon. The focus of the political antislavery crusade, as the cartoon suggests, would henceforward be elsewhere: on the territories.

Conflicts between the fugitive-slave clause and the writ of habeas corpus nevertheless continued, reaching a peak of bitterness on the eve of the Civil War with the issuance of the U.S. Supreme Court decision in *Ableman v. Booth* on March 7, 1859. The opinion, written by the proslavery Chief Justice Roger B. Taney of Maryland, denied the use of any process by state courts to interfere with the enforcement of federal law. By this time, some Northern legislators felt driven to propose laws deliberately aimed at frustrating the constitutional obligation to "deliver up" fugitive slaves. Ohio senator Benjamin F. Wade, referring to a sensational fugitive-slave case that arose after the *Ableman v.*

Booth decision, scoffed characteristically, "If the Supreme Court of Ohio does not grant the habeas corpus, the people of the Western Reserve must Grant it—sword in hand if need be."[11]

Illinois was the only contiguous state of the U.S. not to pass some sort of personal-liberty law before the Civil War, and for that and other reasons, Abraham Lincoln was little involved with the political issues of fugitive slaves and personal liberty. He reluctantly bowed to the constitutional necessity of enforcing the fugitive-slave law and disapproved of state legislation aimed at frustrating the fugitive-slave clause in the Constitution. In 1859, he insisted on Republican adherence to the law and criticized the Ohio party platform, which had called for its repeal. He told Salmon P. Chase and Indiana's Schuyler Colfax that the Ohio position would "explode" the national party and make "the cause of Republicanism hopeless in Illinois," where opposition to the fugitive-slave law would be regarded as "enmity to the constitution itself." Even after his election to the presidency in 1860, Lincoln could say of the Northern personal-liberty laws, "I never have read one." He continued to oppose unconstitutional state laws aimed at blocking the capture of runaways; but he opposed any federal law obligating free-state citizens to aid in their capture and wished to retain "the usual safeguards to liberty, securing free men against being surrendered as slaves"—by which he surely meant the habeas corpus writ.[12]

There is little need to dwell at greater length on the uses of the writ of habeas corpus before the Civil War or on Lincoln's views on the subject, because the war turned everything topsy-turvy. The party associated with the antislavery movement now suspended the writ. The Democratic party, which had encouraged federal intrusions on the power of state courts to issue binding writs, suddenly became the defender of the great writ. The previous uses and abuses of the writ of habeas corpus in the struggle over slavery were no longer of practical interest. The slavery issue was going to be settled by the military power of the executive branch and by the legislature's will, now unobstructed by proslavery Southerners.

The great writ remained a symbol of individual freedom and many knowledgeable Americans in 1861 believed that its suspension would have dire and lasting effects. That feeling would nag the consciences of some Republicans. And it would intensify the sincere anxieties as well as inflate the partisan rhetoric of many Democrats.

But military arrests of civilians never provoked widespread disobedience of the law or violent protests. Only military conscription could do that. Why did military arrests of civilians never arouse such passions as were demonstrated in New York City's draft riots of 1863? The answer surely lies in the identity of the civilians who were actually arrested.

The goal of this book is to settle this question of identity once and for all—not by taking the Republicans' word for it that the victims were spies and traitors, not by accepting without question the Democrats' story that they were the principal victims of partisan excess disguised as zeal for loyalty, and not by listening only to the noisiest of the prisoners themselves, the articulate ones who wrote exposés of prison life and gave speeches about their martyrdom. All of these voices will be heard in this book, but so too for the first time will the mass of political prisoners speak, in their letters asking jailors for release and in their statements taken by interrogating officers in prisons or in military courts. For the first time, history will dig deeper—into the lists and notes and bundles of paper still tied in the now faded red tape of nineteenth-century bureaucracy—to find out who was arrested when the writ of habeas corpus was suspended and why.

The Fate of Liberty

1

Actions Without Precedent

As president-elect, Abraham Lincoln faced the secession of several states from the Union he was soon supposed to govern. He proved ill-prepared to address the challenge with a unifying explanation of the country's national identity. In the previous decade, he had thought constantly of the freedoms and rights denied to slaves. There had been little necessity for thinking about salutary assertions of government power. On the way to Washington for his inauguration, Lincoln betrayed his inexperience in formulating his verbal defenses of government authority. For example, when he stopped in Indianapolis on February 11, 1861, he attempted to describe the ultimately wayward tendency of secession by saying that secessionists thought of the union of the states not as a sacred marriage but as "a sort of free-love arrangement." The vulgar simile surely shocked some people.[1]

By the time of his inaugural address on March 4, Lincoln had found a more sober family analogy, saying that husbands and wives might divorce and separate physically, but the two sections of the country could not do so. Yet when he felt the need for a peroration for this, the most important speech of his political career to date, the right words did not come to mind. He needed a ringing appeal to sentiment for national unity. Though an able speech-maker, Lincoln had to ask for help with this task. He turned to Secretary of State William H. Seward, who wrote the initial draft of the famous conclusion that would evoke the "mystic chords of memory, stre[t]ching from every battle-field, and patriot grave, to every living heart and hearthstone" of the old Union.[2]

Fewer than six weeks after Lincoln made his appeal to these mystic chords of Union, they fell silent, and secession led to civil war. Still lacking a systematic ideology of nationalism to buttress government power, he grasped at any available practical measure that promised to meet the crisis of dissolution. And again he turned to Seward, this time to exert the authority of the national government over disloyal individuals in the North.

3

The First Suspension of the Writ of Habeas Corpus: Maryland

As soon as news of the firing on Fort Sumter arrived on April 14, 1861, worried Northerners thought of restricting civil liberties in Maryland. If unrestrained, the state seemed likely to accompany Virginia, as it followed the more im-petuous Southern states out of the Union.[3] On April 19, when a Baltimore mob blocked the passage of Massachusetts troops en route to guard the nation's capital, the government in Washington began to look seriously into the matter. Lincoln asked Attorney General Edward Bates for an opinion on declaring martial law in Maryland "to the extent of allowing an infraction" of the Fifth Amendment to the Constitution, which required grand-jury indictment for any "capital or otherwise heinous crime."[4] Bates delegated the task to the assistant attorney general, Titian J. Coffey, who prepared the digest of opinions on martial law that was given to Lincoln on the morning of April 20.[5]

The memorandum was not particularly encouraging to a chief executive seeking precedents for decisive action. Coffey excerpted passages from stan-dard British and American authorities that explained the difference between military law and martial law. As commander-in-chief of the army and navy, the president could use military law, but that was clearly prescribed by Congress in the Articles of War of 1806 to apply only to members of the armed services. It was irrelevant to the crisis brewing in Maryland. Martial law, which was applied to civilians in times of war, proved hard to define. Sir Matthew Hale and William Blackstone saw it as no law at all. Thus, the president's powers were unclear but seemed quite limited, for the digest included Supreme Court Justice Joseph Story's opinion that only Congress could suspend the writ of habeas corpus.[6]

The new president's request for the attorney general's opinion suggests that Lincoln did not know the law of this subject upon assuming office. This should come as no surprise, for questions concerning martial law and suspending the writ of habeas corpus were hardly likely to arise in the ordinary practice of a yeoman lawyer in central Illinois. By western standards, at least, Lincoln was a lawyer's lawyer who had argued more than two hundred cases before his state's highest court. But the familiar legal literature read by able practitioners like Lincoln offered no practical ideas about the writ of habeas corpus in times of war. James Kent's *Commentaries on American Law,* for example, stated only that the "privilege of this writ is . . . made an express constitutional right at all times, except in cases of invasion or rebellion, by the constitution of the United States, and by the constitutions of most of the states in the Union." When he dealt with the subject of war in his famous book, Chancellor Kent thought mainly of foreign wars and commented at length only on the inter-national laws of blockade. He said nothing about domestic civil liberties, and the term "martial law" did not appear in Kent's index.[7]

Joseph Story's *Commentaries on the Constitution of the United States,* first published almost three decades before the Civil War, offered somewhat more extensive coverage of the subject. Story assumed that only Congress could suspend the writ of habeas corpus, but he did not explain exactly why. "Hitherto," he said, "no suspension of the writ has ever been authorized by congress since the establishment of the constitution. It would seem as the power is given to congress to suspend the writ of habeas corpus in cases of rebellion or invasion, that the right to judge, whether exigency has arisen, must exclusively belong to that body." Story offered less advice than a president or a judge might want in regard to practical matters like giving notice of the suspension or delegating the power to suspend. Story did show some concern over the consequences if suspension ever occurred:

> It is obvious, that cases of a peculiar emergency may arise, which may justify, may even require, the temporary suspension of the right to this writ. But as it has frequently happened in foreign countries, and even in England, that the writ has, upon various pretexts and occasions, been suspended, whereby persons, apprehended upon suspicion have suffered a long imprisonment, sometimes from design, and sometimes, because they were forgotten, the right to suspend it is expressly confined to cases of rebellion or invasion, where the public safety may require it; a very just and wholesome restraint, which cuts down at a blow a fruitful means of oppression, capable of being abused in bad times to the worst of purposes.[8]

Lincoln probably learned more on this subject from political history than from law books. Having been a Whig politician, he knew, as a useful partisan anecdote if nothing else, that General Andrew Jackson had imposed martial law in New Orleans during the War of 1812 and had subsequently not only ignored a writ of habeas corpus but had also arrested the judge who issued it. Actually, the story might have been more than partisan ammunition in the political war chest of Lincoln's memory. Unlike others in his party, Lincoln had grudgingly admired Jackson's forcefulness and willingness to take responsibility in crises. And to the degree that any episodes from American history were much on Lincoln's troubled mind in this awful time, those in which Jackson had figured were likely prominent. Lincoln had consulted Jackson's proclamation on nullification when he wrote his inaugural address back in January. And he would soon be lecturing Baltimoreans on his duty by recalling memories of Jackson's steadfast devotion to the Union. He told them that there was "no Jackson" in surrendering the government of the United States merely because Maryland did not want to see soldiers march across her soil to the defense of the nation's capital. Still, none of this suggested a Jacksonian imposition of martial law, only a stiffening of the national spine.[9]

Although martial law might be drastic, action in Maryland seemed more imperative than ever. Local authorities burned Baltimore's key railroad bridges on the night of April 19, 1861, maintaining that they acted from fear that other Northern troops on their way to the city might seek revenge for the riot. Bridge-burning looked like plain treason to the government in Washington, which was now defenseless and cut off from the rest of the North.[10]

While anxiety grew in Washington, anger mushroomed in the rest of the land, especially among Republicans. Illinois senator Lyman Trumbull reported to the president from Springfield on April 21 that "everybody" was "greatly excited over the news from Baltimore, & extremely solicitous of Washington." When free passage through the city was refused, the president should have ordered Northern troops to "take possession of Baltimore at once.... Ten thousand troops on the outskirts..., demanding a safe passage through it & prepared to take and hold possession if refused, could have had their demand complied with." Orville Hickman Browning, also from Illinois, a moderate Republican and a friend of Lincoln's, warned him on April 22 that the "fall of Washington would be most disasterous [*sic*]. Communication ought to and must be kept open to Washington. Baltimore must not stand in the way. It should be seized and garrisoned, or, if necessary to the success of our glorious cause, laid in ruin."[11]

Easterners saw eye to eye with Westerners on the Baltimore question. Philadelphia's Andrew H. Reeder told Secretary of War Simon Cameron, *"The administration will be sustained in everything except half way measures.* If Baltimore was laid in ashes the North would rejoice over it." Reeder warned that "when men are told they cannot go to their own Capital for fear of Maryland rebels each man smarts as under a blow or personal insult and a feeling is excited with which the administration dare not trifle." James Watson Webb, the editor of the *Morning Courier and New-York Enquirer,* sent the president similar advice but worded it even more sharply: "For God's sake, if you would not cast a wet blanket upon the Patriotic feeling of the Nation, Order *all* the troops to pass through Baltimore and treat as an insult any proposition to compromise the matter."[12]

Republicans in Washington, understandably, grew anxious. Edward Bates and Salmon P. Chase, cabinet members who represented opposite ends of the Republican-party spectrum on slavery issues, agreed that Maryland required immediate action. The president still refused to act, and the suspense became almost unendurable. Railway executives devised a route to Washington that avoided the Baltimore bridges, from Philadelphia through Perryville via Annapolis. Yet troops did not come.[13]

Baltimore's insult to national dignity seemed minor compared to the possibility of Washington's permanent isolation by the secession of Maryland. In a weak moment, Governor Thomas H. Hicks had called a special session of the Maryland legislature to convene on April 26. Many now feared that the legislators would adopt a secession ordinance, and Massachusetts General Benjamin F. Butler, commanding Union troops near the state capital, threatened to arrest any secession-minded Maryland politicians.[14]

So, too, did General-in-Chief Winfield Scott, but the president would not allow it, ordering him on the 25th to hold off from any such action:

> The Maryland Legislature assembles to-morrow... and, not improbably, will take action to arm the people of that State against the United States. The question has been submitted to... me, whether it would not be justifiable... for you... to arrest, or disperse the members of that body. I think it would *not* be justifiable; nor, efficient for the desired object.
>
> First, they have a clearly legal right to assemble; and, we can not know in advance,

that their action will not be lawful, and peaceful. And if we wait until they shall *have* acted, their arrest, or dispersion, will not lessen the effect of their action.

Secondly, we *can* not permanently prevent their action. If we arrest them, we can not long hold them as prisoners; and when liberated, they will immediately re-assemble, and take their action. And, precisely the same if we simply disperse them. They will immediately re-assemble in some other place.

I therefore conclude that it is only left to the commanding General to watch, and await their action, which, if it shall be to arm their people against the United States, he is to adopt the most prompt, and efficient means to counteract, even, if necessary, to the bombardment of their cities—and in the extremest necessity, the suspension of the writ of habeas corpus.[15]

Lincoln's judgment was vindicated, and the Maryland legislature, though dominated by Democrats, refused to consider a secession ordinance.[16] The letter to Scott, the first document in which Abraham Lincoln, as president, mentioned the writ of habeas corpus, stands as a warning of the difficulties that lie in wait for any reader of the constitutional history of the Civil War. It is an unusual letter in two ways. First, Lincoln did not ordinarily place legal considerations before practical ones, as he did here. Second, Lincoln's letters usually make sense, but the last paragraph of this one does not. How can it be true that suspending the writ of habeas corpus was a remedy for a situation more extreme than one requiring the "bombardment of . . . cities"?

A look at the holograph original of the letter explains the latter problem. The version of the document printed in the standard edition of Lincoln's *Collected Works* fails to describe a change the president made near the end of the last sentence. He originally wrote "if necessary, to the bombardment of their cities—and of course the suspension of the writ of habeas corpus." That made sense. If something as drastic as shelling one's own cities were necessary, then suspending the writ of habeas corpus would be necessary also, as a matter of course. But Lincoln was hesitant to take this unprecedented step toward restricting civil liberties in the North. He reread the order quickly and disliked the casual phrase "of course" immediately preceding the phrase about suspending the writ. So he struck through "of course" and with a caret inserted above the line the phrase "in the extremest necessity." The sentence now sounded properly grave, but the change destroyed its overall sense.[17]

This somewhat rattled order betrayed Lincoln's anxious state of mind in this suspenseful period. Even his admiring private secretaries, John G. Nicolay and John Hay, later admitted in their biography of Lincoln that "this period of interrupted communication and isolation from the North" put him in an obvious "state of nervous tension." Though "by nature and habit . . . calm, . . . equable, . . . undemonstrative," the president paced the floor for nearly half an hour on the afternoon of the 23rd. He then "stopped and gazed long and wistfully out of the window down the Potomac in the direction of the expected ships." Unaware that anyone else was in the room, Lincoln "broke out with irrepressible anguish in the repeated exclamation, 'Why don't they come! Why don't they come!'"[18]

The troops arrived two days after Lincoln's outburst, at noon on Thursday,

April 25, the very day that Lincoln told Winfield Scott not to arrest Maryland's legislators. Celebrations, parades, reviews, and serenades followed.[19] Yet on April 27, two days after the troops came through, Lincoln suspended the writ of habeas corpus for the first time. Why, if the worst danger were now over, did the president take an action appropriate in his mind only for "the extremest necessity"?

Despite the public rejoicing, Scott remained worried about Washington's peril. The general-in-chief drafted an order on the 26th warning that an attack on Washington was possible at any moment and instructing Union pickets not to give way on the bridges leading into the city "till actually pushed by the bayonet"—desperate orders in an age of long-range rifles and artillery. Only 1,600 volunteers had arrived, and the Confederates had 8,000 men at Harpers Ferry. Though spies reported the Southern troops unlikely to attack, Scott would not pronounce the capital safe until April 29. In the midst of continuing peril, then, his daily report of the 26th informed the president that the directors of the railroad between Baltimore and Washington would send an agent the next day "to ask for protection to a reestablishment of trains between the two cities." This agent must have been in Washington on the very day Lincoln gave Scott his authorization to suspend the writ if necessary.[20]

Two versions of Lincoln's historic order exist, one known from the nineteenth century on and another, a draft in Lincoln's hand, apparently discovered by Roy P. Basler and the team that edited *The Collected Works of Abraham Lincoln,* published in 1953. Basler and his co-editors reproduced both but did not explain which document was written first and which was the final order sent to Scott. The sequence is now clear. The unissued draft read thus:

> You are engaged in repressing an insurrection against the laws of the United States. If at any point on or in the vicinity of the military line, which is now being used between the City of Philadelphia and the City of Washington, via Perryville, Annapolis City, and Annapolis Junction, you find resistance which renders it necessary to suspend the writ of Habeas Corpus for the public safety, you, personally or through an officer in command at the point where the resistance occurs, are authorized to suspend the writ.

This language covered only the improvised line to Washington, the one that avoided Baltimore, not a "reestablished" regular route between the cities. The other version of the order permitted suspension "at any point on or in the vicinity of any military line, which is now or which shall be used between the City of Philadelphia and the City of Washington" and omitted the description of the Perryville-Annapolis route. This draft fit the circumstance of the railroad agent's report that day, as it would have protected a reestablished route as well as the makeshift one.[21] We know the order permitting suspension on *any* line was the one actually sent to Scott, because when Congress later asked for copies of the habeas-corpus orders, that was the one the secretary of state sent.[22]

The president had at last taken action appropriate only for the "extremest necessity," but he by no means publicized it. The letter to Scott was not a public proclamation and did not become a general order, printed for circulation among military commands that might be affected by it. Scott himself authorized the commanders of the appropriate departments in Pennsylvania, Delaware, Maryland, and Washington.[23] No one informed the courts or other civil authorities. Lincoln did not say so, but Scott assumed that he could delegate the authority to others.

Some important people did not know about the suspension. When a Maryland judge, William F. Giles, issued a writ of habeas corpus on May 2 for the release of a minor who had enlisted in the army without his parents' consent, the commander at Fort McHenry in Baltimore refused to obey it. The officer said that he acted entirely on his own responsibility; he apparently had no knowledge of the presidential letter or of Scott's orders. When another officer refused to comply with a writ issued later in May by Roger B. Taney, chief justice of the United States Supreme Court, Taney complained, "No official notice has been given to the courts of justice or to the public by proclamation or otherwise that the President claimed this power" to suspend the writ of habeas corpus.[24] Until Taney thus alerted them, the Democratic press outside Maryland remained silent, muffled by ignorance of Lincoln's act, by national outrage over Maryland's obstructionist policies, and by the president's narrow geographical circumscription of the suspension of civil liberties.

The purpose of the initial suspension of the writ of habeas corpus is clear from the circumstances of its issuance: to keep the military reinforcement route to the nation's capital open. It is equally clear that political provocation—the meeting of the Democratic-dominated Maryland legislature—did not cause Lincoln to give Scott the historic authorization. Suspending the writ of habeas corpus was not originally a political measure, and it would never become primarily political.

Florida and Other Early Proclamations and Orders

Abraham Lincoln first suspended the writ of habeas corpus by public proclamation on May 10, 1861:

> Whereas, an insurrection exists in ... Florida, by which the lives, liberty and property of loyal citizens of the United States are endangered ... Now therefore ... I, Abraham Lincoln ... do hereby direct the Commander of the Forces of the United States on the Florida coast, to permit no person to exercise any office or authority upon the Islands of Key West, the Tortugas and Santa Rosa, which may be inconsistent with the laws & constitution of the United States, authorizing him at the same time, if he shall find it necessary, to suspend there the writ of *Habeas Corpus* and to remove from the vicinity of the United States fortresses all dangerous or suspected persons.[25]

Florida had long since seceded from the Union, and this proclamation therefore excited little comment or controversy. Lincoln never bothered to suspend the writ in any other Confederate state or any other Union-held area of a Confederate state. He may not have believed consciously in the formal theory, but Lincoln behaved as though the Southerners who seceded had thereby abdicated their civil liberties under the U.S. Constitution. The lack of public protest may indicate that other Americans thought the same way.

Union officers on the Florida coast had for some time been acting as though the local citizens had abdicated their rights. The commander of the U.S. garrison at Key West, Brevet Major William H. French, resorted to extraordinary measures on his own authority. "My position has required me to take responsibility," he explained.[26] Before the president issued his proclamation, in fact, Colonel Harvey Brown, commanding the Department of Florida, issued his own proclamation suspending the writ of habeas corpus. On May 8, Major French briefed his superiors in Washington on further assumptions of responsibility in Key West:

> There have been no secession flags flying since my peremptory order on the subject. The military organization called the "Island Guards" has disbanded, in consequence of my directing the mayor to furnish me with the muster roll, which he did. The newspaper called the "Key of the Gulf" I suppressed, because it was uttering treasonable and threatening language against the judiciary and other United States officers. I directed the mayor to inform the editor (a Mr. Ward) that he was under military surveillance, and that the fact of his not being in the cells of this fort for treason was simply a matter as to expediency and proper point of time. To enable me to meet such cases with promptitude, I published on the 6th instant Colonel Brown's proclamation suspending the writ of habeas corpus.

The president's proclamation, issued two days after this letter was written, simply recognized a *fait accompli.*[27]

From this point on, theory usually followed fact with President Lincoln in matters having to do with civil liberties. Once he suspended the writ of habeas corpus without suffering dire political consequences, similar actions grew easier and easier.

The president did not go politically unscathed on this issue, however. After the May 25 arrest of John Merryman, a citizen said to be drilling Marylanders in a scheme to take them south and join the Confederate army, Chief Justice Taney, on circuit as a federal district judge, issued a writ of habeas corpus, but he did so as a Supreme Court justice from chambers. When it was disobeyed because of the president's order of April 27, Taney wrote an opinion challenging the legality of Lincoln's suspension of the writ. According to the chief justice, the power to suspend lay with Congress and not with the president. The opinion was quickly published in newspapers and in pamphlet form in Baltimore and Philadelphia. It also received immediate dissemination in the Confederacy—pamphlet versions appeared in New Orleans and Jackson, Mississippi. Following the lead of the aged Democrat Taney, the Northern Democratic press at last took up the issue.[28]

The president offered no impulsive response to the criticism, but he did notice it. On May 30, he asked the attorney general to confer with Maryland

Unionist Reverdy Johnson "and be preparing to present the argument for the suspension of the Habeas Corpus."[29] Lincoln would eventually answer theory with theory, but he would act as needed in the meantime.

While Edward Bates labored to produce "the argument," President Lincoln issued another order suspending the writ of habeas corpus. Addressed to General Scott on June 20, 1861, it stated simply: "You or any officer you may designate will, in your discretion, suspend the writ of *habeas corpus* so far as may relate to Major Chase, lately of the Engineer Corps of the Army of the United States, now alleged to be guilty of treasonable practices against this government." The order was countersigned by Seward and issued from the State Department, which would oversee arrests for disloyalty until February 1862. The origins and significance of this curious order are not at all clear. Presumably Seward wanted Major Chase arrested whether he was found on the Washington-Philadelphia military line, off the coast of Florida, or anywhere else, and this was the only way he could think of to handle the problem. The order bore traces of the confusion wrought by the hectic activity of the perilous early days of the Lincoln administration: the president neglected to supply William Henry Chase's first and middle names. Lincoln surely knew already what he would express openly later, that if he "should suspend the writ . . . , instances of arresting innocent persons . . . are always likely to occur in such cases." In this instance, the president himself invited a possible wrongful arrest.[30]

Less than two weeks later, Lincoln authorized General Scott "personally, or through the Officer in command, at the point where resistance occurs," to suspend the writ of habeas corpus "at any point, on or in the vicinity of any military line which is now, or which shall be used, between the City of New York and the City of Washington."[31] Lincoln had quickly overcome any initial hesitation to use presidential power to suspend the writ. The use of the phrase "military line," for example, was becoming expansive, if not a little deceptive. At first the term had described the threatened route to Washington. Now it referred to no particularly well-described "line" between two places quite far apart on the map.

The President's Defense

Two days after sending to Scott the newest order suspending the writ of habeas corpus, Lincoln presented a formal defense of his actions in a message to the special session of Congress which convened on July 4, 1861. He drafted the document as though it were meant for Congress alone and not, as was inevitably the case, also for the American people. It lacked simplicity, clarity, and homely examples. The president assumed an almost cerebral tone. The libertarian thrust of Lincoln's antebellum political thought still left him unprepared to explain the war powers of the government. Edward Bates and Reverdy Johnson do not seem to have helped him much either. The message was a bit muddled.

Despite his belief in a president's right to employ unused powers in the Constitution, Lincoln proved in his message to Congress to be the intellectual captive of his critics, especially of Taney. The president seemed almost to agree that the legislative branch was the proper body to suspend the writ of habeas corpus. He assured the legislators "that nothing has been done beyond the constitutional competency of Congress" and expressed confidence that they would "ratify" the extraordinary measures taken by the administration in the post-Sumter emergency. "It cannot be believed," Lincoln said, "the framers of the instrument intended, that in every case [of rebellion or invasion], the damage should run its course, until Congress could be called together; the very assembling of which might be prevented, as was intended in this case, by the rebellion." Nothing had kept Congress from assembling before July 4 except Lincoln's decision not to call them into session, but perhaps he was recalling the harrowing period surrounding the first suspension of the writ when troops were having trouble getting through to defend the city, which would have made it impossible for Congress to convene.

He took pains to reassure Congress that the authority he assumed in the crisis "to arrest, and detain, without resort to the ordinary processes and forms of law, such individuals" as the commanding general "might deem dangerous to the public safety," had "purposely been exercised but very sparingly." Yet the president also indulged in idle arguments that unnecessarily betrayed his own uncertainties. "These measures," Lincoln said of his calls to enlarge the army and navy, "whether strictly legal or not, were ventured upon, under what appeared to be a popular demand, and a public necessity." A hostile interpretation of this statement—and Lincoln knew the opposing political party was eager to torture any Republican utterance on this issue—might find in it an admission of an illegal act and uncertainty about public opinion and military necessity. Lincoln stumbled on, assuring Congress in regard to the habeas-corpus suspension that "some consideration was given to the questions of power, and propriety, before this matter was acted upon," but was that enough to erase the doubts he had already expressed?

Without mentioning the source, Lincoln also referred to the *Merryman* decision's "proposition that one who is sworn to 'take care that the laws be faithfully executed,' should not himself violate them." The syntax of Lincoln's response was unusually labored:

The whole of the laws which were required to be faithfully executed, were being resisted, and failing of execution, in nearly one-third of the States. Must they be allowed to finally fail of execution, even had it been perfectly clear, that by the use of the means necessary to their execution, some single law, made in such extreme tenderness of the citizen's liberty, that practically, it relieves more of the guilty, than of the innocent, should, to a very limited extent, be violated? To state the question more directly, are all the laws, *but one,* to go unexecuted, and the government itself go to pieces, lest that one be violated? Even in such a case, would not the official oath be broken, if the government should be overthrown, when it was believed that disregarding the single law, would tend to preserve it?

Having gone that far, the president now shrank from embracing bold theories. He added immediately, "But it was not believed that this question was presented. It was not believed that any law was violated."

The use of the passive voice left Lincoln's own views in doubt, but in a shrewd analysis of the message based on comparing the final version with an earlier draft, historian James G. Randall concluded that "one may read [here], as it were, the President's mental struggling at the time the decision was taken. In this remarkable document may be seen the clearest indication that the appearance of military dictatorship was a matter of deep concern to the nation's war chief and that his action was determined by what he believed to be the imperative demands of the actual situation." Randall denied that Lincoln needed Taney's criticism: "As a matter of fact few measures of the Lincoln administration were adopted with more reluctance than this suspension of the citizen's safeguard against arbitrary arrest."[32]

Lincoln's musings certainly revealed his sincere doubts about the policy, but they are as apparent in the final version as in the first draft. Up to a point, Randall's emphasis on the soul-searching sincerity of the document is proper. In many ways, the message did render the president's experience faithfully. The Bates-Coffey digest, for example, proved prior "consideration . . . given to the questions of power, and propriety." And the president's files contained the letters urging emergency action in Baltimore—proof of "popular demand" for his policies. Winfield Scott's request for exceptional authority and the well-known plight of the capital after the burning of Baltimore's railroad bridges provided adequate proofs of "public necessity." The president could also point to several indications that he exercised this emergency power only "to a very limited extent" and "sparingly," the letter of April 25 ordering Scott not to interfere with the Maryland legislature being by far the most impressive.

Yet, in important ways, Lincoln's message veered away from actual experience toward slippery argument. The "whole of the laws" were, as Lincoln said, "failing in execution, in nearly one-third of the States," but not in Maryland any longer, and they never had failed along the railways and waterways all the way from Washington to New York. A pro-secessionist jury in Maryland might, as the president suggested, relieve more of the guilty than of the innocent and leave disloyal citizens free to harm the Union, but this was surely less likely as one approached New York. Neither popular demand nor public necessity for the suspension above Maryland can be as solidly documented as the vividly threatening events around Baltimore. In short, Lincoln fashioned an able defense of the April suspension of the writ of habeas corpus but left the more recent expansion of the suspension unexplained.

In detecting the sincerity in Lincoln's first message to Congress, Randall uncovered only part of the deeper meaning of the document. What is also apparent in it is the work of a fledgling president, uncertain of his legal ground and his proper audience, nervous, and at once too candid and too unforthcoming. This was not the work of a statesman or of a sure politician. Lincoln would learn fast, but he had not mastered the job by July 1861.

The president ended his message without offering "more extended argument

... as an opinion, at some length, will probably be presented by the Attorney General." The arguments in Edward Bates's 26-page document bore little resemblance to Lincoln's. Bates crafted an essentially Jacksonian defense that asserted the competence of each of the three great branches of government to interpret the Constitution independently. It must not have mattered to Lincoln that his attorney general's argument differed from his own. In fact, constitutional arguments did not interest Lincoln much; he would not himself offer another defense of administration policy on civil liberties for almost two years.[33]

Though Bates's opinion was completed on the day after Lincoln's message, it was not sent to Congress until the House of Representatives passed a resolution on July 12 asking for it. The House also requested "a copy of the General Order suspending the writ of habeas corpus." No such general order existed, of course. Instead, there were three orders and a proclamation. Even Lincoln's cabinet was a little confused about the issue. Secretary of State Seward responded to the House resolution by sending "three general orders upon this subject"—not four. He had apparently forgotten about the strange individual suspension for William Henry Chase.[34]

On October 14, 1861, Lincoln again wrote General Scott, saying that "the military line of the United States for the suppression of the insurrection may be extended so far as Bangor in Maine," and authorizing Scott or "any officer acting under your authority ... to suspend the writ of habeas corpus" anywhere between Bangor and Washington.[35] No obvious specific incidents had occurred to provoke the new order. Troops were not being put on the march to the front from points ever farther north. They were not being blocked by mobs or sabotaged bridges in Massachusetts. "Military line" continued its strange transformation in meaning from a term describing a threatened reinforcement route to the nation's capital to a term according areas far to the rear battlefield status in law. The order differed from previous ones in that the power to suspend was no longer confined to officers at spots where resistance occurred or to officers specially designated by the general-in-chief.

The most important feature of the October 14 proclamation was not its rather loose phrasing or its perhaps threatening content but its authorship. *The Collected Works of Abraham Lincoln* reprints the text as published after the Civil War in the *Official Records,* but the original draft lies in William H. Seward's papers and was written entirely in Seward's hand. Less than six months after his first tentative suspension, Lincoln was delegating the writing of the orders and proclamations suspending the writ of habeas corpus. He wrote so many of his letters and state papers himself that it is a sure sign of indifference for Lincoln to allow such a paper to be ghostwritten.[36]

The Arrest of the Maryland Legislators

While Lincoln continued to stretch the zone of suspended civil liberties up the East Coast, the most important uses of the power thus delegated to military authorities continued to occur in Maryland. Late in the summer of 1861, the president permitted the arrest of several members of the Maryland legislature, though he had previously rejected a proposal to do this. This strange move, five months after the secessionist impulse seemed spent, is difficult to explain.

Documentation is sparse, but the principal testimony that Lincoln may have masterminded the incident came from Frederick Seward, son of the secretary of state and himself assistant secretary of state. In his reminiscences, not published until 1916, Frederick maintained that Lincoln took a carriage ride with him and his father on "a bright summer day in 1861," picked up General George B. McClellan, and then headed to the Georgetown Heights, where, the press assumed, they "had gone up to inspect the camps and fortifications now beginning to cover the hills." The presidential party, in fact, made no inspection stop but proceeded directly to Rockville, Maryland, for a prearranged meeting with General Nathaniel P. Banks, the commander of Union forces in the western district of the state. The secretary of state had already met, "a day or two ago," with the commander in the eastern district, General John A. Dix. In the secret conference that ensued in Rockville, they discussed a meeting of the Maryland legislature supposed to be held on September 17.

The administration believed, according to Frederick Seward, that "a disunion majority" in the Maryland state house would then pass an ordinance to secede. Lincoln had resolved to keep that from happening. Seward recalled:

> As few persons as possible would be informed beforehand. General Dix and General Banks...were instructed to carefully watch the movements of members of the Legislature.... Loyal Union members would not be interfered with.... But disunion members...would be quietly turned back toward their homes, and would not reach Frederick City at all. The views of each disunion member were pretty well known, and generally rather loudly proclaimed. So there would be little difficulty, as Mr. Lincoln remarked, in "separating the sheep from the goats."
> ...When the time arrived...it was found that not only was no secession ordinance likely to be adopted, but that there seemed to be no Secessionists to present one. The two generals had carried out their instructions faithfully, and with tact and discretion.... No ordinance was adopted, Baltimore remained quiet, and Maryland stayed in the Union.

This vague anecdote, written over fifty years after the event by a man who obviously did not keep a diary, is about all that historians have.

Frederick Seward's only indication of the date of the alleged meeting was that it occurred in August. The date may well have been August 23, when the president made a trip reported by the press as an inspection tour of the Georgetown Heights.[37]

The anecdote suffers from other problems. One of them, pointed out by historian Dean Sprague, was that "If Frederick Seward remembered this conversation correctly, the decision to turn back disunion members to their homes was changed radically during the next few days." In fact, the legislators were arrested, and several persons were also arrested who were not members of the Maryland legislature. Another problem is that Secretary of War Simon

Cameron, the principal initiator of the action, was not, according to Frederick, privy to the conference.[38]

No one else involved in the alleged secret meeting left any record of it, but General McClellan's reminiscences throw a little more light on the problem:

> Information from various sources received in Aug. and Sept., 1861, convinced the government that there was serious danger of the secession of Maryland.
>
> The secessionists possessed about two-thirds of each branch of the State legislature, and the general government had what it regarded as good reasons for believing that a secret, extra, and illegal session of the legislature was about to be convened at Frederick [City] on the 17th of Sept. in order to pass an ordinance of secession. It was understood that this action was to be supported by an advance of the Southern army across the Potomac.... It was impossible to permit the secession of Maryland, intervening, as it did, between the capital and the loyal States, and commanding all our lines of supply and reinforcement. I do not know how the government obtained the information on which they reached their conclusions. I do not know how reliable it was. I only know that at the time it seemed more than probable, and that ordinary prudence required that it should be regarded as certain. So that when I received the orders for the arrest of the most active members of the legislature, for the purpose of preventing the intended meeting and the passage of the act of secession, I gave that order a most full and hearty support as a measure of undoubted military necessity.
>
> On the 10th of Sept. Hon. Simon Cameron, Secretary of War, instructed Gen. Banks to prevent the passage of any act of secession by the Maryland legislature, directing him to arrest all or any number of the members, if necessary, but in any event to do the work effectively.
>
> On the same day the Secretary of War instructed Gen. Dix to arrest six conspicuous and active secessionists of Baltimore, three of whom were members of the legislature.
>
> ...On the 10th of Sept. Gen. Dix sent to Secretary Seward and myself marked lists of the legislature. In his letters he strongly approved of the intended arrests, and advised that those arrested should be sent to New York harbor by a special steamer.
>
> The total number of arrests made was about sixteen, and the result was the thorough upsetting of whatever plans the secessionists of Maryland may have entertained. It is needless to say that the arrested parties were ultimately released, and were kindly treated while imprisoned. Their arrest was a military necessity, and they had no cause of complaint. In fact, they might with justice have received much more severe treatment than they did.

McClellan's account added some precision but was also significant for what it did not say. McClellan did not contend that Lincoln ordered the arrests or that there was a secret meeting with the president to plan them; in fact, he did not mention Lincoln's name. It is also noteworthy that such an account came from a leading Democrat, like McClellan, and that he twice justified the arrests as a "military necessity." Thus, the first arrests of leaders of the opposition political party, en route to perform their labors of governance, had bipartisan support. Dix, too, was a Democrat, often chosen by the Lincoln administration for delicate assignments likely to involve the army in conflicts with members of the Democratic party.[39]

Modern accounts generally agree that the secretary of war and McClellan ordered the arrests. Probably most would now disagree with McClellan's assertion that such action was necessary by this late date to keep the state in the Union. At the time, however, Maryland's Unionist governor, Thomas H. Hicks, told Banks: "We see the good fruit already produced by the arrests. We can no longer mince matters with these desperate people. I concur in all you have done."[40]

McClellan's recollections add one element missing altogether from Frederick Seward's story—the idea that the secessionist movement in the state was to have been coordinated with a Confederate invasion. This would support the notion of the arrests constituting a military necessity and explain their timing as well. But the surviving evidence suggests no particular sense of urgency on the part of the federal authorities in Maryland, even among some of those directly involved in the arrests.

Reports that all was quiet on the Maryland front late in the summer discredit McClellan's account. On August 22, for example, Banks reported to Lincoln from his headquarters in Hyattstown, Maryland, "that everything, this side of the Potomac is quiet—promising no serious movement immediately." Reports of Confederate troop strength in Loudoun County, across the river in Virginia, put their number at 5,500 infantry, 600 cavalry, and 200 artillery with 4 cannons. These numbers had not increased in weeks, and the North still wielded a superior force of 9,000 men to guard the river crossings. "In our circle," General Banks went on, "the senior officers do not think it probable the rebels will attempt a passage, but threaten it for various reasons." He added: "There are many intelligent Virginians who entertain the same opinion among our friends—on the other hand, disloyal Virginians declare it to be the purpose of the rebel leaders to enter Maryland, and nearly all intercepted letters speak of an invasion with confidence—some specifying the middle of August as the time fixed for the effort."[41]

Banks's remarks, written the day before the secret meeting described by Frederick Seward likely occurred, betray no real alarm in the field over the possibility of a Confederate invasion. Nor had Banks's letter been elicited by a nervous president asking for confirmation of a rumored Confederate invasion into secession-ripe Maryland. In fact, the general's report was a casual postscript to a letter answering Lincoln's query about the disappearance of a man named Smith.

On August 23, an old political crony of Lincoln from Illinois, Ward Hill Lamon, wrote from Williamsport, Maryland, to say that people there were "unnecessarily alarmed" at the rumor of Confederate invasion. Five days after that, Banks again reported all quiet on the Potomac.[42]

Political alarmism is not readily detectable in the documentary record, either. For example, on August 22, Cameron rescinded an earlier order for the suspension of some Baltimore newspapers, and his action was taken at Seward's suggestion.[43]

Military necessity may seem less imperative today than it did to McClellan, but there have always been some who suspected other motives for the arrests.

Partisan motivation was alleged in some quarters during the war. S. Teackle
Wallis, himself one of the victims of the arrests, put forward the theory to
Republican Senator John Sherman as early as 1863:

> ... I have good reasons for believing that the seizure of the members of the Maryland
> Legislature was not made by Mr. Lincoln's direction or authority at all. . . . [T]he
> whole high-handed proceeding was the work of Mr. Seward, of his own mere
> motion, without the previous knowledge or consent of the President, in any shape.
> That Mr. Lincoln was, afterwards, induced to ratify it, of course, makes his re-
> sponsibility the same as if he had been its author, but I am speaking only of the
> facts as I am confident you will find that they existed when the arrests were made.
> What were the purposes and motives of Mr Seward, you can ascertain more readily
> than I. My only knowledge on the subject is derived from an official despatch of
> Lord Lyons to Lord Russell, bearing date November 4, 1861, and published in the
> Parliamentary Blue Book. I found it in the New York *Times,* of March 1, 1862. His
> lordship reports in it the substance of an interview which he had had, a day or
> two before, with the Secretary of State. "Mr. Seward replied," he says " . . . *that as
> to the recent arrests, they had almost all been made in view of the Maryland
> elections."*

The impact of the arrests on later Maryland elections is difficult to determine,
but they were more likely harmful than helpful to the administration's cause
by supplying an issue to the opposition. In some counties, in fact, the opposition
called itself the "habeas-corpus party."[44]

The reason for the administration's alarm and the consequent arrests of the
Maryland legislators may never be known. Lincoln, for his part, wanted to keep
the reason hidden—for a time, anyway. In a statement printed in the Baltimore
American on September 21, 1861, he said flatly.

> The public safety renders it necessary that the grounds of these arrests should at
> present be withheld, but at the proper time they will be made public. Of one thing
> the people of Maryland may rest assured: that no arrest has been made, or will be
> made, not based on substantial and unmistakable complicity with those in armed
> rebellion against the Government of the United States. In no case has an arrest
> been made on mere suspicion, or through personal or partisan animosities, but in
> all cases the Government is in possession of tangible and unmistakable evidence,
> which will, when made public, be satisfactory to every loyal citizen.

The "proper time" for such public revelations never came.[45]

It is possible that Seward instigated the arrests, perhaps on the strength of
detectives' reports of the intentions of some Maryland political leaders. Neither
the victim of the arrests—Wallis—nor Lincoln's political foe—McClellan—
implicated the president in the planning. And for his part, Frederick Seward,
who was hazy on the details, certainly had an interest in laying responsibility
for the plan on his father's superior rather than on his father. If William H.
Seward did cause the arrests of the Maryland legislators, it was an action fully
in keeping with his reputation for ruthless suppression of civil liberties during
the Civil War.

The Record of William H. Seward

Secretary of State Seward was given control of military arrests of civilians from their inception until the War Department assumed control of them in February 1862. The only lengthy assessment of his tenure in this capacity appears in Dean Sprague's 1965 book, *Freedom under Lincoln,* which denounced the authoritarianism of the Lincoln administration but exempted Lincoln himself from ultimate responsibility. The way Sprague did this was at once to vilify Seward and to magnify his power and ability. Except in the case of Seward's special defender, his son Frederick, this is generally what has happened in the literature on military arrests of civilians during the Civil War.[46]

Sprague insisted that the arrests of civilians so agonized Lincoln that by war's end the president would have readily agreed with a Supreme Court decision preventing future presidents from following his own example. Such an assertion hardly seems to fit the course Lincoln tracked in the first year of the war. If he agonized over the initial decision and awkwardly exposed his doubts in the July 4 message to Congress, by autumn he was comfortable enough to turn over the drafting of habeas-corpus proclamations and orders to someone else. In September, Lincoln assented to the preventive arrests of Maryland legislators, something he had forbidden five months earlier. He readily delegated authority in these matters to Seward, and it would be difficult to argue convincingly that Lincoln was much agonized by the results of Seward's work. Nevertheless, Sprague concluded that this early period was the toughest for civil liberties during the war, and that by the time of the famous Vallandigham arrest in 1863, arrests had become "rare."[47]

Seward has thus been made to shoulder the heavy historical responsibility for military arrests of civilians at their worst. Sprague depicted him as willing "to interfere in the most minute affairs of the smallest community in America." Seward "rapidly expanded his authority until he was ordering nearly all the arrests being made in the free states." This fit the secretary of state's image early in the administration's history as "the real leader of the Republican party" and the power behind the Lincoln throne. Seward carried his power "to incredible lengths, seemingly unrestrained by any factor save naked political expediency."[48]

This image of Seward was not an invention of Sprague but was rooted in the Civil War era, when it was vividly embodied in the widely circulated anecdote about "Seward's little bell." The secretary of state allegedly told Lord Lyons, the British minister in Washington, that he could ring a bell on his desk and arrest a citizen anywhere in the United States. Could even the Queen of England do as much? Seward asked. "Whether the story is true or not," Sprague concluded, "there is no doubt that Seward could honestly have said it.... For six months Seward had more arbitrary power over the freedom of individual American citizens all over the country than any other man has ever had, before or since."[49] All modern treatments of the secretary of state's administration of internal security leave a similar impression of aggressiveness and efficiency.[50]

In truth, Seward and the State Department ordered few arrests. They spent much of their time trying to find out why others had arrested the prisoners who appeared mysteriously in military prisons. The most notable feature of the department's records of "Arrests for Disloyalty," which were published by the U.S. government less than a generation after the Civil War, is their documentation of the government's ignorance of the cause of most arrests.

The State Department's own records show that responsibility for initiating arrest lay with some authority outside the department in at least 43 percent of the cases. Instead of a zealous, active, and efficient operation, Seward's department was rather passive, often baffled and incapable of obtaining accurate information about its prisoners. Many of the records read like this one:

> The first information concerning this man [James Chapin] received at the Department of State was contained in a telegram from John Burt, deputy U.S. Marshal in New York, dated Saratoga, September 3, 1861, as follows: "I have arrested James Chapin, a captain of the Vicksburg (Miss.) Home Guards. What shall I do with him?"

Worse yet, the State Department sometimes received its first news of a prisoner from the prisoner's lawyer or distressed relatives and not from any agent of the government. For example: "The only information received at the Department of State relative to [Bushrod W.] Marriott is contained in a letter from William Price, esq., of Baltimore, stating that Marriott was arrested for giving aid and comfort to the rebels, but the charge was unfounded and he therefore urged the release of the prisoner." Some of the State Department's case files contained only the affidavits of concerned citizens and the protests of prisoners' congressmen gathered to bring about the release of these civilians arrested by military authority.[51]

The curiously one-sided nature of the State Department files caught the attention of Seward's turn-of-the-century biographer, Frederic Bancroft. He attempted an ingenious explanation of the department's arrests:

> Because there was no intention to prosecute, no evidence was collected after the arrest was ordered. Unless the evidence happened to be very strong, the *ex parte* pleas, declarations, and complaints in behalf of the prisoner often indicated that Seward had proceeded without sufficient precaution. The department never made up its case, while that of the defendant is often nearly complete. How few convictions in the criminal courts would seem to be just if one knew only the grounds on which the grand jury based the indictment, and then, at the trial, heard only the witnesses and lawyers for the defence. Yet this is a fair illustration of the disproportion shown in the official records.[52]

Bancroft makes a good point, but it seems ultimately a lame comment on what is really a sign of poor record-keeping, shockingly incompetent organization, and perhaps overzealousness on the part of some local authorities.

Surely, even Bancroft's clever defense could make nothing of such a record as this one: "No papers in the Department of State show at what time the order for Doctor [Alexander C.] Robinson's arrest was issued nor when if ever or on what conditions it was revoked. The correspondence in relation to him begins on the 4th and closes on the 26th of October, 1861. It is almost wholly between parties outside of the Department of State, and it does not appear that any order was made in this Department in relation to the case." Nor should Bancroft have excused the many records consisting only of the prisoner's name, the date he was committed to prison, and the following statement: "There are no papers on file in the Department of State showing upon what charge he was arrested."[53] About the best that can be said of this state of affairs is that a government willing to publish such records for all to see was certainly candid and open.

When the State Department itself did attempt to initiate an arrest, Seward's little bell sometimes rang unheard. On October 4, 1861, for example, the department responded to a federal marshal's report of disloyal activity with an order to arrest, but the recording clerk could later note only, "There is no evidence on file in the Department of State showing that the arrest was made." On other occasions, Seward carelessly abdicated his authority. For example, when Major General Dix asked permission to release a prisoner hitherto unknown to the State Department, Seward told him "to exercise his own judgment."[54]

The secretary of state frequently learned about arrests made on others' authority when the British minister intervened on behalf of civilian prisoners claiming to be British subjects. Dozens of British, Irish, or Canadian citizens were arrested, and still more prisoners claimed foreign citizenship in hopes of being released. For such persons, the State Department was the logical place to inquire, but other distressed relatives and lawyers must have been puzzled about whom to approach. Lincoln never issued a public proclamation giving authority over these matters to the State Department. The War and Navy departments also made arrests on their own, and State's authority over civilian prisoners was never certain or clear, nor necessarily effective. Generals made arrests, and state officials ordered them as well. In this maze of fragmented authority, only the influential, resourceful, or lucky person found the proper channel for an appeal.[55]

The State Department never received any records whatsoever of many prisoners held in federal forts remote from Washington, D.C., like those in St. Louis, Missouri, or Alton, Illinois. At least 30 percent of the total number of arrests of civilians before February 15, 1862, are known, not from State Department records but from fragmentary documents from other sources that never reached the State Department while Seward was overseeing arrests of civilians.

Altogether, then, since some 43 percent of the arrests recorded by the State Department itself were instigated by others, and 30 percent of the total arrests in the period up to February 15, 1862, were never recorded by the State Department, authorities outside the State Department caused the arrest of at least 60 percent of all the civilians detained while Seward was responsible for the program. Such a record can scarcely be reconciled with Seward's tyrannical image.

One reason Seward gained this undeserved reputation was that he did *sound* tough. He was a convinced nationalist whose political ideas contained scant sympathy for doubters or dissenters. To the president, Seward must have seemed perfectly suited for the job of crushing individual opposition to the government in the North.

Though Abraham Lincoln has been compared to Bismarck, there were no echoes of blood and iron in Lincoln's rhetoric about the nation. Seward, on the other hand, had argued as early as 1850 that the Union "was not founded by voluntary choice, nor does it exist by voluntary consent." It was, rather, "the creature of necessities, physical, moral, social, and political." He believed that the "Union ... *is*, not because merely that men choose that it shall be, but because some Government must exist here, and no other Government than this can."[56]

When the Union began to split in the Sumter crisis, Seward's sincere na-tionalism led him to exhaust every avenue of compromise to keep the country together. He proved willing to deal with Southern slaveholders or to provoke a foreign incident that might bring the sections back together for mutual self-defense. After the Civil War began, Seward took little interest in the divisive slavery issue. War itself sealed slavery's doom in his mind, and he looked to the more serious problem of saving the Union. Before the war, he had pro-claimed a "higher law than the Constitution" in denouncing slavery, referring to divinely endowed individual liberties, but as secretary of state during the war, Seward usually saw the highest good embodied in the nation-state.[57]

Seward's willingness to offer the South concessions in the Sumter crisis by no means implied softness on the issue of union. On the contrary, it was rooted in his overconfident faith in the American masses' love of the Union. He as-sumed that by stalling for time in the crisis, he could give these latent senti-ments—these mystic chords of memory—the chance they needed to reassert themselves. Union, he told Lincoln and the rest of the cabinet, was "inestimable and even indispensable to the welfare and happiness of the whole country, and to the best interests of mankind." If the secession crisis tested the question of "Union or Disunion" rather than "slavery, or not slavery," he was confident the crisis would pass.

When Lincoln polled the cabinet about Fort Sumter on March 15, 1861, Seward warned, in a bid for more time, that coercing the seceded states would make "reunion ... hopeless, at least under this administration, or in any other way than by popular disavowal, both of the [resulting] war and of the admin-istration which commenced it." As usual, his syntax was complicated, but the statement revealed Seward's deep fear that war with the South might arouse tremendous opposition to the Lincoln administration in the North. "If this administration ... take up the sword," Seward warned, "then an opposition party will offer the olive branch and will, as it ought, profit by the restoration of peace and Union." In other words, Seward feared that war might arouse so much internal opposition in the North that the Lincoln administration could never win it.[58]

Here was a mentality primed to believe in fifth-column threats. The secretary of state feared dissent and, unlike Lincoln, had long since embraced, along with antislavery sentiments, a tough nationalistic philosophy that prepared him to crush slavery and its Northern sympathizers. Before the war, it is true, he had shown special interest in the writ of habeas corpus. As U.S. senator from New York, he had proposed legislation for the Compromise of 1850 that would have guaranteed habeas-corpus rights to suspected fugitives, and Seward's famous "higher-law" speech was a critique of the compromise that actually became law. But his regard for habeas corpus changed with the advent of civil war. "Let us save the country," Seward told his old friend S. G. Andrews during the war, "and then cast ourselves upon the judgment of the people, if we have in any case, acted without legal authority. The *habeas corpus* will be suspended anywhere, on its being shown that it is necessary to prevent disorganization or demoralization of the national forces."[59]

Yet the statistical record proves that Seward by no means crushed dissent. From a study of the *Official Records,* from the *Democratic Almanac for 1867,* from the *American Annual Cyclopaedia,* from Seward's unpublished papers, and from published reminiscences of political prisoners, one can ascertain that the secretary of state presided over the arrest of only 864 civilians. Most modern estimates of the total number of civilians arrested by federal military authority during the entire Civil War have put the number at over thirteen thousand, and the figure for Seward's period of control certainly looks modest by comparison. Whatever the accuracy of the traditional estimate of some thirteen thousand arrests of civilians for the whole war, one thing seems certain: William H. Seward did not administer the internal security system in its harshest period. That would come later under the supervision of the War Department.[60]

Alarmed though he was at the menace posed by opposition to the war in the North, Seward did little to counteract it. No libertarian ideology inhibited him. He simply lacked the time. Seward's chief responsibility was the conduct of American foreign policy, a task that Abraham Lincoln, without any experience in foreign affairs, willingly handed over to his able secretary of state. Always busy, Seward was especially preoccupied after early November 1861 when he struggled to cope with the worst foreign-policy crisis of the Civil War, the *Trent* affair.

Two other factors impeded Seward's ability to cause the arrest of civilians. First, he lacked any efficient apparatus to manage arrests. He inherited none because the State Department is not an enforcement division of the government. Seward had to borrow arresting authorities and investigative agents from other administrative areas or quickly hire and train them. He employed policemen and marshals as well as soldiers in the work. Second, he was responsible for internal security when few disloyal persons were within reach of Northern military authorities—because so little Southern territory had as yet fallen under United States control.

Seward's reputation stemmed more from what he said than from what he did. Though willing enough to crush dissent, he proved quite unable to do so.

Sailors and Foreigners

Of the 509 cases among the total of 864 before February 15, 1862, in which residence of the prisoner is ascertainable, 159 or 31.2 percent came from states that seceded from the Union. War's outbreak naturally stranded some Southerners in the North, and sailors proved particularly likely to land in the wrong place at the wrong time. Among the Southern sailors in Seward's forts were what might be called the "Southern honor" arrests. When Southern-born crewmen in the United States Navy who happened to be on cruise in April 1861 returned to port, in some instances after months at sea, they landed on hostile soil. Common sailors had no choice in the matter: if they followed their Southern inclinations, they had to desert. Some officers, on the other hand, were allowed to resign their commissions to go south. Other Southern-born naval officers sailed into port, submitted their resignations, and were promptly arrested and thrown into military prisons. A Maryland inmate at Fort Lafayette commented on the outraged honor of these unhappy men: "To me, it appears one of the most barbarous and disgraceful features of this war, to take Southern gentlemen, who have made it a point of honor to bring their ships into Northern ports, and resign their commissions, when they might have with perfect ease, taken them into Southern ports; and immediately on their arrival send them as prisoners to the Forts."[61] At least twenty of the prisoners who appear in the arrest records of the Seward period were sailors of Southern sentiments attempting to leave United States service.

Abraham Lincoln, who in his younger years had proposed absurd terms when challenged to a duel by a feisty Irish-American Democratic politician, possessed a less exaggerated sense of honor than the Southern sailors. He was more a calculating and practical realist, especially where winning the war was concerned, and he wished that the administration had made more such arrests when it had the chance. On one occasion he wrote:

> Gen. John C. Breckienridge, Gen. Robert E. Lee, Gen. Joseph E. Johnston, Gen. John B. Magruder, Gen. William B. Preston, Gen. Simon B. Buckner, and Comodore [Franklin] Buchanan, now occupying the very highest places in the rebel war service, were all within the power of the government since the rebellion began, and were nearly as well known to be traitors then as now. Unquestionably if we had seized them and held them, the insurgent cause would be much weaker. But no one of them had then committed any crime defined in the law. Every one of them if arrested would have been discharged on Habeas Corpus, were the writ allowed to operate.

Most of the Southern sailors detained when they arrived in the North in 1861 were exchanged later in the winter. Like Franklin Buchanan, who eventually commanded the ironclad *Merrimac,* some went on to distinguished careers in the Confederate navy.[62]

Other dislocated sailors presented pathetically innocent cases. William Williams, a twenty-two-year-old common sailor from Liverpool, England, and William Sims, an experienced old salt, forty-nine years old, from Chichester, put to sea on an American bark called the *Susan G. Owen* in the spring of 1861. Her destination was Charleston, South Carolina, where she docked in late April, after the war had begun. The sailors were paid but had no employment and no way home until hired for another voyage. Now they would have to sail through the blockade to return. Late in August, Sims and Williams shipped out in the *H. Middleton,* the first ship they could find heading for Liverpool from Charleston. The U.S. ship *Vandalia* captured the *H. Middleton* shortly after she departed Charleston harbor, on September 7, 1861. The captured crew was imprisoned in Fort Lafayette. The British consul learned of his countrymen's plight and contacted Lord Lyons, the British minister in Washington, who, in turn, made inquiries of the American secretary of state. Seward ordered the British sailors released on September 23.[63]

Although Williams and Sims became innocent victims of world circumstance merely by plying their old and honorable trade, such was hardly the case with most British sailors who attempted to run the Union naval blockade. They knew full well what risks they were taking. Of the 864 arrests made under Seward's administration, reasons for arrest are ascertainable in 638 cases, and of those, 82, or 12.9 percent were civilian blockade-runners.

Most blockade-running men were British, for the South had a weak seafaring tradition and almost no ships, and English mills needed cotton. However, nationality was not consistently noted in the State Department's arrest records, and it is impossible to tell exactly how many blockade-runners were British. Release came quickly for most of these foreign neutrals because international law protected them. Besides, the Lincoln administration wanted to avoid international incidents while its hands were full fighting the Confederacy. An inmate at Fort Lafayette noted enviously in his diary when a new prisoner of this sort arrived, "I suppose as he is a 'British subject' he will be released as soon as the British Consul hears of his imprisonment; lucky thing now-a-days to have been born in England, or anywhere outside the 'Land of the Free and Home of the Brave!'"[64]

Balancing delicate foreign-policy considerations against the potential harm this illicit trade could cause the Union militarily proved difficult. Simply sorting out the relevant facts of each case posed problems. Masters sometimes deceived common sailors as to cargoes or cruise destinations. Non-English-speaking sailors, desperate for work, could easily have been kept in the dark about the political dimensions of a voyage. On October 10, 1861, for example, the U.S. marshal in New York City, Robert Murray, reported to Seward that he had gone to Fort Lafayette to administer the oath of allegiance to several pitiful sailors so that they could be released: "Edward Heinrichs, a Prussian; Erick Brundeen, John Johnson, William Brown, Swedes[;] and George Parker, an Englishman, being foreigners and the first four entirely ignorant of our language I discharged without administering the oath of allegiance, and was obliged from their utter state of destitution to furnish them with sufficient funds to reach the city."[65] Eventually, authorities improvised an oath of neutrality for foreign nationals, who could not take the oath of allegiance to the U.S., but they would never be able to cope adequately with foreign-speaking prisoners.

Sailors were not the only foreign-speaking prisoners, and altogether such minor "international incidents" constituted 11.2 percent of the cases of persons arrested before February 15, 1862. Of these foreign prisoners 76.8 percent were British.[66] Foreigners, especially those not fluent in English, were likely to make innocent mistakes concerning the laws and regulations of American society at war. But since their loyalties were unknown and some, like the blockade-runners, were definitely engaged in activities harmful to the Union cause, the authorities necessarily arrested substantial numbers of them. Most stood an excellent chance of having their cases appealed by the official representatives of foreign governments in Washington and were likely to be released quickly thereafter.

The Nature of Military Arrests of Civilians Before February 1862

Arrests under Seward may not have been as numerous as in later periods, but some of the overall characteristics of Civil War arrests already emerge from the statistics. Many of the 864 arrested were residents of the Confederacy—almost a third (having been trapped in the North at the beginning of the war or arrested in the few areas of the South controlled by the Union army in this early period). Border states provided many prisoners. The largest number of prisoners from any single state in this early period hailed from Maryland: 166 of the 509 cases of known residence, or a substantial 32.6 percent. The border states as a whole supplied 42.8 percent of the prisoners, a figure that grossly underestimates Missourians because place of residence was for some unknown reason not noted in the lists from St. Louis and Alton, Illinois. Probably more than half the prisoners came from the border states. (Their portion can be put as high as 47.1 percent by including prisoners from the District of Columbia, even with Missouri still grossly underrepresented.[67])

Residents of the border states, the District of Columbia, and the Confederacy accounted for most of the 864 arrests under Seward, 78.3 percent. Adding residents of foreign countries will account for 85.6 percent of the persons arrested. Of those arrested above the border states and the District of Columbia, probably no more than 125 were Northerners. Of that estimated number, the records offer firm evidence of only 73 people of Northern residence. The 73 arrests betray no particular geographical pattern. If the estimated 125 people were distributed evenly among the states north of the border states, there would have been at the maximum 7 Northerners arrested in each state in the period. Since Seward was in charge of internal security for ten months, the rate of arrest in the North above the border states and the District of Columbia was less than one person per state per month.[68]

Despite his suspicion of the political opposition and his fear of resistance to the war effort, Seward arrested few of those whose cases would become embarrassingly memorable as symbols of stifled dissent and crushed civil liberty. One famous incident, and perhaps the only one attributable to personal political malice on Seward's part, involved ex-President Franklin Pierce, a sour critic of the administration and a Democrat—but nothing more seditious. His loyalty was challenged on the strength of a hoax perpetrated by a doctor in Michigan who set out to prove how gullible federal authorities could be. Seward and the federal authorities rose to the bait and were on the brink of arresting Pierce, but after an exchange of letters, the administration left the former president alone.[69]

In another curious case, Seward, on the strength of allegations in an unsigned letter written to General Winfield Scott, placed under military surveillance Assistant Judge William M. Merrick of the United States Circuit Court in the District of Columbia. Seward's terse order to Provost Marshal Andrew Porter directed him "to establish a strict military guard over the residence of William M. Merrick." Porter apparently interpreted this to mean something more than surveillance and asked Seward, "Is it desired by the honorable Secretary that the judge should be confined to his house?" Seward replied that such confinement was not expected and that "it may be sufficient to make him understand that . . . when the public enemy is as it were at the gates of the capital the public safety is deemed to require that his correspondence and proceedings should be observed." The same day that he wrote Porter, October 21, 1861, the secretary of state also sent the first comptroller of the treasury this astonishing letter: "I am instructed by the President to direct that until further orders no more moneys be paid from the Treasury . . . on account of the salary of William M. Merrick." Merrick was never arrested, but Seward's order violated the Constitution's stipulation that judges' compensation "shall not be diminished during their continuance in office."[70]

A few cases in this period involved freedom of speech and freedom of the press. "Treasonable language," "Southern sympathizer," "secessionist," and "disloyalty" were standard notations next to prisoners' names on the State Department record books. Even more serious-sounding terms were vague and sometimes denoted offensive words rather than deeds: "aiding and abetting the enemy," "threatening Unionists," or "inducing desertion," for example. A man in Cincinnati was arrested for selling envelopes and stationery with Confederate mottoes printed on them. The display of such items or of sheet music with portraits of Confederate generals on the covers was forbidden by General John A. Dix in Baltimore, though one might display such items in shop windows in New York without fear of arrest. On the other hand, printing blank bills to be used by the Confederate government was more serious; the offense led to the arrest of B. F. Corlies of Brooklyn. The Reverend J. R. Stewart of Alexandria, Virginia, was arrested for omitting a prayer for the president of the United States in a church service. Daniel Cory was arrested in Somerset County, New Jersey, for saying "he would like to put a bullet through" Lincoln, whom he denounced as a Tory and a traitor. Moses Stannard of Madison, Connecticut, became a prisoner for raising a secession flag over his house and saying that he hoped the Confederates would capture Washington and kill the president and his cabinet. Military authorities arrested a candidate for the Maryland legislature, Perry Davis, for making "treasonous" speeches while canvassing for office, but General Joseph Hooker let Davis go when he told the general that he would not vote for a secession ordinance. George J. Jones was arrested in St. Louis for publishing handbills with "treasonous" articles reprinted from British journals.

The prisoners identifiable as newspapermen were Daniel Deckart, publisher of the Hagerstown (Maryland) *Mail;* Reuben T. Durrett, acting editor of the Louisville *Courier;* Francis D. and Joseph R. Flanders, owners and editors of the Franklin *Gazette,* of Malone, Franklin County, New York; F. Key Howard and W. W. Glenn, editors of the Baltimore *Exchange;* Thomas W. Hall, Jr., and Thomas H. Piggott, editors, and Samuel Sands Mills, publisher and proprietor, of the *South,* published in Baltimore; James A. McMaster, editor of New York's *Freeman's Appeal;* Henry A. Reeves, editor of the *Republican Watchman,* published on Long Island; James W. Wall, a contributor of columns to the New York *News;* and several Missouri editors.[71]

Such arrests, however, were exceptional. (Most cases of this sort have already been covered in Robert Harper's *Lincoln and the Press* and therefore need not be discussed at length. Their frequency and significance, of course, will be duly noted.) Lincoln maintained a consistent attitude toward such arrests throughout the war. It was perhaps typified by his response in 1863 to the arrest in St. Louis of newspaper editor William McKee. "I regret to learn of the arrest of the Democrat editor," Lincoln wrote General John M. Schofield. "I fear this loses you the middle position I desired you to occupy."[72] He often learned about such cases only after the fact, if at all, and he usually regretted their occurrence. In general, he suspected that they needlessly polarized opposition to the war by attempting to suppress language from the legitimate party opposition that worked no palpable harm on the U.S. military. "Please spare me the trouble this is likely to bring," Lincoln implored General Schofield. Yet, though critical of individual cases in private and fully aware of their potential to harm the administration politically, he publicly defended a policy that permitted suppressing disloyal papers.

Most arrests had little or nothing to do with the issue of dissent or free speech, as the large numbers of Confederate citizens and blockade-runners among the prisoners suggest. Arrests often dealt with genuinely complicated problems, and most netted prisoners whose liberties most Northerners could not have cared less about—be they Democrats or Republicans. Among the cases that tested the boundaries of international law, for example, were those of James M. Mason, John Slidell, and their private secretaries. After being removed from the British ship *Trent,* these official Confederate emissaries to Europe were held in the forts as civilian prisoners. Whatever the legal merits of their arrests, it can safely be assumed that few loyal Northerners or civil libertarians were much concerned about their rights.

The Seward regime thus was not without its signal successes in arresting dangerous traitors. Among the sinister characters was Rose O'Neal Greenhow, a Washington socialite and Confederate spy who sent General P. G. T. Beauregard the Union plans for the first campaign of the war in Virginia. Thomas A. Jones, too, was arrested on October 4, 1861, without charge and eventually released over the protest of military authorities. He was, in fact, the head of the clandestine Confederate "mail" system in southern Maryland, as he later admitted in an autobiography also describing his role in aiding the escape of John Wilkes Booth after the assassination of President Lincoln.[73]

Few prisoners were as menacing as Jones and Greenhow, but less sinister cases posed tricky problems for harried Union officials. The proslavery Maryland prisoners in Fort Warren, in Boston harbor, took particular glee in describing (in their own mocking proslavery idiom) the cases of Riddick Brooks and William Robinson, the black personal servants of two Confederate army officers captured at Hatteras Inlet in North Carolina. The black men were civilians and wanted to return to their homes in the Confederacy. The federal authorities were willing to accommodate their wishes but required them, like all such prisoners, to take the oath of allegiance before release. One flatly refused, insisting that he was "a secesh nigger." The other asked if his white master had taken the oath. When told that he had not, the black man replied: "I can't take no oaf dat massa George won't take."[74]

Conclusion

In these early months of the Civil War, the Lincoln administration overcame its fears of public reaction to restrictions on civil liberties, instituted a novel internal security system, and came to believe that it worked. Not every historian today would credit it with saving Maryland for the Union, but that conclusion became almost a truism in Lincoln's day. Nathaniel Banks, who commanded the Department of Annapolis in 1861, was a poor general but an astute politician, and he thought the system worked. Indeed, Maryland provided Banks with a model for reconstruction in Louisiana later:

> The secession leaders—the enemies of the people—were replaced and loyal men assigned to ... their duties. This made Maryland a loyal State.... What happened there will occur in North Carolina, in South Carolina, in Georgia, in Alabama and Mississippi. If ... those States shall be controlled by men that are loyal ... we shall then have loyal populations and loyal governments.[75]

The success of the Maryland policy became a political byword and was celebrated, beyond the borders of Maryland, throughout the war. Thus in 1863, a Loyal Publication Society pamphlet on the *War Power of the President* explained the necessity of military arrests rather than reliance on the courts by pointing to that familiar example:

> When the traitors of the loyal state of Maryland were concocting their grand scheme to hurl the organized power of that state against the government, probably not a man of them was known to be guilty of any act for which he could ever have been arrested by civil process. And whatever their offenses against the laws might have been, and whatever the fidelity of the courts in that jurisdiction, the process of civil law would have been far too slow to prevent the consummation of the gigantic treason which would have added another state to the rebellion.... Courts could not have suppressed this unholy work, but the summary imprisonment of those few men saved the state of Maryland to the Union cause.

Republicans would later enjoy substantial bipartisan agreement on the necessity of the early arrests in Maryland.[76]

William H. Seward thought they worked, too. When an old associate of Seward came to Washington to plead for the release of a political prisoner from Kentucky held in Fort Lafayette, the secretary of state readily admitted that no charges were on file against the prisoner. When asked whether he intended to keep citizens imprisoned against whom no charge had been made, Seward apparently answered: "I don't care a d——n whether they are guilty or innocent. I saved Maryland by similar arrests, and so I mean to hold Kentucky."[77]

The earliest days of the Lincoln administration taught the president and his cabinet lessons they never forgot. In fact, these days left fiercely indelible marks on them. This was especially true of Seward. In 1864, when the artist Francis B. Carpenter unveiled his huge historical canvas commemorating the first reading of the Emancipation Proclamation to the cabinet, the secretary of state scoffed at it. He told the artist, at a party given at Gideon Welles's residence, that he had been wrong to choose emancipation as "the great feature of the Administration."

> Seward told him [Welles recalled] to go back to the firing on Sumter, or to a much more exciting one than even that, the Sunday following the Baltimore massacre, when the Cabinet assembled or gathered in the Navy Department and, with the vast responsibility that was thrown upon them, met the emergency and its awful consequences, put in force the war power of the government, and issued papers and did acts that might have brought them all to the scaffold.

The first suspension of the writ of habeas corpus occurred the very week after that fateful Sunday cabinet meeting. Gideon Welles, the secretary of the navy, did not care for Seward, but he remembered those days just as the secretary of state did:

> Few, comparatively, know or can appreciate the actual condition of things and state of feeling of the members of the Administration in those days. Nearly sixty

years of peace had unfitted us for any war, but the most terrible of all wars, a civil one, was upon us, and it had to be met. Congress had adjourned without making any provision for the storm, though aware it was at hand and soon to burst upon the country. A new Administration, scarcely acquainted with each other, and differing essentially in the past, was compelled to act, promptly and decisively.[78]

And act they did.

2

Missouri and Martial Law

Lincoln and his cabinet proved less vigorous in dealing with the West. The president suspended the writ of habeas corpus in places as remote from the Confederacy as Massachusetts six weeks before he made a similar move in the critical border state of Missouri. Yet Missouri surely merited closer attention. Unlike New England or any other state above Maryland, it was the scene of widespread popular revolt, guerrilla violence, and military campaigns from the beginning of the war to its end. But Lincoln, preoccupied with securing the nation's capital in the spring and with battles in Virginia in the summer, at first paid too little attention to the West. He failed to act with sufficient decisiveness to meet Missouri's extraordinary problems, and civil liberties in that state were severely restricted by local military commanders for months before the president did anything. Generals in the Western Department, like French and Brown in Florida, acted on their own.

Martial Law Declared

On May 10, 1861, Captain Nathaniel P. Lyon took as prisoners a large body of Missouri militiamen at Camp Jackson who were allegedly poised to capture the St. Louis arsenal for the Confederacy. Later, Judge Samuel Treat, of the United States District Court, Eastern District of Missouri, issued a writ of habeas corpus for the release of one of the prisoners, Captain Emmett McDonald. General William S. Harney, the area's Union commander, responded to the writ on May 15, saying that despite a sincere desire "to sustain the Constitution and laws of the United States and of the State of Missouri, . . . I must take to what I am compelled to regard as the higher law, [and] even by so doing my conduct shall have the appearance of coming in conflict with the forms of law." This invocation of a higher law than the Constitution would surely have displeased the Republican administration in Washington, had it taken any notice at all. The country at large also paid little attention to this western forerunner of the famous *Merryman* case. For example, the sharply anti-Republican Columbus, Ohio, *Crisis* picked up the McDonald story from the St. Louis *Journal* early in June, but the Douglas Democratic *Illinois State Register* in Springfield, ignored the case.[1]

Missouri quickly became the scene of military operations, not merely an avenue of approach to the South like Maryland nor a sullen occupied area. And after the Confederacy's attempts to help the state failed in the spring of 1862, armed conflict between irregular troops and Unionist forces continued. Erosion of civilian liberties was inevitable under such conditions, whether the writ of habeas corpus was suspended or not. The actions of Ulysses S. Grant, who began his Civil War military career in Missouri, prove the point. Though a West Point graduate and a veteran of the Mexican War, Grant, like most other professional soldiers and sailors, knew little of international law or the so-called laws of war. The military academies did not teach such subjects.[2] At this early point in the war, Grant was only an obscure brigadier, but by virtue of being in Missouri, he was forced to act almost daily on problems with profound implications for the laws of war as well as for the United States Constitution.

On August 25, 1861, for example, he was commanding troops at Jefferson City, Missouri, and sent a captain on a mission with the following orders:

> You will march your men through the country in an orderly manner. Allow no indiscriminate plundering—but everything taken must be by your direction, by persons detailed for the particular purpose, keeping an account of what taken, from whom, its value, etc. Arrests will not be made except for good reasons. A few leading and prominent secessionists may be carried along, however, as hostages, and released before arriving here. Property which you may know to have been used for the purpose of aiding the Rebel cause will be taken whether you require it or not. What you require for the subsistence of your men and horses must be furnished by people of secession sentiment, and accounted for as stated above. No receipts are to be given unless you find it necessary to get supplies from friends.[3]

The next day, Grant sent another officer to see "E. B. McPherson, a true Union man, who will show you a copy of the 'Booneville Patriot.' Bring all the printing material, type &c with you. Arrest J. L. Stevens and bring him with you, and some copies of the paper he edits."[4]

Grant was authorizing the arrest of civilians and the taking of civilian hostages and permitting a mere captain to choose the victims by judging their political loyalty on the hearsay evidence of local Unionists. He was also authorizing the confiscation of a newspaper's press and the arrest of its editor. Grant's superior, John C. Frémont, the commander of the Western Department, had set the tone for his subordinates by closing the *State Journal* in St. Louis in July.[5] All of these things Grant did without orders from above and without the suspension of the writ of habeas corpus in the state. It troubled him a little to invoke such doubtful authority. On August 12, 1861, when he interfered with the delivery of the U.S. mail to persons suspected of disloyalty, he informed his superiors in St. Louis, "I am entirely without orders for my guidance in matters like the above, and without recent Acts of Congress which bear upon them." In October, while commanding the District of Southeast Missouri, he ordered the seizure of goods from boats on the local rivers and reported to headquarters, "I have my serious doubts whether there is any law authorizing this seizure but feel no doubt about the propriety of breaking up the trade now carried on. I respectfully refer this matter to the General Commanding Western Department for instructions."[6]

From headquarters in St. Louis, John C. Frémont tried to respond promptly to such situations. He declared martial law in St. Louis County on August 14 and then throughout Missouri on August 30, 1861. Like Grant, Frémont lacked guidance from his superiors, and he went too far for the War Department or the president when, in his August 30 proclamation, he also attempted to free the slaves of Missouri rebels and threatened to execute Missourians found in arms against the United States.[7]

Lincoln's response to Frémont's proclamation is famous, of course, but more praised than fully understood. The president revoked the emancipation provision on the grounds that it might otherwise scare slaveholding Kentucky out of the Union, and this has, among modern historians, helped Lincoln's reputation. James G. Randall, for example, concluded that Lincoln generally showed himself a man of "shrewdness and tact" with deep "understanding of border-state sentiment." Frémont, on the other hand, "alienated border sentiment, seized functions that belonged to civilian chiefs at Washington, . . . challenged the President's leadership, . . . and precipitated one of those military-and-civil clashes which are always troublesome in a democracy." He proved himself "combative," and his "sensational and unauthorized" proclamation gained support mainly from "that unctuous and impetuous abolitionism which flowed increasingly in anti-Lincoln channels."[8]

In all fairness, it must be said that General Frémont's eagerness to assume responsibility was prompted in part by the failure of Washington authorities to give guidance to its western commanders. Lincoln's reaction to Frémont's proclamation, then, was as much proof of the president's failure to handle this troublesome border state as of any shrewdness or special feeling for these slave societies caught between the warring sections.

Lincoln's objections to parts of the proclamation other than the emancipation provision have been somewhat neglected. He told Frémont, "Should you shoot a man, according to the proclamation, the Confederates would very certainly shoot our best man in their hands in retaliation; and so, man for man, indefinitely. It is therefore my order that you allow no man to be shot, under the proclamation, without first having my approbation or consent." Lincoln did not mention law or the Constitution in this regard. His objections were on entirely practical grounds. Moreover, what Lincoln tacitly permitted Frémont to do was almost as remarkable as what he disallowed. Lincoln did not question Frémont's imposition of martial law. He did not object in principle to the execution of civilian prisoners by the military in a loyal state; he insisted only on his prerogative to review the cases first.[9]

Antislavery advocates criticized Lincoln's revocation of Frémont's proclamation, but so did many other Northerners, among them, Orville Hickman Browning, a conservative Republican and an old friend of Lincoln's. He accused the president of inconsistency in allowing Frémont to execute rebel citizens while disallowing him to free their slaves. Lincoln denied any inconsistency "because I did not also forbid Gen. Frémont to shoot men under the proclamation. I understand that to be within military law." And the president continued to emphasize the practical difficulties: "I also think . . . that it is impolitic in this, that our adversaries have the power, and will certainly exercise it, to shoot as many of our men as we shoot of theirs."

Historians have overlooked a development in this affair that dangerously broadened military authority over civilians in the North. Frémont's proclamation stated clearly: "All persons who shall be taken with arms in their hands within these lines shall be tried by court-martial, and if found guilty will be shot." The novel idea of trying civilians by courts-martial in a loyal state of the Union brought no protest from Lincoln. In September, the very month Lincoln ordered revocation of Frémont's emancipation proclamation, military commissions, essentially courts-martial for civilian defendants, began to try Missouri citizens. From Missouri, they would spread across the nation.[10]

The use of military commissions during the Civil War would prove extremely damaging to Abraham Lincoln's historical reputation. Twentieth-century historians write as though there were some clear legal distinction between the condition obtained by suspending the writ of habeas corpus and that brought about by imposing martial law. The former should permit only imprisonment without charge; the latter would also provide trials by military commission for final disposition of the prisoner. This distinction emerged from the 1866 decision of the United States Supreme Court in *Ex parte Milligan* which declared unconstitutional any military trials of civilians where civil courts were still able to function. At the same time, the Supreme Court did not question the suspension of the writ of habeas corpus during the war or the resulting arrests of civilians without charge.[11]

However clear the Supreme Court may have been by 1866, such distinctions were by no means clear in 1861. When the war began, the Bates-Coffey digest had explained to the president that the Articles of War (and, with them, courts-martial) applied only to members of the United States armed forces. What happened when the writ of habeas corpus was suspended or martial law declared simply was not clear. Years earlier, the U.S. Army had devised a form of court (called a military commission) to try civilians, but Frémont seemed unaware of it, for he used the term "court martial" in his proclamation. Lincoln did not correct the general's terminology or comment in any way on his applying the Articles of War to civilians, even though the Bates-Coffey digest had informed him this was not allowed. In fact, in his letter to Browning of

September 22, 1861, Lincoln himself had used the term "military law" instead of the correct term, "martial law."

The *Milligan* decision five years later essentially created this distinction between suspension of the writ of habeas corpus and martial law. Before that, trials by military commission had come into use during the Civil War without any particular notice by the officials of the Lincoln administration. It was almost a year after Frémont instituted them that a War Department order first dictated their use. Washington officials often used the terms "suspension of the writ of habeas corpus" and "declaration of martial law" interchangeably.

Lincoln may never have grasped such legal distinctions, but his feel for the practical was unerring. Even as the president prepared his letter warning Frémont against executions of civilians that might lead to a sordid war of atrocity and retaliation, Confederate General M. Jeff Thompson of Missouri threatened to take the very actions against which Lincoln warned:

> Whereas, Maj. Gen. John C. Frémont... has seen fit to declare martial law through-out the whole State and has threatened to shoot any citizen soldier found in arms within certain limits..., I... solemnly promise that for every member of the Missouri State Guard or soldier of our allies the armies of the Confederate States who shall be put to death in pursuance of said order... I will hang, draw and quarter a minion of... Abraham Lincoln.... I intend to exceed General Frémont in his excesses.[12]

Frémont's star sank quickly after his brash proclamation, and new reports of corruption and incompetence led to his replacement on November 18, 1861, by Henry W. Halleck. Halleck was among the few Union generals with extensive knowledge of international law, having written a textbook on the subject published shortly before the war began. This made him somewhat more careful about legal niceties than other generals. Besides, he needed to distinguish himself quickly from Frémont, who had too boldly assumed authority. Instead of issuing any emancipation proclamation, Halleck disallowed escaped slaves in his lines. And rather than continue Frémont's exercise of martial law, Halleck *asked* for permission to impose it. On the day he assumed command, he telegraphed the general-in-chief in Washington: "No written authority is found here to declare and enforce martial law in this department. Please send me such written authority and telegraph me that it has been sent by mail."[13]

George B. McClellan had by this time replaced Winfield Scott as general-in-chief, and the new commander proved more cautious. He did not reply promptly, and Halleck had to write again. Assuming that Washington was reluctant to grant him written authority for fear of its abuse in the field, Halleck now explained:

> It is not intended to either declare or enforce martial law in any place where there are civil tribunals which can be intrusted with the punishment of offenses and the regular administration of justice. But in some places there are no such tribunals, and it devolves upon the military to arrest and punish murderers, robbers, and

thieves, and martial law already exists in these places. In this city [St. Louis], for example, it has existed for months, but by what legal authority I am unable to ascertain. In the absence of the proper civil tribunals it is impossible to entirely dispense with it, but I intend to restrict it as much as possible.[14]

Before this letter reached Washington, an aide to McClellan replied to Halleck's first telegram: "The general-in-chief desires you to give your views more fully as to the necessity of enforcing martial law in your department, and if you think the necessity is sufficiently pressing for such a step to mention the names and addresses of the officers to whom you think the power should be given."[15] McClellan had already proved himself willing to order military arrests of civilians in Maryland, but now he needed to know more before he acted. The president, on the other hand, was already satisfied. Lincoln had noted on Halleck's original telegram: "If General McClellan and General Halleck deem it necessary to declare and maintain martial law at Saint Louis the same is hereby authorized."[16]

By observing the chain of command and essentially delegating the decision, the president permitted more delay, and Halleck had to write a third letter on November 30:

...the enemy is moving north with a large force and...a considerable part of Northern Missouri is in a state of insurrection. The rebels have organized in many counties, taken Union men prisoners, and are robbing them of horses, wagons, provisions, clothing, &c....

To punish these outrages and to arrest the traitors who are organizing these forces and furnishing supplies it is necessary to use the military power and enforce martial law. I cannot arrest such men and seize their papers without exercising martial law for there is no civil law or civil authority to reach them. The safety of Missouri requires the prompt and immediate exercise of this power, and if the President is not willing to intrust me with it he should relieve me from the command. It is and has been for months exercised here by my predecessors but I cannot find any written authority of the President for doing so. I mean to act strictly under authority and according to instructions and where authority will not be granted the Government must not hold me responsible for the result.[17]

When Halleck's request finally squeezed through the McClellan bottleneck, the president, whose mind had been effectively made up long before, did not bother to write the requisite order himself. He delegated the task to Seward, whose handwritten draft went unamended by Lincoln and became the text of the order to Halleck, dated December 2, 1861:

General: As an insurrection exists in the United States and is in arms in the State of Missouri, you are hereby authorized and empowered to suspend the Writ of Habeas Corpus within the limits of the military division under your command and to exercise martial law as you find it necessary in your discretion to secure the public safety and the authority of the United States.[18]

Halleck had never mentioned the writ of habeas corpus and neither had Abraham Lincoln in this Missouri exchange. The correspondence leading up to the order of December 2 proves that the administration recognized no particular distinction between imposing martial law and suspending the writ of habeas corpus. Halleck formally reinstituted trials by military commission in General Order No. 1 of January 1, 1862.[19]

Application of Martial Law in Missouri

General Halleck and his subordinates urged vigorous action against the disloyal populace in Missouri while attempting at the same time to restrain overzealous commanders from encumbering the makeshift federal prisons with people arrested for trivial offenses. In order to restrict the application of martial law, Halleck had declared it in force only in St. Louis and "in and about all railroads in this State."[20] He reiterated the point in a general order on March 13, 1862, saying that "Martial law has never been legally declared in Missouri except in the city of St. Louis and on and in the immediate vicinity of the railroads and telegraph lines." Actually, Halleck had not mentioned telegraph lines in his original order back in December, and he was thereby unconsciously expanding the suspension while urging others not to.[21]

Ulysses S. Grant, like other generals in Missouri, continued to encounter unanticipated problems after Halleck and Lincoln acted. While still in Cairo, Illinois, commanding the District of Southeastern Missouri, Grant learned that four Union pickets had been shot on the morning of January 11, 1862. With his usual decisiveness, he gave Brigadier General Eleazer A. Paine, commanding at Bird's Point, Missouri, this order:

> If...the assassins were citizens, not regularly organized in the rebel Army, the whole country should be cleaned out, for six miles around, and word given that all citizens making their appearance within those limits are liable to be shot. To execute this, patrols should be sent out, in all directions, and bring into camp at Bird's Point all citizens, together with their Subsistence, and require them to remain, under pain of death and destruction of their property until properly relieved.
>
> Let no harm befall these people, if they quietly submit[,] but bring them in, and place them in camp below the breastworks and have them properly guarded.
>
> The intention is not to make political prisoners of these people, but to cut off a dangerous class of spies.
>
> This applies to all classes and conditions, Age and Sex. If, however, Woman [sic] and Children, prefer other protection than we can afford them, they may be allowed to retire, beyond the limits indicated, not to return until authorized.

General Paine moved swiftly, bringing in perhaps a hundred citizens. Indeed, he showed an alacrity for this sort of unpleasant work that would get him into deep trouble later in Kentucky. "I think," Paine replied to Grant, "I shall find out who shot the picketts and when I do I shall shoot the guilty parties on very short notice." Grant was tough, but he wanted nothing to do with summary justice. He told General Paine, "If you have reason to believe, that the parties guilty of shooting our pickets, are discovered, inform me, and I will order a Court of Commission, that will act without delay." The use of military commissions, officially introduced under that name by Halleck's order of January 1, remained so unfamiliar to Grant that he could not readily come up with the technical term.[22]

Confusion concerning military authority over civilians in Missouri was not confined to incidents involving aggressive generals like Grant or ruthless ones like Paine. A more typically ambivalent communication, beginning with caution but ending in excess, came from Provost Marshal General Bernard G. Farrar on March 8, 1862. He told a major in Hannibal:

> As a general rule release those who are not guilty of irregular warfare or other violations of rules of war, burning bridges, &c., upon taking oath and giving bond with good security. Such as are clearly guilty of robbing and other offenses against law where the evidence is clear, turn them over to the officers of the law and make them do their duty. Do not hesitate to assume any responsibility your judgment may dictate as necessary to thwart any plan of secessionists or to wrest from them any power they have civil or otherwise. They have first discarded law and have appealed to force. It is now purely a question of power not one of law. Do not hesitate to seize and hold their property. Where there is no law there is no property. If they deny the power of the Government they are without law and let them feel the consequences.[23]

Arrests for mere disloyal sentiments in Missouri often seemed futile, and Farrar's letter was one among many that urged the release of minor offenders. Back in the time of Frémont's command, a staff officer told a zealous colonel in Rolla that the commander of the department would "hold as prisoners those men taken by you bearing arms against the United States; others the charges against whom are not more serious than entertaining secession feelings he has discharged." In another letter, more exasperated in tone but written to the same man the same day, the staff officer added, "If entertaining secession feelings constitutes a grave offense, one sufficient to imprison a man on, the Government would have two-thirds of the State to feed at its expense."[24] Yet, restricting the freedom of Missourians because of their opinions never came to an end. Farrar himself gave orders in March 1862 not to allow a particular Protestant minister, a prisoner on parole, to preach in certain counties of the state.[25]

Despite his legal-mindedness and timid image, Halleck usually defended drastic enforcement of martial law. Bridge-burning, he explained to one critic, "is not usually done by armed and open enemies but by pretended quiet citizens living on their farms. A bridge or building is set on fire and the culprit an hour after is quietly plowing or working in his field. The civil courts can give us no assistance as they are very generally unreliable. There is no alternative but to enforce martial law. Our army here is almost as much in a hostile country as it was when in Mexico."[26]

The First Trials by Military Commission

The war in Mexico, fifteen years before the Civil War, had witnessed the first military commissions convened in American history. General Winfield Scott, who introduced their use, explained in his *Memoirs* (published late in the Civil War) that "by the strange omission of Congress, American troops take with them beyond the limits of their own country, no law but the Constitution of the United States, and the rules and articles of war. These do not provide any court for the trial or punishment of murder, rape, theft, &c., &c.—no matter by whom, or on whom committed." Scott's "martial-law order" for Mexico enumerated and limited punishments for "offenses, any one of which, if committed within the United States or their organized Territories, would, of course, be tried and severely punished by the ordinary or civil courts of the land." Among them were "assassination, murder, poisoning, rape, or the attempt to commit either; malicious stabbing or maiming; malicious assault and battery, robbery, theft; the wanton desecration of churches, cemeteries or other religious edifices and fixtures; the interruption of religious ceremonies, and the destruction, except by order of a superior officer, of public or private property." No such specific catalog of crimes was established for the use of martial law in the Civil War.[27]

One of the striking features of the offenses enumerated by Scott is that many of them could have been perpetrated *only on the enemy in occupied territory by Scott's own troops.* The very model of a professional soldier himself, "Old Fuss and Feathers" thought of military commissions mainly as a way of restraining unruly U.S. volunteer soldiers. He despised such troops and told the secretary of war, William L. Marcy, on January 16, 1847:

> ...our militia & volunteers, if a tenth of what is said be true, have committed atrocities—horrors—in Mexico, sufficient to make Heaven weep, & every American, of Christian morals *blush* for his country. Murder, robbery & rape on mothers & daughters, in the presence of the tied up males of the families, have been common all along the Rio Grande. I was agonized with what I heard—not from Mexicans, and regulars alone, but from respectable individual volunteers—from the masters & hands of our steamers. Truly it would seem unchristian & cruel to let loose upon any people—even savages—such unbridled persons—free-booters, &c., &c. The respectable volunteers—7 in 10—have been as much horrified & disgusted as the regulars, with such barbarian conduct. As far as I can learn, not one of the felons has been punished, & very few rebuked—the officers generally, being as much afraid of their men as the poor suffering Mexicans themselves are afraid of the miscreants. Most atrocities are always committed in the absence of regulars, but sometimes in the presence of acquiescing, trembling volunteer officers.

With "no legal punishment" available for the "atrocities" of "the wild volunteers," Scott took the initiative to develop trials by military commissions and martial law for the United States Army. He was proud of this work. "Without it," Scott wrote, "I could not have maintained the discipline and honor of the army, or have reached the capital of Mexico."

The reception given Scott's order in its day provided a foretaste of the controversy eventually stirred up by military commissions during the Civil War. As a Whig general under a Democratic administration, Scott's relations with his superiors were never very good, but the dangerous novelty of this order stunned Washington officials. The secretary of war appeared startled, refused comment, and "soon silently returned" the order "as too explosive for safe handling." The attorney general was "stricken with legal dumbness." "All the authorities," Scott recalled, "were evidently alarmed at the proposition to establish martial law, even in a foreign country, occupied by American troops."[28]

When martial law and trials by military commission were applied to Missouri during the Civil War, they certainly stretched Scott's precedent. Obviously American troops operating in Missouri were not "beyond the limits of their own country," and "murder, rape, theft" and other similar crimes were punishable under Missouri law—whoever the perpetrators. Further, martial law and military commissions did not serve to restrain the United States Army. They were used to restrain the civilian populace; they operated mostly to deprive Missourians of rights they would otherwise have enjoyed in the absence of U.S. troops.[29]

Trials by military commission restrained United States forces in the Civil War mainly by imposing systematic record-keeping and an atmosphere of legality on the army's dealings with a hostile populace. Military commissions played a role in preventing martial law from degenerating into what the cynical Duke of Wellington had called it, "the will of the general."[30] No Union general in Missouri could, according to Winfield Scott or Henry W. Halleck or Ulysses S. Grant, order summary justice for a civilian offender and march on without so much as recording the event.[31] The records, most of them today still in tight bundles of crisp paper tied by red tape, occupy rows and rows of shelves at the National Archives.

Military commissions dictated not only record-keeping but also system, regularity, review, and some safeguards for defendants' rights. The commissions were composed of a minimum of three officers and often many more. The accused had to be present at the trial and was allowed counsel, though lawyers could not speak for their clients in the courtroom. Witnesses had to confront the defendant; *ex parte* evidence by affidavits was not permitted. Sentences were reviewed by the departmental commander, and capital sentences had to be reviewed by the president.[32]

The actual workings of military commissions were nevertheless quite imperfect. The early trials in Missouri often brought before these tribunals persons who were disloyal in heart, mind, and outward behavior, but these did not necessarily correspond to the crimes of which they were officially accused.

For example, Joseph Aubuchon, of Ironton, Missouri, perhaps the first civilian to be tried by a military commission in the United States, was charged with "Treason against the Government of the United States." The specification, which was supposed to identify the time, place, and specific act, stated that

Aubuchon "did assume an attitude of open rebellion against the Federal Government by taking up arms against the same, by assuming and exercising the functions and office of lieutenant in the rebel army within the limits proper of the state of Missouri from and after about the 20th day of August, 1861." The military commission, which met in the St. Louis arsenal on September 5, 1861, found him guilty of the specification "except the words 'By taking up arms against the same, by assuming and exercising the functions in the rebel army.'" That left only assuming "an attitude of open rebellion" as a hardly definitive specification. Nevertheless, the commission found him guilty of the charge, confiscated his property, and sentenced him to imprisonment at hard labor for the duration of the war.[33]

Joseph Aubuchon did not commit treason, a crime carefully defined in the United States Constitution by men who, in the eyes of some Englishmen, had themselves committed treason. Treason must consist of an "overt Act" of levying war against the U.S. or adhering to its enemies in war by giving them aid and comfort. And it must be confessed to in open court or testified to by two witnesses. Aubuchon's "attitude of open rebellion" did not fit the constitutional definition.[34]

Though far from legally correct, military commissions nevertheless provided something more than show trials and sham justice. Aubuchon, for example, was released when General Frémont reviewed his sentence, because "the offense charged occurred previous to the proclamation" which had established martial law in Missouri. In another case, that of George W. Higginbotham, tried September 29, 1861, at Pilot Knob, Missouri, the commission itself found the accused innocent and released him unconditionally with the request "that the commander of the post issue an order forbidding the arrest of persons without evidence of their guilt."[35]

Especially in these earliest trials, the legal footing of the military authorities was unsure. The army particularly sought to punish bridge-burners and other saboteurs under a charge of violation of the laws of war.[36] But the commissions often added to that charge a charge of treason, as evidenced by the defendant's taking up arms against the government and exercising the functions of a rebel soldier. The accused parties were allowed to employ lawyers, and it did not take such men long to devise an obvious defense strategy: plead not guilty to bridge-burning but guilty to the treason charge of taking up arms and serving as a rebel soldier. The latter, as specified, appeared to qualify the defendant for status as prisoner of war and, hence, not liable to execution as a civilian would be. Thus, early Missouri defendants confessed to one form of disloyalty to avoid execution for another form.

The trial of George M. Pulliam provides an example. A military commission meeting in Palmyra, Missouri, charged him with bridge-, railroad-, and car-burning; with giving aid and comfort to bridge- and railroad-burners; and with treason specified in the customary way. Pulliam pleaded not guilty to the first two charges but guilty to the third—the pattern followed by other defendants facing similar charges. Pulliam admitted having served as rear-guard for a party of bridge-burners, but he maintained that it was a military action under officers' orders.

A crucial exchange in the trial occurred when a witness from the party that arrested Pulliam testified that the defendant "when we met him ... had a gun that was loaded. He looked at us a bit and said would we treat him as a prisoner of war? Major Linder said we would." Pulliam seized on the point at the trial: "I would like to ask if he did not hear me say to the major if he did not treat me as a prisoner of war I would not surrender?" When Pulliam took the witness stand afterward, he insisted that he had been a member of Confederate General Sterling Price's army, that he did not go voluntarily to burn bridges, and that he did not instigate plots to burn them. He was simply obeying orders: "Privates knew nothing at all about what was going on." Despite this apparently telling defense, the military commission found Pulliam guilty on all three charges and sentenced him to death. General Halleck approved the sentence.[37]

Later, Halleck overruled verdicts of treason on the grounds that "such charges were not triable by a military commission." Afterward, Missouri's military commissions typically charged defendants accused of bridge-burning with violating the laws of war by taking up arms against the U.S. while *not* being soldiers in any lawfully organized military force at war with the U.S.[38]

Under the new charges, defendants caught in the act had little choice but to plead guilty to bridge-burning but not guilty to violating the laws of war. Thus John Bowles, tried in Danville, Missouri, on February 13, 1862, admitted that he had destroyed railroad and telegraph lines, but he asserted that his captain had ordered it. Though recruited by the captain for Confederate General Sterling Price's army, Bowles had been unable to reach his unit. The captain assured him that Price had ordered the destruction of the Union communications. For its part, the court held it illegal for Price to recruit behind Union lines. Bowles denied knowledge of its illegality but was sentenced to be shot, and Halleck sustained the sentence.[39]

Even after military commissions in Missouri ceased trying "treason" cases, irregularities in their work continued. In Cape Girardeau, for example, a commission found defendant Joseph Bollinger guilty of a charge but not guilty of the only specification listed under that charge. Higher authorities reviewing his sentence overturned the guilty verdict.[40]

The justice meted out by military commissions was of a rough sort, but justice was usually their goal. They did not victimize innocent and guilty alike in kangaroo courts or in show trials of predetermined outcome. In fifty-four trials by military commission that occurred before June 1862 but after the treason-charge difficulty had been cleared up, thirteen, or 24.1 percent, of the accused parties pleaded guilty to all or some of the charges or specifications. Leaving aside these cases involving admissions of guilt by the defendants, six of the remaining forty-one, or 14.6 percent, were acquitted. Of the remainder, sixteen had their sentences mitigated upon routine review by higher officers, and ten who were sentenced to be executed had their sentences approved.[41]

Thus military justice, though rough-and-ready, could be tempered with mercy. Review was a regular part of the system of trials by military commission. The commissions themselves sometimes recommended mercy because of the youth, stupidity, or previous loyalty of the convicted person.

Trials by military commission did not often serve the purpose of repression for political opinion. Many of the defendants had, in fact, been in arms against the United States. Of the 101 cases tried in Missouri from September 1861 to June 1862, the reports of whose trials are printed in the *Official Records,* only one involved mere political beliefs or freedom of expression. Edmund J. Ellis, editor and proprietor of the *Boone County Standard,* was tried in Columbia, Missouri, in February 1862. He was accused of publishing information for the benefit of the enemy and of encouraging resistance to the U.S. government. The army cited as evidence certain articles and communications that had appeared in Ellis's newspaper. The military commission found him guilty, banished him from the state, and confiscated the press and equipment of the newspaper.[42] Military commissions remained always capable of such practices, but their attention was usually focused on more serious cases.

The Nature and Extent of Military Arrests of Civilians in Missouri

The substantial level of disloyalty—actually taking up arms against the U.S.— revealed in the records of the early military commissions in Missouri helps explain the extraordinary numbers of civilians arrested by the U.S. Army in that state. More such arrests occurred in Missouri than in any other state by far.

The precise number is unascertainable. Records are extremely fragmentary, especially for the period when William H. Seward oversaw military arrests of civilians. Extant lists of prisoners of state in St. Louis and Alton account for only 160 detainees in that period, and though most of these were from Missouri, not all of them were.[43] Even these partial figures account for a sizeable portion of all civilian arrests made under Seward's supervision: 18.5 percent.

Numbers of civilian prisoners from Missouri reported later in the war, available in the records from the provost marshal general's office, are much higher. Surely, similar rates of arrest prevailed in 1861 and early 1862, for policies and personnel varied little between the periods, and Washington never exercised much control over Missouri.[44]

Although more records are available for the period after February 1862, the jumble of papers is unsystematic to the point of chaos. No comprehensive lists of prisoners in Missouri for that period remain. The records consist mostly of receipts exchanged when guards brought prisoners into the city from the hinterland, when prisoners were taken from their cells to the provost marshal's office for interrogation, or when they were transferred from one prison to another. Fragmentary as the records are, they are also frustratingly repetitious, as the same detainee may be mentioned in more than one record as he was moved from field to prison to interrogation to another prison. Counting becomes exceedingly difficult under such conditions, as every name on a list must be checked against every other list. With no rules for writing such receipts and no standard way to write names, problems in obtaining accurate figures are compounded by the need to know whether "W. Smith," "William Smith," "W. F. Smith," and "William Franklin Smith" were one, two, three, or four persons.

Many of the records, often hastily scrawled by harried officers of low rank, were not the equals of the work of professional clerks or copyists in the employ of institution-minded prison administrators. Captains serving as local provost marshals or lieutenants commanding detachments of prisoners in transit—even corporals—drafted or signed some lists.

Moreover, the Missouri records almost never mentioned the charge against and only rarely the residence of the prisoner. Citizens were not always clearly differentiated from prisoners of war. M. Jeff Thompson, for example, a general in the pro-Confederate Missouri State Guard, shows up on a list of civilian captures and is distinguishable only because of his fame.[45] The Missouri records are now arranged in a huge series in the National Archives called the Union Provost Marshal's File of Papers Relating to Two or More Civilians.[46] A few lists of soldiers identified by rank and unit appear amidst these records, and some soldiers' names, like General Thompson but not as famous, undoubtedly appear on lists purportedly containing the names only of civilian prisoners. For example, some lists show men designated as military officers but contain no prisoners identified as enlisted men; the remaining names on such lists likely include citizens and common soldiers indiscriminately mixed.

Federal authorities early on developed terms to describe the three important classes of prisoners held in military prisons: "prisoners of war" were captured Confederate soldiers and sailors; "United States prisoners" were members of the U.S. armed services held for crimes committed in camp, like theft or murder; "prisoners of state" were civilians. Consistent use of the terms, however, was confined to those who administered military prisons and to other high-ranking officials. In the field, the usage was inconsistent at best.

These crucial terms were used too carelessly to make the Missouri records clear to historians. When, for example, a military commission reported early in 1862 on the petitions of various St. Louis prisoners seeking release, one group "to be retained as prisoners of war" included a list of nine persons, identified by residence rather than rank or regiment. They were almost certainly civilians and could not be held as "prisoners of war."[47] Besides, discriminating soldiers from civilians was perhaps the central problem of Union authorities in this state torn by guerrilla warfare.

Intelligent soldiers at the time recognized the inadequacies of the records the army was creating. For example, W. J. Masterson, commanding the Gratiot Street prison in February 1863, commented thus on eleven men sent in by Captain James Call of the Third Missouri Volunteer Cavalry: "There being no descriptive list sent with these prisoners it is impossible for me to give any explanatory notes concerning these men. There was no roll of *names* even, except a list made by the officer bringing them in. The Provost Marshal at Rolla who sent the prisoners forward deserves *great credit* for his business talents in this line at least." And an exasperated officer in the provost marshal general's office in St. Louis filled in a printed form this way:

> Prisoners received at the *Gratiot* Street Military Prison the *2* day of *Feb* 1863, from *Corporal Willis Knight* sent forward from *Dont know where* on the *ditto* date of *Feb* 1863, by order of *Don't know who, as no papers came with guard or prisoners.*[48]

Problematic notations further debased the Missouri records—for example, "citizens but prisoners of war, bushwhacker, etc."; "claims to be a Confederate captain"; "charged with belonging to C.S.A."; and "citizens and recruits."[49] But the greatest problem in the records of Missouri's political prisoners is neither imprecision nor duplication. It is lack of information altogether. Only a little over 14 percent of the names listed show any record whatever of cause of arrest.

However limited, these are the only records left to the historian. Carefully sifted, the ones for Gratiot Street prison for the period from April 1862 through October 1863 indicate that at least 2,014 different civilian prisoners entered, passed through, or remained in that military prison in that nineteen-month period. That yields an average of 106 prisoners per month. At that rate, from the imposition of martial law by Frémont in September 1861 to the end of the war, Gratiot Street prison would account for 4,770 political prisoners. Union authorities thus locked up as prisoners of state well over one out of every one hundred males in the state of Missouri in Gratiot Street prison, not to mention the thousands of Missourians held as prisoners of war. These figures are conservative because the available records are only fragmentary and Gratiot Street was not the only prison in use.[50]

St. Louis's other military prison, on Myrtle Street, was smaller and operated for a shorter period of time. It left records of 244 civilian prisoners committed between February 3, 1862, and November 22, 1864.[51] Other civilian prisoners were detained in stockades at remote military posts and were never transferred to St. Louis. The exact number of civilian prisoners taken in Missouri will never be known, but any extrapolation from the available records will produce a formidable if not staggering figure.

General Order No. 11, August 25, 1863

The statistics fit Missouri's matchless reputation as the scene of cruel guerrilla warfare and desperate military repression. A symbol of her sad fate is General Order No. 11, regarded by some as "one of the cruelest and most unusual orders issued by a general during the Civil War." This document, issued by Brigadier General Thomas Ewing, commanding the District of the Border, after

Confederate guerrilla chief William C. Quantrill's raid on Lawrence, Kansas, ordered the evacuation of four counties in western Missouri, which were said to have sustained Quantrill's guerrillas. Independence and a few other settlements were exempted, and part of one county fell outside the boundaries of the military district; otherwise, every resident had to move. Those who could establish their loyalty to the satisfaction of the commanding officer of the nearest military post would be issued certificates allowing them to move to military posts in the state. Everyone else was supposed to leave the state. Hay and grain found after that date would be destroyed or, if located near a military post, taken into the post and credited to loyal owners.[52] The order may have created as many as twenty thousand refugees from the western Missouri counties. Though it did not directly create any political prisoners, many of these homeless refugees must have wandered eventually into Union lines and were doubtless arrested.

Guerrilla warfare constantly threatened to break down the customary distinction between soldiers and civilians. Well before Quantrill's raid, General Ewing had been considering actions that would severely test the boundaries between soldier and civilian in this technically loyal state. On August 3, 1863, he described the situation in his district to his superiors in St. Louis this way:

About one-half of the farmers in the border tier of counties of Missouri in my district, at different times since the war began, entered the rebel service. One-half of them are dead or still in the service; the other half...have returned to those counties. Unable to live at their homes if they would, they have gone to bushwhacking, and have driven almost all avowed Unionists out of the country or to the military stations. And now, sometimes in squads of a dozen and sometimes in bands of several hundred, they scour the country, robbing and killing those they think unfriendly to them, and threatening the settlements of the Kansas border and the towns and stations in Missouri.

So large a portion of the troops under my command are held fast, guarding the Kansas border and the towns and stations in Missouri, which are filled with refugees, that I cannot put in the field numbers equal to those of the guerrillas.... The country is rich and supports them well, but it is so rugged and heavily timbered, and full of places of concealment and ambuscade, that these bands could not possibly be expelled from it with forces in the field less than three times their own.

About two-thirds of the families on the occupied farms of that region are kin to the guerrillas, and are actively and heartily engaged in feeding, clothing, and sustaining them. The presence of these families is the cause of the presence there of the guerrillas. I can see no prospect of an early and complete end to the war on the border, without a great increase of troops, so long as those families remain there. While they stay there, these men will also stay, if possible. They know they cannot go home and live peaceably because of the fierce feeling against them among the loyal men of the border, who have suffered at their hands. Against these loyal men no amnesty now or hereafter can protect them. They will, therefore, continue guerrilla war as long as they remain, and will stay as long as possible if their families remain. I think that the families of several hundred of the worst of these men should be sent, with their clothes and bedding, to some rebel district south, and would recommend the establishment of a colony of them on the Saint Francis or White Rivers, in Arkansas, to which a steamboat can carry them direct from Kansas City. About one-half of them could take with them no provisions or money of any consequence, and would have to be temporarily supplied by Government.

From St. Louis came approval of the plan for the guerrillas' families, along with the suggestion that Ewing "collect them at one or more points on or near the Missouri River, where they can be temporarily guarded and quartered." Headquarters also urged, "On account of the expense and trouble necessarily attendant upon the carrying out of this plan, and also the suffering it may cause to children and other comparatively innocent persons, the number to be transported should be as small as possible, and should be confined to those of the worst character."[53]

A week before General Order No. 11, Ewing issued General Order No. 10, which included this drastic provision:

> ...officers will arrest, and send...for punishment, all men (and all women not heads of families) who willfully aid and encourage guerrillas, with a written statement of the names and residences of such persons and of the proof against them. They will discriminate as carefully as possible between those who are compelled, by threats or fears, to aid the rebels and those who aid them from disloyal motives. The wives and children of known guerrillas, and also women who are heads of families and are willfully engaged in aiding guerrillas, will be notified by such officers to remove out of the district and out of the State of Missouri forthwith. They will be permitted to take, unmolested, their stock, provisions, and household goods. If they fail to remove promptly, they will be sent by such officers, under escort, to Kansas City for shipment south, with their clothes and such necessary household furniture and provision as may be worth removing.[54]

This order predated Quantrill's sack of Lawrence by three days and proves the tendency of guerrilla warfare to cause the breakdown of distinctions between soldier and civilian even without massacres or other sensational incidents to arouse passionate hostility. The potential for creating civilian prisoners in such a country was virtually limitless.

President Lincoln approved of the notorious General Order No. 11—far more than he did of interfering with freedom of speech or political organization. Thus, he wrote General John M. Schofield, commanding the Department of the Missouri, on October 1, 1863, with this broad advice:

> Under your recent order, which I have approved, you will only arrest individuals, and suppress assemblies, or newspapers, when they may be working *palpable* injury to the Military in your charge; and, in no other case will you interfere with the expression of opinion in any form, or allow it to be interfered with violently by others. In this, you have a discretion to exercise with great caution, calmness, and forbearance.
>
> With the matters of removing the inhabitants of certain counties *en masse;* and of removing certain individuals from time to time, who are supposed to be mischievous, I am not now interfering, but am leaving to your own discretion.

Sixteen more months passed, Union military success came nearly everywhere, and there was "no organized military force of the enemy in Missouri." Yet "destruction of property and life" remained "rampant every where" in the state. Exhausted and remedyless, Lincoln could only counsel Union forces there in 1865 to encourage "neighborhood meetings" where "old friendships will cross the memory; and honor and Christian Charity will come in to help."[55]

Missouri: A Failure in Policy for the Lincoln Administration

When questioned about military repression in the border states, Republican contemporaries of Abraham Lincoln pointed with pride to the role of the administration's vigorous measures in saving Maryland for the Union. Even Democrats in the North frequently admitted the necessity of the early arrests in Maryland. Likewise, historians have long emphasized the crucial place of Kentucky on the Union map, and many of them have praised the administration for keeping that state in the Union by a shrewd mixture of tough policies and sympathetic treatment.

Nobody bragged about Missouri. Secession was avoided in that border slave state also, but otherwise it proved to be the locus of many serious problems for the Lincoln administration. The president could devise no better way to state his policy goals there, once regular Confederate forces had been driven from the state, than that the army should "compel the excited people there to leave one another alone."[56] From Missouri would come incidents and practices that would dog Lincoln's historical reputation ever after. One cannot be certain without public opinion polls, but it seems likely that the most unpopular measure taken by President Lincoln in the first year of the war was his revocation of Frémont's emancipation proclamation for Missouri. At the time, his policy seemed to reveal a reluctance to hit the offending South hard, and thus Lincoln appeared a weak leader. After the war, the revocation damaged Lincoln's reputation in another way. It seemed inconsistent with any claim to being a "great emancipator." Eventually historians managed to find in it a shrewd feeling for the public opinion of the border states, but certainly public opinion in Missouri did not rally to the Union in any remarkable way after the incident.

Missouri also saw the origin of trials by military commission, and the use of these by the Lincoln administration would lead shortly after the war to sharp condemnation by the United States Supreme Court in *Ex parte Milligan.* That decision, in turn, dealt a blow to Lincoln's reputation as a steward of the Constitution from which he never recovered in the constitutional history books. And the disproportionately large numbers of civilians arrested in Missouri point also to a dismal record of failure of administration policy in this state.

If Maryland and Kentucky somehow became early "success stories" for the

Lincoln administration, Missouri proved from start to finish to be a sorry blemish on the administration's record. It became a nightmare for American civil liberties. What a different story this book would tell if Missouri and its thousands of political prisoners could be left out.

3

Low Tide for Liberty

By the autumn of 1862, when President Lincoln announced the Emancipation Proclamation, the writ of habeas corpus remained technically secure in most of the North. It was officially suspended only around Florida, in St. Louis and near the railroad and telegraph lines of Missouri, and along a "military line" stretching from Washington, D.C., to Bangor, Maine. In fact, the writ of habeas corpus was a dead letter in many other areas of the North as well: in the seceded South, in much of the border states, and in some far-flung and carelessly governed territories.[1]

Only administration insiders knew how hardened Abraham Lincoln's attitudes on such constitutional issues had already become and thus how shaky was the legal status of the writ of habeas corpus everywhere in the country. Most of the cabinet members learned about a new proclamation, the first to suspend the writ throughout the nation, by reading the newspapers. "The President has issued a proclamation on martial law," Secretary of the Navy Gideon Welles wrote indignantly on September 25, 1862, "suspension of *habeas corpus* he terms it, meaning, of course, a suspension of the *privilege* of the writ of *habeas corpus.*" Welles concluded that he was "not sorry" to have known nothing of this proclamation until it was issued: "I question the wisdom or utility of a multiplicity of proclamations striking deep on great questions."[2]

Suspending the Writ of Habeas Corpus Nationwide

Frederick Seward, who had seen the events firsthand as assistant secretary of state, maintained in his biography of his father that the "form" of the Florida suspension of May 10, 1861, "was afterward made general, and proved adequate to its purpose during the war. Applying the suspension merely to exposed points and actual cases of disloyalty, it did not interfere with the rights of the general public in the courts."[3] The level of confusion on the subject is again apparent in this misleading statement. The writ was suspended broadly, and not merely at "exposed" geographical "points," and to have confined it to "actual cases of disloyalty" would have required an omniscient government. What Frederick Seward wrote in this instance bordered on nonsense.

Lincoln suspended the writ of habeas corpus in certain kinds of cases throughout the nation on September 24, 1862, with a proclamation that at the time attracted surprisingly little attention. Although the president called a special meeting of the cabinet on Wednesday the 24th (they customarily met on Tuesdays and Fridays), they did not discuss the habeas-corpus proclamation. Instead, Lincoln invited the opinions of his advisors on an Indian treaty and on treaties with foreign governments desiring colonies of American blacks. Colonization may have been an important subject to Lincoln, who had issued his preliminary Emancipation Proclamation only two days before, but Indian affairs were always marginal to the main work of the administration. The writ of habeas corpus had definitely lost much of its importance as a delicate issue for Lincoln.[4]

The origins of this new habeas-corpus proclamation lay in the Militia Act of July 17, 1862, which empowered the secretary of war to draft for nine months the militiamen of states that failed to upgrade their militias. This rather technical-sounding law that smacked of routine military housekeeping proved to be a disguised conscription law, authorizing the first national military draft in American history. (It came to be called the Militia Draft of 1862.) Congressmen, sensing the potentially explosive unpopularity of conscription in the individualistic United States, had obscured its real purpose as much as possible. The president intended his habeas-corpus proclamation to enforce this provocative conscription law.[5]

The reason for the quiet indifference of most of the cabinet members was that the president, once again, was putting a gloss on a *fait accompli.* Six weeks before Lincoln's proclamation, the War Department, anticipating that trouble would arise from the Militia Act, had written a series of orders "to prevent evasion of military duty and for the suppression of disloyal practices":

1. By direction of the President of the United States it is hereby ordered that until further order no citizen liable to be drafted into the militia shall be allowed to go to a foreign country. And all marshals, deputy marshals and military officers of the United States are directed, and all police authorities especially at the ports of the United States on the seaboards and on the frontier are requested, to see that this order is faithfully carried into effect. And they are hereby authorized and directed to arrest and detain any person or persons about to depart from the United States in violation of this order and report to Maj. L. C. Turner, judge-advocate at Washington City, for further instructions....

2. Any person liable to draft who shall absent himself from his country or State before such draft is made will be arrested by any provost-marshal or other United States or State officer wherever he may be found within the jurisdiction of the United States and conveyed to the nearest military post or depot and placed on military duty for the term of the draft, and the expenses of his own arrest and conveyance to such post or depot and also the sum of $5 as a reward to the officer who shall make such arrest shall be deducted from his pay.

3. The writ of habeas corpus is hereby suspended in respect to all persons so arrested and detained and in respect to all persons arrested for disloyal practices.

Stanton issued the orders on August 8, 1862, thus making the secretary of war rather than President Lincoln the first official to suspend the writ of habeas corpus across the whole United States. Stanton later swore under oath that Lincoln gave him "verbal direction" to issue the order and that the president read Stanton's draft of the order before the secretary signed it in his presence.[6]

A similar order published the same day added the threat of trial by military commission:

Ordered:
1. That all U.S. marshals and superintendents or chiefs of police of any town, city, or district be, and they are hereby, authorized and directed to arrest and imprison any person or persons who may be engaged, by act, speech, or writing, in discouraging volunteer enlistments, or in any way giving aid and comfort to the enemy, or in any other disloyal practice against the United States.
2. That immediate report be made to Maj. L. C. Turner, judge-advocate, in order that such persons may be tried before a military commission.

This was the first War Department order explicitly prescribing military commissions for trials of civilian offenders. When the orders of August 8 were published as General Order No. 104 on August 13, Stanton also sent this explanatory note to the press:

The Recent Orders to Prevent the Evasion of Military Duty
These orders are designed to operate on two classes of persons, viz., those who contemplate leaving the United States for the purpose of evading their military duty, and those who leave their own State or place of residence and go into other States for the same purpose. The object is to compel every citizen of the United States subject to military duty to bear his share in supporting the Government. Instructions have been prepared, and will be issued on Monday, to military commandants, marshals, and police officers respecting the mode of executing the orders so as to interfere as little as possible with individual pursuits and business, and limit the operation of the order to cases of evasion.[7]

The tone of the War Department press release seems to suggest a fear of public reaction to the bold orders, but after the August 8 orders were discussed by Seward in that day's regular cabinet meeting, Chase wrote in his diary, "Nothing proposed and nothing done of any moment."[8]

In fact, the orders of August 8 had momentous effect on civil liberties in the United States. The brief period of sweeping and uncoordinated arrests that followed their issuance constituted the lowest point for civil liberties in the North during the Civil War, the lowest point for civil liberties in U.S. history to that time, and one of the lowest for civil liberties in all of American history. It showed the Lincoln administration at its worst—amateurish, disorganized, and rather unfeeling. The War Department proved incapable of controlling or coordinating so vast a system. The secretary of war could not even draft an order that would fit the circumstances of the whole nation. Thus D. W. Alvord

of Greenfield, Massachusetts, had to write the War Department for clarification
on August 11:

> The High Sheriff of this County, and the acting constable of this town, have con-
> sulted me (I being District Attorney) in the question whether they, or either of
> them, is charged, under your order, with the duty of arresting persons using
> seditious language and discouraging enlistments.... The order, literally construed,
> does not charge either of them with this duty; and I have so advised them. They
> now request me to write to you for your opinion, or instructions.
>
> In the country towns of Massachusetts there is no such officer as "Chief of
> Police." There is no officer belonging to any one of the classes of officers named
> in your order, resident within forty miles of this place.... The High Sheriff
> of a county may in some sense be said to be Chief of Police, but he has no such
> title. That title is confined in this State to the Chiefs of the municipal police of char-
> tered cities.

Judge Advocate Levi C. Turner replied that all sheriffs, constables, and deputy
sheriffs were regarded as authorized to arrest.[9]

The overall effect of the orders of August 8 was to allow a horde of petty
functionaries to decide without any legal guidelines one of the highest matters
of state: precisely who in this civil war was loyal or disloyal. Reporting the
arrests to Major Turner provided a measure of restraint but usually only after
the arrest had been made and the victim's local reputation stained. Nor is it
clear from the original wording of the order that the intent was to have Turner
act as watchdog. In fact, it sounds as though Stanton saw Turner's role as one
of scheduling trials by military commission rather than riding herd on the
Dogberrys unleashed across the land.

As it turned out, the judge advocate wrote dozens of letters explaining who
could and could not be arrested, demanding affidavits to prove allegations
against the victims, and restraining the overzealousness of small-town consta-
bularies or federal marshals too far from the War Department to appreciate
the real desires of the Lincoln administration. But he by no means acted always
and solely to restrain others. Sometimes he goaded the timid into action,
reminding overcautious officials who wrote for advice in doubtful cases that
it was their duty to arrest.

Perhaps the poorest sense of judgment in the field belonged to David L.
Phillips, United States marshal for the Southern District of Illinois. Historian
Frank L. Klement's sharply critical account of this man's work is essentially on
the mark. Phillips tended to find traitors lurking in almost any gathering of
Democrats.

The marshal's most sensationally wrongheaded arrest was that of Dr. Israel
Blanchard of Murphysboro in Jackson County, Illinois, who was dispatched to
the Old Capitol prison in Washington on August 24. Blanchard, a native of
New York State but an Illinois resident since 1852, was arrested on the affidavit
of a man who maintained that the doctor had attended a meeting of the
mysterious Copperhead organization called the Knights of the Golden Circle

in Pinckneyville, Perry County, Illinois, on August 10 and had there made disloyal remarks.

On August 30, John A. Logan, a Democratic politician born in Murphysboro and at the time a twice-wounded Union brigadier on leave between military campaigns, wrote President Lincoln about the case. Dr. Blanchard was General Logan's brother-in-law. Logan pointed out that Blanchard had frequently begged for help in obtaining a commission in the army and had once tried to raise part of a company so he could enter the army as a lieutenant. Moreover, many respectable local residents could testify that on the day of the alleged disloyal meeting, Blanchard was tending a sick child and the day after had attended a meeting near town of a company of the 81st Illinois Volunteers. "From my own knowledge of the country," Logan wrote, "I know that the place designated as the place of the meeting of the K.G.C. is twenty-eight or thirty miles from Murphysboro and is not connected by rail-road or steamboat communication. In other words he could not have been here the evening of one day and the morning of the next and yet attend a meeting of any kind in that or any other part of Perry County."

The president eventually reviewed Logan's letter, endorsing it on October 9, "Submitted to the Secretary of War, with the remark that I strongly incline to discharge Dr. Blanchard." The prisoner was released and returned to Illinois, where he was elected the next year as a state senator on the Democratic ticket.[10]

Blanchard went free in part because his arrest had been glaringly wrongful but also because he was lucky. His case, among many other doubtful ones, gained the president's attention because the prisoner was well connected. His brother-in-law, John A. Logan, though a Democrat, was important to the Lincoln administration. He played a key role, especially after the death of Stephen A. Douglas in 1861, in rallying Illinois Democrats to enthusiastic support of the Northern war effort. A lifelong resident of southern Illinois, he symbolized the loyalty of even the geographically southernmost areas of the North. Lincoln did not want to offend Logan if it could be helped.

Blanchard's was perhaps the most embarrassing among many useless arrests made under the orders of August 8, 1862. The total number of arrests was extraordinarily high. Between August 8 and September 8, at least 354 civilians in the North became prisoners as a direct result of the War Department orders. Other arrests prompted by these careless orders—likely dozens and dozens of them—also occurred but are too poorly documented to count. During the same period, still other arrests for the usual reasons—for contraband-trading, blockade-running, communication with the enemy, traveling to or from the Confederacy, and other offenses—continued to occur. But such arrests would have occurred no matter what the August 8 orders specified.

If arrests had occurred throughout the Civil War at the rate they took place in the first month following the August 8 orders, then at least 16,992 Northerners would have been imprisoned. The early operation of military arrests of civilians under the orders of August 8, 1862, in fact, resembled what the bitterest critics of the Lincoln administration have had in mind when they envisioned the internal security system as a whole. Many historians gained the impression, as Allan Nevins did, that arrests became "commonplace" events in Northern life.[11] They imagined that throughout the war thousands of arrests occurred from the Canadian border to the margins of the Confederacy, that they were usually timed in the late summer or early autumn before elections, and that they frequently involved Democratic politicians, editors, and partisans. Historians imagined that thousands of these arrests were conceived and executed by small-minded, vindictive, and narrowly partisan local office-holders— sheriffs, deputies, and constables—throughout the Civil War.

This traditional picture fits the arrests in the first month following the August 8 orders better, perhaps, than those from any other period in the war. For example, a 58-year-old New Hampshire physician and Democratic politician named Nathaniel Bachelder was arrested on the recommendation of the state's Republican governor, Nathaniel S. Berry, for having said at a recruiting rally that three-fourths of the men would be killed and go to hell. When the state's chief justice issued a writ of habeas corpus in the case, higher authorities became involved. Secretary of War Stanton made inquiry and Judge Advocate General Joseph Holt replied, "If any case of disloyalty, short of taking up arms against the government, could justify the arrests and imprisonment of the offender, this is one." He urged that a "formal order ... be made suspending the writ of Habeas Corpus in this case which should be communicated to the Marshal as his warrant for refusing obedience to the writ." The need for any such formal order as Holt described is hardly clear because the orders of August 8 suspended the writ for "any person ... discouraging volunteer enlistments." But Holt was not the only member of the Lincoln administration unsure of the practical effect of suspension in any specific case.

No such special order was issued, and cooler heads soon prevailed. New Hampshire Senator Daniel Clark sized up the situation quickly: "It is not desirable that there should be a collision between the State and Federal authorities, in this matter." He did not, however, want to see Bachelder released by state authority and thought it best that he be imprisoned out of the state under War Department authority. Then the marshal could tell the justice he did not have the prisoner. Clark preferred this course, but in any event thought that Bachelder "should either be held, or set free by the *Government,* without let or hindrance of local offices."

As it turned out, the chief justice relented, deciding he had not power enough to enforce the writ if refused by the U.S. He thus modeled his behavior on Taney's when he had been faced with defiance of his writ in the *Merryman* case. Bachelder was not eager for a protracted martyrdom and wanted to be released on bond. The arresting marshal thought the "effect of his arrest & his imprisonment have been widely and beneficially felt.... Its influence would not be increased by longer confinement." The judge advocate general then relented, and Bachelder was ordered released on September 26 on taking the oath of allegiance, promising future loyal behavior, and posting $10,000 bond.[12]

In some similar cases, the local authorities failed to show enough initiative to suit Washington authorities. When Democrat and ardent white supremacist C. Chauncey Burr and others "broke up" a recruitment rally "with hooting and noise" in Newark, New Jersey, on August 11, the local U.S. Attorney, A. Q. Keasby, reported it to Levi Turner. The judge advocate pronounced the evidence "abundantly sufficient" to justify arresting the leaders of the mob. "They should have been arrested, at once," Turner admonished, "as is authorized & directed by the order of the 8th, & a report thereof made to this office." Turner also ordered the arrest of the editors of the Newark *Evening Journal,* copies of which Keasby had sent along to the judge advocate.[13]

Whatever such government interference accomplished, it did not include beneficial effects on the future conduct of Burr. He had founded *The Old Guard,* a monthly journal sharply critical of the administration's war effort, in June 1862. Though production was interrupted after August 1862, the journal resumed publication, as bitterly critical as ever, in April 1863.[14] Burr also contributed to a notorious New York Copperhead newspaper called *The Day Book or Caucasian.* He remained one of the noisiest of the nation's peace Democrats and continued until 1869 to publish *The Old Guard,* the pages of which contained some of the most extreme anti-Lincoln views available in periodical literature.[15]

The orders of August 8 caused problems for Pennsylvania's Republican governor when the chief of police in Wilkes-Barre, a man named Ricketts, arrested Ezra B. Chase, George B. Kulp, and Ira Davenport for discouraging enlistments. Chase was the town's district attorney, Kulp a minor functionary, and Davenport a merchant. The day after the War Department orders, Ezra Chase delivered a speech at a meeting in which, Ricketts alleged, he "advised his hearers not to go to the war but to stay at home and go to the polls, and that they could more readily settle the present difficulties of our country, by electing men who would . . . put affairs in such a state as would bring the Southern Confederacy back into the Union without any fighting." After the arrests, Ricketts was served with a writ of habeas corpus. The judge, Ricketts pointed out soberly, "while a good and true man cannot take official notice of a suspension of this writ simply published in the newspapers." Legal commentators would say that the writ still issues even when it is constitutionally suspended, though the arresting officer does not have to obey it. Still, many judges sympathetic to the Republican administration would not have issued writs had they been certain the privilege was suspended. The Lincoln administration would have done well to publicize its suspension more systematically. Notifying the judiciary would also have robbed hostile judges, like Taney, of an issue.

The writ did issue in the Wilkes-Barre case, and Judge Advocate Turner instructed Ricketts to ignore it. But on September 2, Governor Andrew G. Curtin, a Republican, telegraphed Turner to urge parole of the prisoners, as "everything necessary had been accomplished by the arrest." Washington ignored the governor's plea. For his part, Ricketts said that he had risked his reputation to arrest these men for the government, and he felt they should be tried. "What my motives were in making these arrests you may judge," he explained, "when I tell you that I am a pro-slavery democrat that my intercourse with two of the parties has always been friendly and the other I did not know." To top it all, Ricketts was acting chief of police at the time of the order, serving the community as a favor, without pay. Later, he would be sued for false arrest, and the action would drag on into the spring of 1863, when by an Act of Congress such suits were transferred to the United States courts.[16]

In the Wilkes-Barre case, though active Democrats were the victims, their arrests were made by a proslavery Democrat who wanted the men tried by military commission, while a Republican governor attempted to gain their release on parole. Partisan motives are not always easy to sort out in the records of military arrests.

A Republican federal marshal, H. M. Hoxie of Iowa, made perhaps the most damaging of all the arrests immediately subsequent to the August 8 orders when he seized the editor of the Dubuque *Herald* for publishing material to discourage enlistments. The editor was Dennis A. Mahony, an able and energetic Catholic who would do much to harm the reputation of the Lincoln administration from 1862 on. Once Mahony became a prisoner, he mustered an impressive array of affidavits to prove his loyalty. He had even helped raise a Union regiment of Irish-Americans in Iowa. After he landed in Washington's Old Capitol prison, Mahony learned that Archbishop John Hughes, the highest-ranking Catholic prelate in America, was visiting the city and wrote to him for help. Hughes was on intimate terms with Secretary of State Seward and wrote him about the matter on August 27:

> I have known him for some thirty years. At first I thought him an idiot—or what the Irish people, in their strange blendings of charity & poetry together, would have called *"an innocent."*
>
> He is not a traitor, though he may have been foolish. Let him have a chance to take the oath of allegiance . . . and let him off with as little delay as possible.
>
> I have written to him to this effect; but if he should refuse to take the oath of allegiance then let him be dealt with according to a merciful but yet vigorous interpretation of the law.[17]

The prisoner was released on November 11 but was far from silenced. The outraged Iowa editor hastily composed a book, *The Prisoner of State,* published in New York in 1863. In February of that same year, he also founded the Prisoners of State Association, and under that group's auspices, after the war John A. Marshall would write *American Bastile* [*sic*], which became one of the most widely circulated anti-Lincoln books of all time.[18]

In Archbishop Hughes's eyes, Democrat Charles Ingersoll of Philadelphia seemed a more likely ringleader of Northern disloyalty than Mahony. In a speech delivered at a Democratic mass meeting on August 23, Ingersoll was reputed to have said that a more corrupt government than Lincoln's could be found only in the older regions of Asia. And he asked whether any government with so much power had ever accomplished so little on the battlefield. The "whole object of the war . . . [has been] to free the nigger," he complained. On August 27, William Millward, the U.S. marshal in Philadelphia, arrested Ingersoll for thus discouraging enlistments and then informed Stanton that Democratic Judge John Cadwalader would surely issue a writ of habeas corpus and "not recognize the suspension of it under your order." Millward, incidentally, was so unsure who held ultimate authority for internal security in the Lincoln administration that he sent telegrams about the arrest to both the secretary of state and the secretary of war. After Cadwalader issued the expected writ of habeas corpus in the case, the government relented. On August 30, Turner ordered Ingersoll's release.[19]

Such arrests seem not to have alarmed or dissuaded the authorities in Washington; indeed, Turner occasionally urged authorities to arrest first and ask questions afterward. But other incidents dismayed even Turner. Unlike Bachelder, Mahony, and Ingersoll, who were men of influence, some victims were obscure, and their outbursts hardly threatened national security. Charles Anderson, who gave a cheer for Jefferson Davis in Buffalo, New York, seemed to Turner to be only a "worthless vagabond." When Peter Anthony's supposedly treasonable language was reported from Monroe County, New York, Turner dismissed him as a man of no social or political influence who was drunk when he spoke the offending words. Alcohol played a role in numerous cases. John D. Bond, arrested by the Chicago superintendent of police, excused his boisterous language on the grounds of intoxication and quickly enlisted in the army. When Bentley J. Goheen told a recruiting officer in Philadelphia that he would rather take an oath to Jefferson Davis than to Lincoln, he too was drunk; Turner decided he was a worthless vagabond of no influence, anyway. Samuel Strantzenheimer, arrested by a federal marshal in Indianapolis, was drunk when he hurrahed for Jefferson Davis and was a "poor-looking specimen of humanity" to boot. There were still other causes of bizarre behavior besides disloyalty and drunkenness. The provost marshal in Chicago arrested Lewis Bobson for saying that Lincoln was a damned fool and the South's cause was just, but friends quickly informed authorities that Bobson was subject to fits of temporary insanity.[20]

Many incriminating remarks were heard not in boisterously intoxicated or noisily insane public displays of poor judgment but in private conversations. For example, Charles J. Bush, of Fond du Lac County, Wisconsin, was said to have told someone at the end of a recruiting rally that if he were drafted, he would fight for the South. Nicholas Grier of Philadelphia apparently said to a new recruit, "God damn your heart—what did you go into this damned abolition war for?" But his brothers maintained that he was insane, and Turner instructed the authorities in Philadelphia to release Grier, reminding them that "the insane are objects of pity." George Slagel, an unnaturalized German himself, was held for twelve days for having said that he wished all German and Irish soldiers in the U.S. Army had their throats cut.[21]

What lay at the heart of aggressively antiwar sentiment seems apparent after examining a number of the remarks reported by arresting authorities. Though

the arrests under consideration here occurred weeks before Lincoln announced the Emancipation Proclamation on September 22, many of the president's vociferous opponents already assumed that emancipation was the real
object of the war. In Vermont, M. V. Barney was arrested for calling it a damned
abolition war. Two brothers in Frederick, Maryland, James and John Fraley,
threw stones at the officer who came to enroll them for the draft and said that
Lincoln was a damned abolition son of a bitch and ought to be shot. Norman
H. Gray was arrested in Hunter, New York, allegedly for having asked in a
speech at a recruiting rally what the 200,000 Union dead had been sacrificed
for. "The nigger," he answered.

Marshal Phillips reported from Illinois that David Lyon had said: "Anyone
who enlists is a God Damn fool, besides being a God Damn black abolition
son of a bitch." William H. Palmer, a candidate for sheriff in Jackson County,
Ohio, was arrested for discouraging enlistments by writing articles in the *Iron
Valley Express,* in one of which he said that not fifty soldiers would fight to
free Negroes. Another Ohioan, Dr. C. Hamilton Peters of Stoutsville, called the
national conflict a "nigger and abolition war." Still another in Hamilton, Ohio,
spoke of the war as a "nigger-freeing" affair and urged letting the abolitionists
fight it themselves. He proved to be "a mere boy without influence," twenty-
one years old. And in New Jersey, Joseph Wright was alleged to have said that
anyone who enlisted was "no better than a goddamned nigger." If the offending
language reported by arresting authorities contained specific social or political
content rather than mere exhortations or expletives, race and abolition were
the issues most often alluded to.[22]

And yet, did these blustering scoffs, insults, and curses really threaten the
existence of the nation? And how often were denunciations, uttered in private
conversations, reported to authorities because of grudges or old feuds? Turner
saw plenty of this. J. M. Kershaw, for example, assured Attorney General Edward
Bates that his arrest for discouraging enlistments had been the work of a
personal enemy. The undoing of New York City's John C. King had been
witnesses to his remarks who had "an ill feeling against him." And W. W.
Meredith of Kent County, Delaware, produced his own countervailing affidavit
asserting that the charges against him were contrived. To be sure, Turner did
not take all such arrests for "irresponsible outbursts" lightly; he occasionally
ordered trials by military commission for men who cursed the "damned abolition war" or cheered for Jefferson Davis and the Confederacy.[23]

The situation immediately following Stanton's orders of August 8, 1862,
proved a nightmare for civil liberties in the North. Nevertheless, the authorities
were in many ways sincerely attempting to end resistance to enlistment and
the draft. Of the 354 well-documented cases under the orders reported to
Turner from August 8 to September 8, 181 involved young men suspected of
attempting to escape the draft by leaving the country or the state in which
they lived, and 10 more involved persons accused of aiding such attempts.
Thus, a majority—54 percent—of the arrests under the August 8 orders involved men who appeared to be fleeing the draft. Most of them were young

men picked up in towns along the Canadian border or in trains or boats heading for the border.

Nor was the system notably corrupt. Whereas local grudges caused some foolish arrests, in none of the 354 cases were the arresting authorities themselves parties to the feud, avenging personal wrongs or continuing old quarrels. Mahony, for example, was taken from his bed between three and four o'clock in the morning by shadowy figures who did not explain their purpose. He cried "murder" and feared the worst, but when Marshal Hoxie and his deputy emerged from the shadows, Mahony was, by his own admission, "relieved" to see them. Their presence seemed to mean that he was to be arrested rather than assassinated for opposing the war. Mahony was already acquainted with Hoxie and they were able to engage in a reasonable conversation. There was no grudge here; this was not an old enemy come to avenge a private wrong under the guise of public purpose.[24]

Although largely free of personal motivation, the arrests were not always exempt from political prejudice. Most of those arrested for expressing sentiments that discouraged enlistments were surely Democrats, and the arresting authorities were often Republicans. Such facts touch on one of the fears associated with all internal security systems: that they operate in partisan ways. Historian Frank Klement repeatedly suggests in his books and articles on Copperheads that the system was instituted for political gain, that allegations of disloyalty were excuses trumped up by Republicans to eliminate political enemies at convenient times of the year—near elections.

Such was not the case with the orders of August 8, 1862. They were issued on the eve of an autumn election campaign, and their victims were often Democrats, but their purpose was to prevent draft resistance. When they could no longer serve that purpose, the Republican administration revoked the orders. On September 8, 1862, Levi C. Turner circulated the following printed order:

> The quota of volunteers and enrollment of militia having been completed in several States, the necessity for stringent enforcement of the orders of the War Department in respect to volunteering and drafting no longer exists.
>
> Arrests for violation of these orders, and for disloyal practices, will hereafter be made only upon my express warrant, or by direction of the military commander or Governor of the State in which such arrests may be made; and restrictions upon travel, imposed by those orders, are rescinded.[25]

Turner and the War Department would tolerate the busy authoritarianism of the marshals, policemen, and sheriffs only in the face of what they feared was a grave national emergency.

And if the purpose was sincere, enforcement was correspondingly tough. Levi Turner ruled late in August that persons who advertised in newspapers that they would guarantee exemption from the draft for a fee of one dollar should be arrested for discouraging enlistments. What the misleading advertisers apparently offered the young men who answered their ads was a list of substitutes, names of men available at a price to serve in another's place. Substitution was a legal and accepted part of nineteenth-century conscription, and so, too, must have been advertising about substitutes. These ads would hardly have been as dishonest as other customary ads of the period—for patent medicines, for example—but the War Department and Turner despised both the advertisers and those who answered the ads. They thought that the whole substitute business not only discouraged enlistments but was also "tainted with *cowardice.*" The provost marshal in New York City arrested ten men who took out such ads in various city newspapers.[26]

In the past, historians have assumed that the first ten months of the Civil War, when Seward controlled military arrests of civilians, were the worst for civil liberties in the North. The arrests under the August 8 orders suggest that is untrue. Stanton proved much tougher on civil liberties because he was more efficient and, unlike Seward, wielded from his post at the War Department a genuine enforcement apparatus. After Stanton assumed control of the arrests in February 1862, there followed a period of prisoner releases and reviews of cases in the winter and spring of 1862. But the secretary of war was faced with establishing conscription in the summer, and the special measures taken to enforce that novel program, along with the continuing arrests for the same offenses Seward had punished, rendered the highest monthly arrest rate during the war. In fact, August and September 1862 were the only months during the war actually marked by numerous arrests north of the border states. And the rate under Stanton's orders of August 8, 1862, outstripped anything Seward and the State Department ever managed to produce.

At least 354 persons were arrested by the orders of August 8 in one month, the only month the orders operated stringently. By contrast, only 864 civilians were arrested under Seward in the first ten months of the war, a rate of 86 per month, less than a fourth of the rate at which Stanton's juggernaut crushed civil liberties.

The 354 arrests inspired by the orders of August 8 were not the only arrests of the August 8–September 8 period. Counting only cases brought to Levi C. Turner's attention in Washington, one would have to add at least another 97 persons to the figure. In fact, that would approximate the figure for the East alone; few cases of Missourians or Kentuckians were reported to Turner. Nevertheless, the resulting sum, 451 persons in one month, equals over half the 864 persons arrested in the ten months that Seward oversaw internal security (the exact percentage is 52.2). In addition, St. Louis's Gratiot Street prison received at least 191 civilian prisoners arrested between August 8 and September 8 who were never reported to Turner. Myrtle Street received one. Altogether, then, at least 643 arrests occurred in the month following the orders of August 8. In just one month, Stanton presided over arrests equalling in number at least 74.4 percent of the total reached under Seward's entire tenure. The secretary of state's record was quite modest, and the shocking effect of the August 8 orders at last becomes vividly apparent.

Apologists for this internal security system in the past have insisted that

Abraham Lincoln's humane, charitable, and lenient personal character colored the whole system, or at least ameliorated its harsher mistakes. This was never true, and it is especially evident after examining the August 8–September 8 period. Lincoln himself intervened to correct a wrongful arrest under the orders of August 8 in only one case: that of Dr. Blanchard.

In the same period, Lincoln also asked the secretary of war to respond to a telegram from a Kentucky official protesting "indiscriminate arrests" of "quiet law abiding men holding state rights dogmas ... required to take an oath repulsive to them or go to prison." Stanton dutifully told General Jeremiah T. Boyle to exercise caution and to arrest "persons in civil life" only "where good cause exists or strong evidence of hostility to the Government." And he explained to the Kentuckians that the arrests complained of—including those of two men over seventy years old—had "not been directed by this Department," an excuse often used to minimize Washington's responsibility for military arrests of civilians.[27]

The only historian who has previously stressed the importance of the orders of August 8, Robert J. Chandler, failed to notice Turner's order of September 8 ending stringent application of the orders, so he was unable to see the administration's sincere focus on the problem of draft resistance. Chandler therefore succumbed to the temptation to try to find more in the orders than was really there:

> The secretive Lincoln probably had additional motives: The two orders [of August 8] were a necessary prelude to emancipation of Southern slaves. Following the collapse of negotiations with the border states for compensated emancipation, Lincoln presented a draft of the Emancipation Proclamation to his cabinet on July 22, 1862. He postponed issuing the document until after the Union army won a battle, but in the meantime he worked to build public acceptance.
>
> The orders for arrest would deter opposition to emancipation, allow the administration to move against antiwar Democrats, and even if unsuccessful in operation, show Lincoln's firmness. He publicly illustrated the tie between both policies two days after the preliminary Emancipation Proclamation appeared. On September 24, Lincoln issued a second proclamation following the language and provisions of the two August orders, and Stanton established a network of provost marshals to enforce it.[28]

In fact, the War Department relaxed enforcement of the orders of August 8 two weeks before announcement of the Emancipation Proclamation on September 22.

Chandler has done careful work on Civil War dissent, and his argument is therefore especially symptomatic of more fundamental problems with the historical literature on civil liberties in the Civil War. Historians have long tried to unearth hidden meanings in the orders and proclamations suspending the writ of habeas corpus, while neglecting their straightforward meanings and true intent. They have been too willing to take the literature written by the opposition at face value, while searching for hidden motives behind the arguments in favor of suspending the writ. An early forerunner of Chandler's interpretation was a speech by James Brooks, who addressed the Democratic Union Association of New York on September 29, 1862, on "The Two Proclamations." Brooks told his audience: "The proclamation is a corollary of Proclamation No. 1. It substantially says to the free white people of the North, if you discuss and agitate this subject of emancipation, if you make war against the Administration upon this subject, you shall be incarcerated in Fort LaFayette."[29]

The orders of August 8, 1862, were aimed at enforcing the first national conscription in American history. Their issuance was unrelated to the Emancipation Proclamation, the timing of which was premised on the unpredictable: a military victory. Though flawed in execution in predictably partisan ways— Republican functionaries tended to imagine too much threat to military recruitment in Democratic criticism of the president—the arrests, nevertheless, were meant to reduce obstacles to military recruitment. The best proof of that is the 54 percent of the arrests that netted young men (and their accessories) heading for Canada or other areas to escape conscription.

The sort of arrests that occurred under the orders of August 8 did not end with the September 8 order. They merely diminished in frequency as higher authorities took control. (Indeed, some such arrests had occurred before the orders of August 8. For example, when a Missourian named Thomas C. McDowell said on August 6, "[I] wouldn't wipe my ass with the stars and stripes," General Halleck had him arrested.[30]) One of the most laughably wrongheaded arrests occurred in Baltimore in the second week of October, when David T. Shaw, a hotelier of British origins, was arrested for making disloyal remarks. He was accused of saying that he "would be the first man to take a rope to hang Abe Lincoln" and that he hoped the Union would lose the war and Union men in Baltimore be poisoned. Shaw was drunk at the time, and when in more sober condition he appealed for release, he vehemently protested his loyalty by insisting that he had written the patriotic song, *Columbia, the Gem of the Ocean,* as indeed he had![31]

The nature of military arrests of civilians began to change gradually after September 8, 1862, in large part as a result of changing War Department policy. In December, for example, Turner instructed an overeager Ohioan to release a prisoner arrested for having been overheard to say he had voted the "secession ticket." "Persons now in custody for disloyal utterances are discharged," Turner wrote, "besides the affidavits show that he was intoxicated and obviously 'game not worth the candle.'" In a more exasperated tone, he instructed another Union officer: "To hurrah for Jeff Davis simply will not warrant the order of arrest by this department."[32]

The dramatic outcome of the orders of August 8 makes clear why Lincoln's proclamation of September 24, 1862, the first one to suspend the writ across the whole nation, was an anticlimax. The system had already been in effect for six weeks. It had made headlines and brought forth habeas-corpus writs and Democratic ire. The damage had already been done when Lincoln issued the essentially redundant proclamation:

Whereas, it has become necessary to call into service not only volunteers but also portions of the militia of the States by draft in order to suppress the insurrection existing in the United States, and disloyal persons are not adequately restrained by the ordinary processes of law from hindering this measure and from giving aid and comfort in various ways to the insurrection:

Now, therefore, be it ordered, first, that during the existing insurrection and as a necessary measure for suppressing the same, all Rebels and Insurgents, their aiders and abettors within the United States, and all persons discouraging volunteer enlistments, resisting militia drafts, or guilty of any disloyal practice, affording aid and comfort to Rebels against the authority of the United States, shall be subject to martial law and liable to trial and punishment by Courts Martial or Military Commission:

Second. That the Writ of Habeas Corpus is suspended in respect to all persons arrested, or who are now, or hereafter during the rebellion shall be, imprisoned in any fort, camp, arsenal, military prison, or other place of confinement by any military authority or by the sentence of any Court Martial or Military Commission.[33]

The restriction of civil liberties in the North was now geographically complete. From the autumn of 1862 to the end of the war, persons who discouraged enlistments, impeded the draft, or afforded aid and comfort to the enemy were theoretically subject to martial law: arrest, trial, and punishment by the U.S. Army. Once imprisoned, these persons might have no recourse to civil courts, for the writ of habeas corpus was suspended for any such person. Trials by military commission were fully established. Moreover, martial law applied to all the cases mentioned in the proclamation and the categories of offenses were vague enough, in effect, to have placed the whole of the United States under martial law.

Yet the administration felt no more secure, no better equipped to cope with disloyalty in the North than before. Though couched in broad and sweeping language of long-range significance, the proclamation of September 24, 1862, in the minds of the administration, was really only an ad hoc remedy for an immediate problem—resistance to the draft. As soon as there occurred a crisis of another sort involving alleged disloyalty, the Lincoln administration acted as though the proclamation of September 24 had never been issued.

The Vallandigham Case

Union military fortunes did not improve over the winter of 1862–1863, and war weariness shortened partisan tempers. On May 1, 1863, General Ambrose Burnside caused the most famous arrest of a civilian during the whole Civil War, that of ex-Congressman Clement L. Vallandigham, a Democrat from Ohio. A military commission the very next day sentenced him to imprisonment for the duration of the war. The president and his cabinet learned of the arrest and trial from the newspapers. They did not much like what Burnside had done, but they saw no alternative to backing up the general afterward.[34]

Vallandigham had been courting such martyrdom and promptly applied for a writ of habeas corpus. Rumor had it that he might file his application in the Ohio federal district court with Justice Noah H. Swayne, a Lincoln appointee to the U.S. Supreme Court who would be sitting as a circuit judge. Stanton feared that Swayne might issue the writ. Therefore on May 13, he drafted this general order:

> Whereas C. L. Vallandigham ... is now held ... as a prisoner by Major General Burnside ... ; and whereas, in the judgment of the President, public safety requires the suspension of the privilege of the writ of habeas corpus in the case of said Vallandigham. ...
>
> Now, therefore, it is hereby ordered by the President that the privilege of the writ of habeas corpus, and any writ which has been or may be hereafter issued during the present rebellion in the case of said Vallandigham be and the same is hereby suspended.

In a companion order, Stanton instructed Burnside to inform the court of the general order suspending the writ of habeas corpus for Vallandigham.[35]

Both orders were left on the president's desk, but Lincoln never signed them. The proposed special suspension in Vallandigham's case was surely legally superfluous. The allegation that "Valiant Val" had publicly expressed "sympathies for those in arms against the Government of the United States" and declared "disloyal sentiments and opinions, with the object and purpose of weakening the ... Government in its effort to suppress the unlawful rebellion" surely qualified him as an "aider and abettor" of the insurgents who gave "aid and comfort to Rebels against the authority of the United States," as Lincoln's September 24 proclamation stated. However, Lincoln's reason for not signing the orders had nothing to do with their redundant nature. Instead, the president consulted Seward and Chase, who advised him not to "issue the special suspension of the Writ of Habeas Corpus." The secretary of the treasury had heard that the case was before Judge Humphrey Leavitt, not Justice Swayne, and thought "that the writ will probably not issue, whichever the application may be before." Chase was an Ohioan who read the newspapers from home, and in one of these, he had seen that Judge Leavitt "stated that Judge Swaine & he refused a similar application last year."[36]

Stanton's proposed special suspension constituted nothing more nor less than bullying the judge and was thus typical of the secretary's overbearing and confrontational style. Chase, Seward, and Lincoln proved more politic. In typical fashion, the president heeded the practical advice that neither judge was likely to issue a writ. Lincoln may still have been unsure of the legal ground, but he recognized good political advice when he got it.

Nevertheless, the Vallandigham arrest, which led to widespread Democratic protests throughout the North, so dramatized the habeas-corpus issue that the president was at last compelled to explain his policy again to the American people. This time he utilized homely examples and minimized the tone of self-doubt, writing "in language," as Nicolay and Hay described it later, "so terse and vigorous that it is difficult to abridge a paragraph without positive mutilation." The occasion was a formal reply sent on June 12, 1863, to Erastus Corning and other Democrats who had protested Vallandigham's arrest at a mass meeting in Albany, New York.[37]

The president's vigorous tone and able style differed from his earlier attempt to defend the policy in July 1861, though the constitutional argument had not changed much. The military arrest of civilians came "within the exceptions of the Constitution" specifically permitted for rebellions. Lincoln had acted "without ruinous waste of time" to do what Congress would have done had it been in session. He had nevertheless been "slow to adopt" these admittedly "strong" measures and had done so only "by...degrees." Among new elements in his letter was an accusation of conspiracy aimed at his civil libertarian opponents. The rebels, Lincoln argued, had planned all along "to keep on foot amongst us a most efficient corps of spies, informers, supplyers and aiders and abettors of their cause" under cover of specious cries for liberty of speech, liberty of press, and habeas corpus. They knew "they had friends who would make a question as to *who* was to suspend" the writ of habeas corpus. Lincoln's accusations appeared more believable in that day, when the electorate had been shaped by Republican charges of a "Slave Power conspiracy" in the middle and late 1850s.[38]

The peculiar nature of the Vallandigham arrest dictated other fresh elements in the president's letter. Although Lincoln did not say so, this prominent Ohio Democrat was neither the sort of person meant to be arrested nor the sort who usually was. All along, the policy had been aimed at dangerous Confederate sympathizers like John Merryman, at bridge-burners in Missouri, and at the legions of draft dodgers throughout the land. All along, the policy threatened freedom of speech and the press as well, but only incidentally. Because Vallandigham, a bona fide political activist, was arrested for a speech he gave, the president here had to defend the excesses of the internal security policy rather than its normal operations. Lincoln admitted that some innocent people were bound to be arrested under such a system and argued that the rebels had counted on this: their friends in the North would raise "a clamor...in regard to this, which might be...of some service to the insurgent cause."[39]

The New York protest also forced Lincoln to address an issue he had avoided in his 1861 message to Congress. The Albany Democrats had insisted that arrests should not be made "outside of the lines of necessary military occupation, and the scenes of insurrection," to which Lincoln replied:

> Inasmuch...as the constitution itself makes no such distinction, I am unable to believe that there is any such constitutional distinction. I concede that the class of arrests complained of, can be constitutional only when, in cases of Rebellion or Invasion, the public Safety may require them; and I insist that in such cases, they are constitutional *wherever* the public safety does require them—as well in places to which they may prevent the rebellion extending, as in those where it may be already prevailing—as well where they may restrain mischievous interference with the raising and supplying of armies,...as where the rebellion may actually be.

The president said it would have been wrong to arrest Vallandigham merely because his speech "was damaging the political prospects of the administration." But such was not the case. Vallandigham was hurting the army by discouraging enlistments and encouraging desertion. Then Lincoln devised what had been conspicuously lacking in his 1861 defense, a homely example that made his point without Latinate distinctions or close constitutional reasoning. "Must I shoot a simple-minded soldier boy who deserts, while I must not touch a hair of a wiley agitator who induces him to desert?" Lincoln asked. "I think that in such a case, to silence the agitator, and save the boy, is not only constitutional, but, withal, a great mercy." Burnside's unfortunate act caused Lincoln to fight on ground not of his own choosing, but he fought exceedingly well. "Few of the President's state papers," Nicolay and Hay recalled, " . . . produced a stronger impression upon the public mind than this."[40]

The Proclamation of September 15, 1863

Well before the Vallandigham arrest in May, the United States Congress had acted to eliminate doubt about the legality of the suspension of the writ of habeas corpus. Its Habeas Corpus Act of March 3, 1863, was a model of legislative ambiguity that left it unclear whether the presidential suspensions before the act had always been legal or were legal only now because of congressional approval.[41] Whatever the case, suspension was surely legal now. Anyone who doubts Lincoln's profound lack of interest in constitutional theory should contemplate this act of Congress and the president's defense of his habeas-corpus policies in the Corning letter three months later. In some three thousand words of justification in the public letter, President Lincoln never once mentioned the Habeas Corpus Act. He apparently felt justified without it.

In fact, he made his own views clear in a letter written on June 29, 1863, to answer the Ohio State Democratic convention's protest of Vallandigham's arrest. Their question, Lincoln wrote,

divested of the phraseology calculated to represent me as struggling for an arbitrary personal prerogative, is either simply a question *who* shall decide, or an affirmation that *nobody* shall decide, what the public safety does require, in cases of Rebellion or Invasion. The constitution contemplates the question as likely to occur for decision, but it does not expressly declare who is to decide it. By necessary implication, when Rebellion or Invasion comes, the decision is to be made, from time to time; and I think the man whom, for the time, the people have, under the constitution, made the commander-in-chief, of their Army and Navy, is the man who holds the power, and bears the responsibility of making it.

Nevertheless, the official proclamations suspending the writ of habeas corpus—and more were to come despite the seeming completeness of suspension by the summer of 1863—would hereafter be based both on the U.S. Constitution and the Habeas Corpus Act. But this was the result less of Lincoln's own thinking than of his reliance on cabinet members to draft routine proclamations.[42]

The later proclamations had similar origins to the ones already on the books of the Lincoln administration. They arose as practical responses to immediate problems and seemed unconscious of legal groundwork previously laid.

Conscription continued to be the most provocative issue. On the same date that the Habeas Corpus Act passed in March 1863, Congress also passed a genuine conscription act, which it disingenuously called the Enrollment Act. As its previous evasive militia draft had provided for only nine months' service by the draftees, Congress had to act by May; more important, Union manpower was flagging. Conscription was a necessity, but Congress understandably feared its consequences and shied away from calling the law a conscription act.

When the first New York draftees under the new law were called up on July 11, 1863, the greatest incidence of domestic violence in all of United States history ensued: the New York City draft riots.[43] Afterward, the unpopularity of conscription became a political byword, a serious practical problem for the Lincoln administration, and a genuine impediment in many people's eyes to the war effort. Lincoln's mind boggled at government officials who opposed the draft and at political obstructionism that impeded the draft. He could comprehend the reluctance of the common man, but resistance from responsible men infuriated him.

After the war, the provost marshal general recalled that in the 1863 draft "the practice of serving writs of *habeas corpus* on the officers of the bureau," which was charged with enforcing conscription, "became so prevalent as to interfere seriously with the progress of the business." War Department solicitor William Whiting advised that such writs, if issued by state courts, should be ignored. But the provost marshals were to obey writs issued by United States courts and confusion reigned. The worst problem arose in Pennsylvania. There, the writs of state courts "threatened for a time, in several districts to defeat, or at least to suspend, the business of raising troops and arresting deserters, and either to throw the officers of this bureau into custody, or keep them so constantly before the courts as to prevent their attendance upon the duties for which they were appointed."[44]

This practice greatly provoked President Lincoln. Indeed, one must look back to the anxious April days of 1861 to find Lincoln so apparently exasperated. In the midst of the crisis in Pennsylvania, he called a special cabinet meeting on Monday morning, September 14, 1863, to discuss the problem. Attorney General Bates found the president "more angry than I ever saw him." "The course pursued by certain judges," Lincoln is reported to have said, "is defeating the draft." The courts were "discharging the drafted men rapidly under *habeas corpus,*" and Lincoln was "determined to put a stop to these factious and mischievous proceedings" if he had the authority. According to Bates, the president even "declared that it was a formed plan of the democratic copperheads, deliberately acted out to defeat the Govt., and aid the enemy. That no honest man did or could believe that the State Judges have any such power."[45]

Bates and Seward both expressed belief that the president already possessed requisite authority to end the practice. The secretary of war remained silent, despite the obvious relevance of the subject to his department, and that curious circumstance led Gideon Welles to surmise that Stanton had instigated the president's action. The secretary of the navy, though often a conservative critic of Seward and Stanton, in this instance agreed wholeheartedly that something must be done. He had himself introduced the subject on previous occasions, for the navy had been "suffering constant annoyance—vessels were delayed on the eve of sailing, by interference of State judges who assumed jurisdiction and authority to discharge enlisted men . . . on *habeas corpus.*" A "factious and evil-minded judge—and we had many such holding State appointments—could . . . stop armies on the march." He thought that "United States judges were the only proper officers to decide on these naval and military cases."[46]

No one expressed opposition to the president until Salmon P. Chase raised some doubts about taking a headlong course toward confrontation with the judges. As Chase described it in his diary,

> . . . I had always been accustomed to regard the Writ of Habeas Corpus as a most important safeguard of personal liberty. It has been generally conceded, . . . or at least such has been the practice, that State Courts may issue Writs of Habeas Corpus for persons detained as enlisted soldiers, and to discharge them. Several cases of this kind have occurred in Ohio, and the proceeding of the State Court was never questioned, to my knowledge. Of course, a proper exercise of the power does not justify its improper exercise. If the Writ is abused with a criminal purpose of breaking up the Army, the persons who abuse it should be punished as any other criminals are. But before taking any action, which even seems to set aside the writ, a clear case should be made, which will command the concurrence of the people and their approval. I suggest, therefore, that the Secretary of War should make a statement of the number of persons discharged from military service under the Writ, and such notes of the circumstances as will show the abuse of it.

Chase feared that if a confrontation took place civil war might break out in Pennsylvania and elsewhere in the North.

The president seemed unwilling at first to accommodate Chase's doubts. He would not only enforce the law, he said, but send the interfering judges after Vallandigham if they continued their course. The discussion proceeded, with Blair and Interior Secretary John P. Usher more or less supporting Chase. Blair said that as a state judge in Missouri he had himself discharged soldiers on writs of habeas corpus. Bates, Seward, and Stanton insisted on action. Bates took the rather astonishing view that the president was above all process as commander-in-chief of the army and navy and could himself disregard the judges and instruct his subordinate officers to do so as well—all without even suspending the writ of habeas corpus. Stanton, apparently responding to Chase's question about numbers of men released under the writ, pointed out that state judges were not the only or even the prime offenders. Judge John Cadwalader, of the U.S. District Court for Eastern Pennsylvania in Philadelphia, and Judge Wilson McCandless, of the U.S. District Court for Western Pennsylvania in Pittsburgh, both appointees of James Buchanan, had released more soldiers than all the state judges put together. Finally, the president promised to prepare a paper on the subject for discussion by the cabinet at a meeting the next morning.[47]

Chase may have cared more about the law and probably knew more about it than Lincoln or anyone else in his cabinet. It was true that state courts had jurisdiction to issue the writ of habeas corpus for federal soldiers held by federal officers. State court decisions had been all but unanimous on that point, and Chancellor Kent had eventually come around to the view as well in his famous *Commentaries.*[48] Confronted with such precedents, Lincoln, had he known his law better, might have responded by citing Roger B. Taney's own opinion in *Ableman v. Booth,* in which he had argued the necessity of federal power to do its constitutional duty without interference from writs issued by state judges. But, like Welles, Lincoln knew only the practical necessity of forbidding such interference with mobilization in a major war. Besides, federal judges were behaving no better than state judges.

On the morning of September 15, at a second cabinet meeting on the subject, Lincoln read his proposed order, substantially as follows:

> Please order each Military officer of the United States, that, whenever he shall have in his custody any person, by the authority of the United States and any writ of *habeas corpus* shall be served upon him, commanding him to produce such person before any court or judge, he, the said Military officer, make known to the court or judge issuing such writ, by a proper return thereto, that he holds such person by the authority of the President of the United States, acting as the Commander-in-Chief, of the Army of the United States in time of actual rebellion and war against the United States; and that, thereupon he do not produce such person according said writ, but that he deal with him according to the orders of his military superiors; and that, having made known to such court or judge, the cause of holding said person as aforesaid, if said court or judge, shall issue any process for his, said officer's, arrest, he refuse obedience thereto; and that, in case there shall be any attempt, to take such person from the custody of such officer, or to arrest such officer, he resist such attempt, calling to his aid any force that may be necessary to make such resistance effectual.[49]

Here was the very provocation to civil war in the North that Chase had feared: an order that actually contemplated the use of armed force by the executive to resist officers of a court enforcing judicial orders. Lincoln's directive seemed to raise the spectre of armed conflict between two branches of the government.

Chase felt impelled to speak again. He pointed out that the draft order "does not suspend the writ in terms, though it probably does in effect." This would leave "the question of suspension open to debate, and will lead to serious collisions probably, with the disadvantage on the side of the Federal authority." Chase suggested that "instead of this order there should be a Proclamation

distinctly suspending the Writ of Habeas Corpus so far as may be necessary
to prevent the great evil of virtually disbanding the Army.... By this bold and
direct action, I think you will command the confidence of the public, avoid
collisions on uncertain grounds, and secure most completely the great objects
you have in view."[50] Everyone agreed, perhaps recognizing these cases as
something different. Lincoln turned the routine proclamation-writing over to
Seward.

The secretary of state came up with a draft which, somewhat modified by
a cabinet discussion that afternoon, became the habeas-corpus proclamation
of September 15, 1863:

> Whereas the Constitution... has ordained that the privilege of the Writ of Habeas
> Corpus shall not be suspended unless when in cases of rebellion or invasion the
> public safety may require it, And whereas a rebellion was existing on the third
> day of March, 1863, which rebellion is still existing; and whereas by a statute
> which was approved on that day, it was enacted by the Senate and House of
> Representatives of the United States... that, during the present insurrection, the
> President..., whenever, in his judgment, the Public safety may require, is au-
> thorized to suspend the privilege of the Writ of Habeas Corpus in any case through-
> out the United States or any part thereof; and whereas in the judgment of the
> President the public safety does require that the privilege of the said writ shall
> now be suspended throughout the United States in the cases where, by the authority
> of the President..., military, naval and civil officers of the United States or any of
> them hold persons under their command or in their custody either as prisoners
> of war, spies, or aiders or abettors of the enemy; or officers, soldiers or seamen
> enrolled or drafted or mustered or enlisted in or belonging to the land or naval
> forces of the United States or as deserters therefrom or otherwise amenable to
> military law, or the Rules and Articles of War or the rules or regulations prescribed
> for the military or naval services... or for resisting a draft or for any other offense
> against the military or naval service. Now, therefore, I, Abraham Lincoln, President
> of the United States, do hereby proclaim... that the privilege of the Writ of Habeas
> Corpus is suspended throughout the United States in the several cases before
> mentioned, and that this suspension will continue throughout the duration of the
> said rebellion, or until this proclamation shall, by a subsequent one to be issued
> by the President... be modified or revoked.[51]

After this reiteration of public purpose, the president quietly issued the provocative orders to the provost marshals two days later on September 17.

The dates of these documents are quite deceiving, and only by considering
the real order of their composition can one measure the evolving feelings and
intentions of President Lincoln. The habeas-corpus proclamation had become
a ritual exercise, a gesture—one might almost say a public relations gambit.
The real intention—defiance of judicial writs of habeas corpus that interfered
with the draft and the punishment of deserters—was embodied in tough language afterward. Form followed function in Lincoln's mind but, in this instance,
not in the way in which the policies were announced to the American public.

Chase was clearly on to something, though Gideon Welles, characteristically,
pooh-poohed it. "He feared," the secretary of the navy said of the secretary of

the treasury, "if the President acted on Executive authority a civil war in the Free States would be inevitable;—fears popular tumult, would not offend Congress, etc. I have none of his apprehensions, and if it is the duty of the President, would not permit legislative aggression, but maintain the prerogative of the Executive." On the 18th, Welles reported that "the proclamation suspending the privilege of the writ of *habeas corpus* has been generally well received. I have never feared the popular pulse would not beat a healthful response even to a stringent measure in these times, if the public good demanded it."[52]

Chase, on the other hand, saw the president on the 17th, the day the tough order embodying his real intentions went to the provost marshals, and in a brief discussion of the issue, Lincoln said that Chase had been "quite right" to recommend the proclamation suspending the writ of habeas corpus rather than the issuance of the raw order to the provost marshals. According to the secretary of the treasury, Lincoln said that "he had been convinced of it as soon as he heard my statement of the law."[53]

Despite the unusually extensive documentation of the origins of this habeas-corpus order, it is still difficult to sort out what really happened. Very likely Welles was wrong to think that Lincoln bowed to the concern over congressional rather than executive authority for this dangerous act. Surely that issue had been settled in Lincoln's mind long before.

Likewise, Chase's assertion that his legal arguments had swayed Lincoln was at most only partly true. What Lincoln took to heart was probably less the legal than the political advice from Chase. If the president simply took the bold action he was determined to take without any ceremony of suspension, it would leave matters "open to debate" and would "lead to collisions." He needed to "command the confidence of the public." Though the Pennsylvania problem was novel, proclamations suspending the writ of habeas corpus were becoming not so much official legal notices as government propaganda.

Questions of legal and constitutional form still took a back seat in the Lincoln administration. At the end of the affair of the Pennsylvania judges, Chase commented in his diary, "I was surprised to find that in a matter of this importance, no one but myself seemed to have read the Act of March 3d with reference to the subject under discussion, and that its provisions were unfamiliar to all."[54]

If law and the Constitution were not the first considerations, neither was politics in the narrowly partisan sense.[55] No Northern Republican governor treated the habeas-corpus issue more gingerly than Pennsylvania's Andrew G. Curtin, a candidate for reelection in the fall of 1863. He had issued a somewhat embarrassed message the preceding February asking the Pennsylvania legislature to pass a joint resolution about military arrests of civilians. Curtin wanted the legislators to urge the U.S. Congress to provide "proper legislation" on the matter:

Whether such legislation should include a suspension of the writ of habeas corpus in any, and what part of the country, is a question which belongs exclusively to the legislative authorities of the United States, who under the Constitution have the right to determine it. That great writ ought not to be suspended, unless, to the wisdom of Congress, the present necessity shall appear urgent.

The Habeas Corpus Act of March 3 had not been passed yet, and this letter condemned the acts of the Republican president without mentioning his name. Governor Curtin had decided that only Congress could suspend the writ— the heart of the Democratic position on the habeas-corpus question until mid-1863. The governor's message ended with a plea for "laws defining and punishing offenses" that were dangerous but stopped "short of the technical offense of treason." And he asked that those laws provide "for the fair and speedy trial by an impartial jury of persons charged with such offenses, in the loyal and undisturbed States, so that the guilty may justly suffer and the innocent be relieved."

Less than two weeks before the Pennsylvania judges provoked Lincoln's September 15 suspension, Curtin told the president, "If the election were to occur now, the result would be extremely doubtful, and . . . my impression is, the chances would be against us. The draft is very odious in the State, and unfortunately is not producing more than one-sixth of the men anticipated for the public service."[56] Curtin did not ask for local relief from the conscription law, but he did list it first among factors working against the Republicans in Pennsylvania. Any presidential actions aimed at toughening enforcement of the draft would hardly constitute a political asset.

Lincoln ignored Governor Curtin's political fears. And after Lincoln's proclamation and orders of mid-September, Curtin wrote again, saying, "The proclamation suspending the writ of *habeas corpus* is a heavy blow but as it is right we can stand it."[57]

4

Arrests Move South

As the number of military arrests of civilians in the North fell from the forbidding heights of August 1862, the most important new developments occurred not in Washington cabinet meetings but in the field. The president and his advisors appeared to view suspension of the writ of habeas corpus as a means of enforcing the odious draft in the North, but the military prisons filled with civilians from the South arrested for other kinds of offenses.

The "Southernization" of Civilian Arrests

Though nothing matched the frenzy of August 1862, the overall rate of military arrests of civilians rose in 1863 to a level above that of the first year of the war. The increase can be judged by comparing the period when William H. Seward oversaw internal security (15 April 1861–15 February 1862), with the same dates in 1863–1864, when Edwin M. Stanton controlled the program. Whereas 864 civilians had been arrested throughout the country in the first period, 1,111 citizens were held in the later period in Washington's military prisons alone, not to mention the other forts (numerous by 1864) where civilian prisoners were kept. The D.C. prisoners were meticulously reported in Levi C. Turner's manuscript Record of Prisoners of State, a handwritten ledger that included essentially only those persons arrested in Maryland, the District of Columbia, and Virginia.

Although military arrests of civilians remained a border-state problem, the number of Marylanders arrested had fallen from the early days. In 1861, 166 of the 509 cases of known residence involved Marylanders, that is, 32.6 percent. In 1863–1864, the number of arrests in Maryland constituted but 136 of 1,001 cases where place of arrest is known, or only 13.6 percent. The absolute number of Marylanders arrested had fallen only a little, but the tempo of arrests had increased significantly elsewhere.

They had burgeoned across the Potomac, where Union forces at last controlled some Confederate territory. As of January 1, 1863, Lincoln proclaimed "the counties of Berkley, Accomac, Northampton, Elizabeth-City, York, Princess Ann, and Norfolk" and "the fortyeight counties designated as West Virginia" fully under Union military control.[1] And contested areas of the state naturally produced civilian prisoners as well. Of the 1,001 cases in the 1863–1864 period where place of arrest is known, 709 took place in Virginia. Military arrests of civilians were moving south.

Conversely, military arrests of civilians were not increasing in the North. Secretary of War Stanton managed to hold on to Maryland, for example, while making arrests there at a slower rate than Seward had. Stanton was arresting, on average, about 13 persons per month in Maryland.[2] And not all of the persons arrested in Maryland were Marylanders. Of the 120 citizens arrested in the state in this period, for whom cause of arrest can be established, 25 were, or claimed to be, refugees from the South—20.8 percent. This influx of citizens attempting to escape to the Union was bound to increase as the fortunes of the Confederacy declined.

Most refugees wanted to go someplace where they could resume their ordinary occupations. Despite its reputation as a state straining at the bonds of Union, by 1863 Maryland had more the appearance of a society going about its ordinary business, or attempting to do so, while situated on the edge of the greatest war in the country's history. In fact, the commonest infraction causing Marylanders in 1863–1864 to run afoul of Union authorities involved commerce. Variously described in the prison records as violations of the blockade, involvement in contraband trade, or smuggling, these activities, no doubt, represented in substantial part only an attempt by entrepreneurs to resume business as usual. Now, however, old patterns of trade to the South were no longer legal, and attempting to follow in the familiar groove might land a merchant in prison.[3]

The greatest number of arrests came in St. Mary's and Charles counties, south of Washington and directly across the Potomac from Confederate Virginia; 49.6 percent of the arrests in Maryland in this period took place in those two counties. They were less the breeding grounds of active political disloyalty than thresholds of trade and travel between the warring sections of the old Union. Of sixty arrests in those two counties for which the charge is known, twenty-six, or 43.3 percent, involved running the blockade, smuggling, or carrying contraband goods. Another twenty-two, or 36.7 percent, netted refugees from the Confederacy. Thus fully 80 percent of the persons arrested in Charles and St. Mary's counties were traveling and trafficking rather than plotting or sniping. Political loyalty was not exactly at issue in these arrests. At least four of the contraband traders, for example, were black men, surely more interested in the economics of trade across the Potomac than in aiding the Confederacy. Only four were arrested specifically for disloyalty. Other charges included: one man arrested for having formerly been enlisted in J. E. B. Stuart's cavalry; three, as suspicious characters; one, to be held for information; one, as a carrier of "rebel mail"; one for attempting to cross the Potomac without a pass; and one, as a deserter from the U.S. Army. Altogether, that made twelve arrests for offenses other than traveling or doing business across a forbidden border.

The pattern for the rest of the state was about like that in St. Mary's and Charles counties. In addition to the 20.8 percent who were refugees, 35.8 percent of the persons arrested in Maryland in this period for whom cause of arrest is known were violating the blockade in some way. Throughout the state, then, over half the people arrested ran afoul of travel and trade restrictions—56.6 percent. A dozen men were arrested as "suspicious characters" and another eleven for "disloyalty." The rest had allegedly served time in the Confederate army, aided someone to cross the Potomac illegally, attempted to cross the river themselves without a pass, or defrauded the government while working in the large army corral located in Maryland.

A majority of the offenses had little to do with political questions. True, blockade-running could be a deliberate attempt to aid the Confederacy with vital war supplies, but most often it was nothing more than a petty attempt to make money—or to make ends meet. Edward Downs, for example, came across the Potomac from Westmoreland County, Virginia, to obtain supplies for his family. He was arrested in St. Mary's County, Maryland, on November 14, 1863, as a "blockade-runner." Interrogators reported that Downs, a poor man with eight children, admitted his crime but maintained he had no other way to obtain food for his family.[4]

The Case of the Disappearing Noncombatants

Civilian refugees and petty contrabandists like Downs were the sorts of prisoners routinely created by war. They are relatively unfamiliar to Americans because so few of our wars have been fought on U.S. land. But wherever armies have fought, civilians have gotten in the way. In the American Civil War, such ordinary citizens were arrested and, some of them, sent back to prisons where they swelled the rolls of civilian prisoners and were later misinterpreted by historians as "political prisoners" or victims of "arbitrary arrest." Members of the Lincoln administration often used the term "political prisoner," which seems almost a moral category today, to designate any civilian in military custody (though the more customary official term was "prisoner of state"). The poor Virginia farmer Downs and the self-conscious political martyr Vallandigham alike fell under that category. To understand the prison statistics of the Lincoln administration, one must give such terms as "political prisoner" or "prisoner of state" their proper historical definition rather than their morally charged one of modern times.

Though many civilian prisoners were of a commonplace type by the standards of Continental European warfare between contiguous nations in the nineteenth century, the American Civil War did in other small ways anticipate the wars of the twentieth century, which have increasingly victimized civilians.

By the fourth year of the war, the logic of military events was moving some generals toward ideas that undermined ages-old distinctions between soldiers and civilians. These changes were not dictated by innovations in military technology. The ballistic and logistical nature of land combat in the American Civil War—despite rifled artillery, railroads, and the telegraph—bore a close resemblance to the wars of the eighteenth century and the Napoleonic wars of the early nineteenth.[5] Weapons, as Civil War soldiers used them, were not so much more indiscriminately destructive nor the speed with which armies could invade enemy territory so much greater that most civilians could not get out of harm's way. The principal agencies of change were conscription and guerrilla tactics.

Civil War generals sometimes reacted to guerrilla warfare and conscription by the enemy with sweeping judgments and harsh orders that, if obeyed by their subordinates, would have created tens of thousands of civilian prisoners on every front. While these developments in military thought are interesting, they took place largely only in the realm of ideas. Practical realities usually intervened to prevent their systematic application to the American Civil War, which therefore was fought, by and large, within the customary restraints of civilized warfare on the Western European model. Nevertheless, some generals began at least to think in newly destructive ways.

Among the first to recognize the changes and to respond to them with remorseless logic was Ulysses S. Grant. He kept his eye on large factors that determined the numbers of combat-ready men the enemy could put in the field. He also formulated pithy and articulate statements that made his arguments clear and persuasive to superiors and inferiors alike in the chain of command. Happily, his views were usually informed by a practical wisdom that halted his logic before it marched toward excess or fanaticism.

The logic of Confederate conscription worked to break down the distinction between combatants and noncombatants in Grant's mind. By the spring of 1864, he and President Lincoln understood that Robert E. Lee's army was the major objective in the eastern theater. Grant told Benjamin F. Butler that "Lee's Army, and Richmond," were "the greater objects toward which our attention must be directed in the next campaign," but the sequence in which he listed the objectives was significant: Lee's army, then Richmond. Even more significant was the fact that he assigned Butler, whom Grant regarded as a military incompetent, to attack Richmond with thirty thousand men. The much larger Army of the Potomac, under the fully competent George Gordon Meade, was given these orders: "Lee's Army will be your objective point. Wherever Lee goes there you will go also."[6]

The ultimate limits of Southern manpower figured ever larger in Grant's strategic thinking. It seemed clear by 1864 that when Lee's ranks received no replacements, the war in the East would be over. Manpower shortages in the underpopulated Confederate states had led their Congress to embrace conscription even before the North had—on April 16, 1862, males eighteen to thirty-five were made liable to draft. By February 1864, the age limits had been expanded to seventeen and fifty. Grant realized as early as the summer of 1863 that Confederates were conscripting everyone they could lay their hands on. So, when Confederate General Edmund Kirby Smith suggested that a prisoner who had served as a volunteer aide-de-camp, without commission or rank, on a Confederate general's staff should be treated as a citizen rather than a soldier, Grant expressed skepticism: "The Conscript act is so rigidly enforced in the South that everyone, to be secure, must enroll themselves in some capacity. Many whose interests... incline them to remain at home enroll themselves as a Volunteer Aid on the Staff of some Gen."[7]

Confederate conscription seemed fast on its way to turning every mobilizable male into a proto-combatant. The categories of noncombatants shrank accordingly. When guerrilla or partisan warfare further exasperated him, Grant proposed radical measures. To stop the pesky Confederate cavalry leader John Singleton Mosby, Grant eventually suggested a blistering raid by General Phillip Sheridan:

> If you can possibly spare a Division of Cavalry send them through Loudo[u]n County to destroy and carry off the crops, animals, negroes, and all men under fifty years of age capable of bearing arms. In this way you will get many of Mosby's men. All Male Citizens under fifty can farely be held as prisoners of war and not as citizen prisoners. If not already soldiers they will be made so the moment the rebel army gets hold of them.

Mosby provoked another uncharacteristically savage order from Grant to Sheridan at the same time:

> The families of most of Moseby's [sic] men are know[n] and can be collected. I think they should be taken and kept at Ft. McHenry or some secure place as hostages for good conduct of Mosby and his men. When any of them are caught with nothing to designate what they are hang them without trial.

Grant seemed to be acquiring an unwholesome taste for summary execution, a policy he had rejected in his early days of command in Missouri, when he was likewise bedeviled by partisan and guerrilla warfare.[8]

The partisan cavalry under Mosby along with the unit commanded by Loudoun County's Elijah White had troubled Union authorities for months and months. Phillip Sheridan willingly testified in his *Memoirs* to their effectiveness in depleting his "line-of-battle" strength by forcing him to provide large escorts for his supply trains. Nevertheless, Sheridan was at first too busy with a campaign against Jubal A. Early in the Shenandoah Valley to order the special operation against Mosby that Grant desired. When the Confederate partisans subsequently killed Sheridan's chief quartermaster and his medical inspector, he decided to turn his attention to them after the campaigning slowed in the late autumn. His orders of November 27 to cavalry commander Wesley Merritt embodied the scorched-earth aspects of Grant's suggestions, but the wholesale rounding up of civil population seems not to have been attempted:

> This section has been the hot-bed of lawless bands, who have, from time to time, depredated upon small parties on the line of army communications, on safe guards

left at houses, and on all small parties of our troops. Their real object is plunder and highway robbery. To clear the country of these parties that are bringing destruction upon the innocent as well as their guilty supporters by their cowardly acts, you will consume and destroy all forage and subsistence, burn all barns and mills and their contents, and drive off all stock in the region.... This order must be literally executed, bearing in mind, however, that no dwellings are to be burned and that no personal violence be offered to the citizens. The ultimate results of the guerrilla system of warfare is the total destruction of all private rights in the country occupied by such parties.

In this instance, the ordinarily fierce Sheridan retained a greater sense of distinction between guilty and innocent civilian populations than Grant, but the logic of military events was driving him to similarly ruthless-*sounding* conclusions. More than likely, Sheridan spared the civilians less out of consid- erations of conscience than practical military necessity. Thousands of civilian prisoners in tow would hardly have made the Union cavalry's task of rounding up Mosby's men easier. After all, the partisans were, as Merritt ruefully reported, "mounted on fleet horses and thoroughly conversant with the country."[9]

In the end, both Grant and Sheridan stopped short of obliterating the dis- tinction between noncombatants and soldiers, even while fighting the aggra- vating Mosby. Levi C. Turner's Record of Prisoners of State makes this clear. The number of civilian arrests directly attributable to fear of Mosby and White, made between December 1862 and October 1864, was quite small. Only 36 out of 2,187 arrested were accused of association with the famous Confederate partisan rangers. Many high-ranking Union officers, like Sheridan, spoke loosely of Mosby's forces as a gang of thieves, but the federal authorities in fact treated his men when captured as prisoners of war. One man, John Taylor, captured in September 1863, was discovered to be a deserter from White's cavalry, at which point he was transferred to the prisoners of war list from the list of prisoners of state. The other thirty-five were civilians accused of aiding Mosby or White in some way—often as "scouts." For all their vaunted knowledge of the country, Confederate partisan cavalry leaders needed scouts and guides, men whose work was closely related to spying.

Many of those arrested for aiding Mosby were indeed suspicious persons and not innocent victims of Union hysteria. Benjamin H. Hatton, for example, arrested on November 30, 1863, in Fairfax County as a dangerous guerrilla, was referred to many years later in John Singleton Mosby's cynically humorous *Memoirs:*

> I received information that there was a pretty strong outpost on a certain road in Fairfax, and I was determined to capture it.... Near the post lived a man named Ben Hatton, who traded in the camps and was pretty familiar with them. So, around midnight, we stopped at his house ... and he told us that he had been there that evening—I suppose to get coffee and sugar. Ben was impressed as a guide to conduct us to the rear of the enemy. When we reached that point, I determined to dismount, leave our horses, and attack on foot. Ben had fully discharged his duty and, as he was a non-combatant, I did not want to expose him to unnecessary

danger. . . . So the horses were tied to the trees, and two of my men . . . stayed with
Ben as a guard over the horses. . . . We were soon ready to start back with our
prisoners and their horses, when a fire opened in our rear, where we had left the
guard and horses. . . . To our surprise we found the horses all standing hitched to
the trees, and Ben Hatton lying in a snowbank, shot through the thigh. . . . Ben was
lifted on a horse behind one of the men, and we started off with all the horses
and prisoners. . . . We left him at home, curled up in bed, with his wife to nurse
him. He was too near the enemy's lines for me to give him surgical assistance, and
he was afraid to ask any from the camps. The wound would have betrayed him to
the Yankees had they known about it, and Ben would have been hung as a spy!
He was certainly innocent, for he had no desire to serve any one but himself. His
wound healed, but the only reward he got was the glory of shedding his blood
for his country.[10]

It is characteristic of Mosby's flippant style not to take notice of poor 46-year-
old Ben Hatton's real reward—nearly three months in federal prison. Hatton,
incidentally, told the Yankees that he had been "conscripted" by Mosby.[11]

Other civilians arrested as guides, independent scouts, spies, and suppliers
of information for Mosby and White were surely as guilty as Hatton. But not
all were, and perhaps the federal forces grew less discriminating as the Confederate rangers grew more irritating. By July 1864, prison records could carry
this notation: "lives in Mosby's counties and must therefore cooperate with
him."[12] Ordinary soldiers were beginning to follow the logic of Grant—a logic
that could have broken down distinctions between soldiers and civilians.

Arrests in Fauquier and Loudoun counties were frequent throughout the
Civil War, but these were by no means all attributable to the Union authorities' anxieties over the partisan cavalry. Of the 2,187 prisoners in this same
record series (of persons arrested between December 1862 and October
1864), 316 were arrested in Loudoun or Fauquier, a substantial 14.4 percent. But many were arrested in areas of the counties outside Mosby's control, and many were arrested for reasons not having anything to do with
guerrilla or partisan warfare. Only forty can be identified as having been arrested in the areas controlled by Mosby. The Union authorities talked tough
about Mosby, even to each other, but their actions usually fit the traditional
standards of civilized warfare.

It required the extreme provocation of the frustrating campaigns of the
summer of 1864, knowledge of the relentlessness of Confederate conscription,
and the embarrassing irritations of Mosby and White to drive Grant to declare
on August 16, 1864, that essentially all Southern males between the ages of
seventeen and fifty be treated as combatants. And afterward, no one really
acted on the new declaration. The Union armies never gathered all Southern
males, aged seventeen to fifty, from any area and took them as military prisoners.
Sheridan, who was specifically told to do so, did not.

The Confederates heard about Grant's idea, however, and those whose job
it was to keep the war within civilized bounds, expressed alarm. On November
1, 1864, Robert Ould, the Confederate agent for exchange of prisoners of war,

wrote a troubled letter to Secretary of War James A. Seddon, at the end of
which he said:

> The enemy still continues the arrest of non-combatants. I have been notified by
> the Federal authorities that "all white persons between the ages of seventeen and
> fifty, residents of the Confederate States, captured by the U.S. forces, will be held
> and deemed to be soldiers of the Confederate Army, and will be treated as prisoners
> of war and held for exchange." In view of their practice and this declaration, the
> course to be pursued by us toward non-combatants who are residents of the United
> States, or who, being citizens of the Confederate States, are hostile to our cause,
> becomes a subject of the gravest importance. After much reflection, I am fully
> convinced that the only effectual method of preventing the outrages which are
> being daily perpetrated upon our loyal non-combatant citizens is to cause the
> arrest of every citizen of the United States who may be within our reach and of
> such citizens of any one of the Confederate States as are known to be inimical....
> If the plan suggested worked no other result, it would furnish us, in the event of
> an exchange, with more material. I know there are very many grave objections to
> this course, but yet I think...that the horrors under which our non-combatant
> population are now suffering can hardly be increased. When we have resorted to
> such arrests as are made by the enemy, there is some chance that the whole system
> will break down by the sheer weight of its gigantic misery.[13]

Wholesale military arrests of citizens never came about, in fact, for the war
was fought mostly by practical men of action and not by theoreticians, fanatics,
or ideologues. Although Grant, among other Union generals, has been said to
have devised a forward-looking military doctrine of "annihilation" and warred
on populations rather than armies, such is hardly the case.[14] Whatever the
declared policy, men like Ulysses S. Grant tailored their actions to accommodate day-to-day realities. In its most extreme formulation, in his letters to
Phillip Sheridan, for example, Grant included only draft-age males and the
known relatives of guerrillas in his new broad definition of belligerent population, not women, children, or the aged. Within days, as new information
came to him, he was forced to modify his own drastic orders to suit the political
realities in the field. "I am informed by the Asst. Sec. of War," Grant told
Sheridan, "that Loudo[u]n County has a large population of Quakers who are
all favorably disposed to the Union. These people may be exempted from
arrest." He qualified the order again two weeks later, instructing Sheridan to
exercise his own judgment "as to who should be exempt from arrest and...
who should receive pay for their stock grain &c. It is our interest that that
County should not be capable of subsisting a hostile Army and at the same
time we want to inflict as little hardship upon Union men as possible." Two
months later, Grant relented still further in his thinking on the civilian population of Mosby's Confederacy:

> Do you not think it advisable to notify all citizens living East of the Blue Ridge to
> move out North of the Potomac all their stock, grain and provisions of every
> description? There is no doubt about the necessity of clearing out that country so
> that it will not support Mosby's gang and the question is whether it is not better

that the people should save what they can. So long as the war lasts they must be prevented from raising another crop both there and as high up the valley as we can controll.[15]

As a practical matter of fact, Ulysses S. Grant never applied a unitary military philosophy to the South, though as lieutenant general late in the war he wielded power to set policy for all theaters and armies. Rounding up civilians and destroying the crops and livestock by which a local army could live—these were strategies Grant ordered only in bitterly disloyal areas infested with guerrillas. Where the political complexion of the local populace appeared different, Grant's orders took a different tone. After Vicksburg's fall in 1863, for example, he issued General Order No. 50, counseling restraint on the part of U.S. forces, which now controlled the entire western third of Mississippi. He called upon the people of the state "to pursue their peaceful avocations in obedience to the laws of the United States," and assured them that if they did so, the occupying forces would be "prohibited from molesting in any way the citizens of the country." "In all cases," he added, "where it becomes necessary to take private property for public use a detail will be made, under a Commissioned officer, to take specified property, and none other. A staff officer of the Quartermaster or Subsistence Dept. will . . . give receipts for all property taken, to be paid at the end of the war, on proof of loyalty or on proper adjustment of claim under any regulation hereafter established." Even proof of loyalty was not always necessary to receive humane treatment from Grant's armies. For Warren County, which had been "laid waste by the long presence of contending Armies," he made provision to "issue articles of prime necessity to all destitute families" who called on the Union armies for them. The local generals could make "such restrictions for the protection of Government as they deem expedient," but for his part, Grant did not stipulate proof of loyalty as a requisite for receiving U.S. aid.[16]

Political judgment more than humanitarian sentiment dictated Grant's differing policies. Where the potential for reconstructing civilian loyalty appeared high, he treated the local populace gently. Grant told Halleck that Mississippi and Louisiana "would be more easily governed than Kentucky or Missouri if armed rebels from other states could be kept out." When General William T. Sherman sent a cavalry expedition toward Memphis, Grant instructed him to "impress upon the men the importance of going through the State in an orderly manner, abstaining from taking anything not absolutely necessary for their subsistence while travelling. They should try to create as favorable an impression as possible upon the people." Sherman, who would soon become famous for taking the opposite approach to the enemy populace, was at the time exasperated by recent depredations by U.S. troops and responded enthusiastically: "It will give me excessive pleasure to instruct the Cavalry as you direct, for the Policy you point out meets every wish of my heart."[17]

Later, Grant urged tough treatment for Loudoun County, Virginia, the playground and logistical base for John Singleton Mosby, but he showed moderation elsewhere. In 1864, for example, after he was given overall command of the

Union armies, he learned from an old general, Henry Price, a veteran of the Seminole and Mexican wars, that General Eleazer A. Paine, whom Grant had known in Missouri, was oppressing the people of Kentucky. Price protested "in the name of God and of all my countrymen who respect the rights of mankind." Grant ordered Paine removed from command in Paducah:

> He is not fit to have a command where there is a solitary family within his reach favorable to the Government. His administration will result in large and just claims against the Government for destruction of private property taken from our friends. He will do to put in an intensely disloyal district to scourge the people but even then it is doubtful whether it comes within the bounds of civilized warfare to use him.

Paine was later court-martialed and reprimanded.[18]

Grant was not growing soft; he simply believed the commander ought to be tailored for the district he commanded. Thus, he thought Benjamin F. Butler worthless as a soldier, but in "taking charge of a Dept.mt where there are no great battles to be fought, but a dissatisfied element to controll no one could manage it better than he." As late as the summer of 1864, Grant contemplated a restructuring of military districts that would put Butler in command of Kentucky or Missouri. These areas Grant had seen himself, and he regarded them as harder to control than Mississippi. Butler, whose notorious treatment of civilians in occupied New Orleans earned him the nickname "Beast" and made him an outlaw in the Confederacy, seemed ideal for the intractable western border states. Grant adapted his policies to the situation at hand, but he usually remained "within the bounds of civilized warfare."[19]

The Problem of Confederate Deserters

The rules of civilized warfare dictated that the Union armies would take mostly military prisoners, but it was not always easy to distinguish soldier from civilian. Men who left the South and gave themselves up to Union pickets, maintaining that they were deserters from the Confederate armed services, posed an awkward problem. General Order No. 100, the rules drafted by the Lincoln administration in 1863 to govern the Union armies, did not spell out the status of enemy deserters, but the principles governing such prisoners were well understood throughout the Union army by 1863. For example, William Hoffman, the commissary general of prisoners in Washington, told the provost marshal general in Louisville early in 1863, "Deserters from the rebel army who are really such cannot be held as prisoners of war and they should be released; but to insure their future loyalty they should be required to take the oath of allegiance with the penalty of death for its violation."[20]

It must have offended the nineteenth century's unwritten code of military honor to treat enemy deserters in such lenient fashion, while captured soldiers who had fought bravely, had obeyed orders, and had performed their duties languished as inmates in notoriously ill-equipped prisoner-of-war camps. Yet officers had a clear grasp of the unheroic reasoning that lay behind the policy. Major General Horatio G. Wright, commanding the Department of the Ohio in 1863, explained it this way to a subordinate in Kentucky, where many such prisoners fell into Union pickets' hands:

> Deserters from the rebel ranks, recognized as being such, are to be treated according to the laws of war with all the leniency compatible with our own safety, it being an established principle to weaken the enemy as much as possible by encouraging desertion from his ranks. Under this rule the practice has been to permit the deserter to remain at home on his taking the oath of allegiance and giving proper bonds for its observance.[21]

The policy may seem simple and reasonable, but it proved difficult to implement. As General Hoffman put it, "If professed deserters come within our lines they may be spies and every commander should judge of each case after careful inquiries according to the circumstances." General Wright instructed his Kentucky subordinate, "Should any doubt exist ... of the reliability of the individual then such other steps should be taken, such as sending him to Camp Chase [in Ohio] as a political prisoner, as will give reasonable security against his doing harm." How could there be anything but doubt about the reliability of such professed deserters? To be reliable in Union eyes would be to prove genuine unreliability as Confederate soldiers. Documents, such as enlistment papers, might prove only that the person in question really was a soldier. Witnesses were unlikely to be found, and they were equally unable to look into the alleged deserter's heart. Simple rules could hardly work justice in such inherently treacherous matters.

Carelessness, distrust, or dislike of these deserters led occasionally to substantial periods of imprisonment. When the British minister inquired about one Confederate deserter named John Spencer, a subject of the queen who had been imprisoned for some time in Cumberland, Maryland, local authorities explained to Levi Turner that "Deserters are often held for the purpose of eliciting facts in regard to their cases, to satisfy the military authorities that they are acting in good faith and have deserted with the intention of not returning again to the Rebel Service." Early in the war, C. W. Nelson, a prisoner in Washington, wrote a letter of protest to the editor of the *Intelligencer* which he called "How we Live at the Old Capitol":

> There are men in this very room who have been confined for three long, weary months—*for what?* For the unpardonable crime of desertion from the rebel ranks. ...Does not every deserter from the ranks of the enemy weaken the rebellion? They cross the lines and deliver themselves to the Federal pickets, who welcome them with open arms and generous hearts, for they know that they are no longer enemies, but loyal. They are furnished with an escort and sent to Washington, where the Old Capitol doors yawn to receive them. They are kindly told that they will be released in a few days, and sent to Philadelphia. They patiently wait, but nothing is heard from the outside....We are aware that some men who have

deserted the enemy should be regarded with suspicion. But we must say that it looks very unjust to make an indiscriminate confinement, without giving a man a hearing.

The wonder is, not that many of these men were held briefly as "political prisoners," but that so many were quickly released to return home.[22]

Perhaps even more surprising was the lenient treatment of discharged Confederate soldiers who fell into Union hands. General Wright advised:

> Persons returning home after serving their term in the rebel service but who are believed to be truly penitent may if deemed reliable be permitted to remain on similar terms [to the Confederate deserters']. They are not entitled to the same consideration as deserters, but as they might if prevented by dread of imprisonment from returning remain in the rebel ranks it is good policy to encourage their return home.... If any doubt exists as to their reliability they should be sent with proper written charges and proofs to Camp Chase as political prisoners. They cannot be considered as prisoners of war and are not entitled to the immunities granted that class by the cartel agreed upon by the United States with the rebel authorities.[23]

As Confederate desertion grew to epidemic proportions, and discharged soldiers refused reenlistment, such Southerners occupied cells, however briefly, as civilian prisoners in the military prisons of the North. They were only passing through, so to speak, but the marks they left unfairly swelled the indictment against the Lincoln administration, accused of holding thousands of "political prisoners."

Selling Liquor to Soldiers

The army at times invoked its power to arrest civilians more out of convenience than grim necessity, and the practice tended to spread to cover military eventualities unthought of before the war. Arrests of liquor-sellers provide a vivid example. With no state or federal prohibitions on the sale of alcoholic beverages to soldiers, generals attempted to remedy the perennial problem of a drunken soldiery by arresting liquor-sellers. No such arrests occurred while Seward oversaw internal security, but they commenced in 1862 and continued well into Reconstruction. Among the 5,108 civilian arrests in the period from 1862 to 1865 recorded in the Turner-Baker Papers in Washington, for example, 14 were made for selling liquor to soldiers. The earliest instance occurred in Upton Hill, Virginia, on October 2, 1862, and another arrest for the practice occurred on Rikers Island, New York, that same month. By 1863, in the Washington, D.C., area, standard punishments of thirty to sixty days' imprisonment had been devised.[24]

This category of arrest accounts for few of the civilian prisoners in the North, but it reveals one of the tendencies of the program of military arrests in the Civil War. Arrests eventually were made, for convenience's sake, for causes never dreamed of when the president took his first hesitant steps toward suspending the writ of habeas corpus and imposing martial law. And, as administered by the Republicans, the arrest program took on a somewhat puritanical, moralistic, or purifying cast.

Military arrests of civilians who sold liquor to soldiers, though heretofore largely unnoticed by historians, did not altogether escape public notice in the Civil War, and one case, in fact, led to a sweeping state supreme court decision on the war powers of the president. On June 8, 1863, the chief provost marshal of the military District of Indiana and Michigan told Indianapolis provost marshal Captain Frank Wilcox to "issue an order prohibiting the sale of liquor, by any party, to enlisted men." The "many intoxicated soldiers in our streets" constituted an "evil" that "must be stopped." Captain Wilcox issued this order on the same day:

> All persons engaged in the traffic and sale of spirituous and intoxicating liquors within this city are notified that they are strictly prohibited, from and after this date, from selling the same to any enlisted soldier. A violation of this order, by any person whomsoever, will be visited with severe punishment.

When Joseph Griffin, properly licensed to sell liquor to all except minors and intoxicated persons, was subsequently arrested for violating the order, he sued Wilcox in the Marion County Common Pleas Court for false imprisonment. The case reached the Indiana Supreme Court on appeal, and Justice Samuel E. Perkins handed down a decision in January 1864. Perkins was a Democratic journalist as well as a judge, and despite protestations to the contrary, he may have been eager to cripple military arrests of civilians. He almost did.[25]

Perkins proved himself a shrewd jurist, and his decision was notable for making careful legal distinctions in the sort of partisan controversy that usually led to carelessly sweeping pejoratives. He noted, for example, that the decision did "not invoke the question of the right, in any person, or body of men, to suspend the writ of *habeas corpus*" because "Griffin did not apply for that writ in order to effect his discharge from imprisonment." Perkins's sharp mind thus enabled him to probe the Lincoln administration's policies where they were weakest: in applying military justice to civilians, not in merely arresting them and holding them for a time without charge. The habeas-corpus suspension clause appeared in the U.S. Constitution in plain English, but that document said nothing about prohibiting the sale of liquor to soldiers or "severe punishment" of civilians by the army. Perkins noted "that the suspension of the writ of *habeas corpus* does not legalize a wrongful arrest and imprisonment; it only deprives the party thus arrested of the means of procuring his liberty, but does not exempt the person making the illegal arrest from liability to damages, in a civil suit, for such arrest, nor from punishment in a criminal prosecution."[26] Perkins was on firm legal ground here. The Habeas Corpus Act of Congress of March 3, 1863, provided money for counsel for officers sued for false arrest and protected them from state courts, but it did not immunize them from legal accountability.[27]

Perkins asserted that the "real question, lying at the bottom of this case, involves the *war power* of the President of the United States, that is, his power to act upon martial law without its having been first declared by the sovereign power of the state." The judge distinguished carefully between military law, established by Congress and applicable only to members of the United States armed forces, and martial law, the application of military authority to civilians. Had Captain Wilcox issued his order to apply to soldiers only, they were lawfully obliged to obey. The situation was different in this case:

> ... when martial law supercedes the civil, or is exercised concurrently with it, the civil being permitted by mere military sufferance, or as a matter of convenience, where it does not interfere with, or is subservient to the war power, the military assume the government of the citizens to just the extent they please. The assuming to prohibit the sale of liquor to soldiers in Indianapolis was upon this theory. So were the military orders prohibiting the sale of arms and ammunition to citizens in contravention of their constitutional right to procure and keep them. So were the arbitrary arrests for pretended disloyal opinions....

With no state or federal law prohibiting the sale of liquor to soldiers, the question became "when and where can the citizen be subjected to martial law?"

Without an act of Congress, Judge Perkins answered, the citizen could be subject to that law only "upon *necessity*—occasioned by force, actually existing or immediately threatened, at the time and place where martial law is exercised." He did not enquire whether Congress could impose martial law, but he did attempt to describe carefully the conditions under which it could be imposed by whatever authority. American soldiers could impose martial law in a war with a foreign country when they drove "the governing power from the whole or part of it." Also, when "force may expel the civil authority from a part or the whole of our own territory; or, perhaps, it might be said, martial law is exercised in our country, the military being on the spot to execute it, where no civil authority exists. But where the civil authority exists, the Constitution is imperative that it shall be paramount to the military." Perkins's theoretical understanding of martial law may well have been as deep as that of any American jurist at the time. Seizing on a distinction made by New York Democratic leader Horatio Seymour, Perkins explained that martial law

> is exercised precisely upon the principle on which self-defense justifies the use of force by individuals. Robbers and burglars, in some cases, rioters, may be resisted and even slain, in self-defense, by private individuals.... This is the doctrine expressed in the maxim, *"inter arma silent leges."* ... This corresponds with Lord Coke's idea of Cicero's maxim. He says: "When the courts of justice be open and ministers of the same may by law protect men from wrong and violence, and distribute justice to all, it is said to be time of peace. So when by invasion, insurrection, rebellion or such like, the peaceable course of justice is disturbed and stopped, so as the courts be as it were shut up, *et silent inter leges arma,* then it is said to be time of war."

The right, Perkins argued, "temporarily and locally to exercise martial law, in case of necessity, is the war power of the Governor of a State or of the President of the United States, and it is all the war power that either possesses by virtue of which he can assume to govern independently of the civil law; and this war power, each executive usually exerts through his subordinate military officers." Thus Perkins concluded:

> Where force prevails, martial law may be exercised. But in all parts of the country where the courts are open and the civil power is not expelled by force, the Constitution and laws rule, the President is but President, and no citizen, not connected with the army, can be punished by the military power of the United States, nor is he amenable to military orders.

The power to impose martial law within the nation, if it existed at all, Perkins argued, was derived solely from the president's power as commander-in-chief. It did not come from his oath to support, protect, and defend the Constitution. And it did not come from the power to suspend the writ of habeas corpus. This argument was aimed directly at the Lincoln administration, which was always a little confused about the distinction between declaring martial law and suspending the writ of habeas corpus.

> Simply because the *habeas corpus* is suspended, is it right to destroy every man's liberty and property? The right, in a case of emergency, to exercise the war power, temporarily and locally, supposing that power to exist at all, under the Constitution, does not depend upon the fact of the *habeas corpus* being suspended, or not suspended.

In the end, Perkins could not restrain his partisanship. After explaining the sectional nature of the current conflict, which aimed not at overthrowing the old government in the North but at setting up a new one in a section of the old Union, Perkins argued that opposition to the administration in the North was peaceful. He went on:

> And the question now is, does such peaceful conflict of opinion and argument justify the Administration in subjecting those who differ with it to the military power? For the case at bar, though perhaps not of that description in its facts, yet rests entirely upon that principle . . . of governing by martial law; as it would not be pretended that the military could make such arrest of the citizen as that involved in this case, in time of peace. We have found no legal principle that will justify such a course. *We know of no precedent for such an exercise of the war power as that above propounded,* viz: of subjecting opponents, simply in political opinions, to martial law for expressing those opinions; for such opinions are not force, nor is the expressing of them force, nor is it a crime by any law of the land.

Yet Mr. Griffin had no known political differences from the Lincoln administration; he merely wished to sell a drink to a thirsty enlisted man. Such legal opinions, while they brilliantly sounded out the problems most feared in wartime arrests of civilians by military authority, left a legacy of historical distortion as to the actual reasons for most arrests. As legal principle, however, Perkins's opinion, as has been accurately pointed out in the past, looked forward to the *Milligan* decision of the U.S. Supreme Court in 1866. Moreover, Perkins fought

on the same ground Lincoln had chosen to defend in the Corning letter of
June 1863—the high ground of political opinion and not the low ground of
saloon-keepers or Confederate deserters.[28]

As for the tendencies revealed not by Judge Perkins's decision but by the
arrest of the saloon-keeper Griffin, they are suggested by an editorial in the
most influential Democratic newspaper in Indiana, which referred to Griffin's
arrest "for violating the Maine Law enacted by the Provost Marshal." The
"Maine Law" was the earliest law in the U.S. banning the sale of alcoholic
beverages and had caused numerous bitter party struggles in the 1850s. Indeed,
Judge Perkins himself had ruled against a prohibition law in the state of Indiana
in the 1850s. Thus, the Griffin case recalled the familiar issues of antebellum
party contests, when Democrats defended the German's and Irishman's drink
against the puritanical Republican desire to regulate Americans' lives according
to the code of evangelical Protestantism.[29]

The Last Proclamation

President Lincoln and his cabinet proved more than willing for the American
people to see how the administration would regulate their lives. The number
of habeas-corpus proclamations continued to grow, and the proclamations
grew longer as well. By the end of 1863, Lincoln had issued eight orders and
proclamations suspending the privilege of the writ of habeas corpus and had
come close to issuing another for Vallandigham. It has generally been assumed
that the president used the successive orders and proclamations to bring ever
greater areas of the North under the suspension. Historians were no doubt
encouraged to adopt such an explanation by Lincoln's own statements, for
example, in the Corning letter, that he adopted the measures "by slow degrees."
James G. Randall, who published in 1926 what remains the fullest work on
the subject, *Constitutional Problems under Lincoln,* mentioned only half the
documents in his book and suggested vaguely that the president moved from
a "limited" suspension on April 27, 1861, to a "general" suspension on September 24, 1862. Fifty years later, Don E. Fehrenbacher, both a constitutional
historian and a Lincoln specialist, wrote that after the first suspension, the
president "extended these orders geographically, and in 1862 ... made them
nationwide for specified offenses tending to impede the prosecution of
the war."[30]

Yet even after the proclamation of 1862 making the offenses "nationwide,"
three more habeas-corpus documents were drafted, and among those characterized as spreading the suspension geographically was one that applied to
only one man, the army engineer named Chase. By now it should be obvious
that the habeas-corpus orders and proclamations did not share the logic of
either development by slow degrees or steady geographical expansion. On the
contrary, the administration lurched from problem to problem drafting hasty
proclamations and orders to meet the objective of the moment.

The last of these somewhat redundant proclamations appeared on July 5, 1864. Over the years, the suspensions had added new legal justification in their preambles and had grown generally longer. This final one applied only to Kentucky:

...whereas, the Congress...by an act approved on the 3d. day of March 1863, did enact that during the...rebellion, the President...is authorized to suspend the privilege of the writ of Habeas Corpus...;

And whereas the said insurrection and rebellion still continue...especially in the States of Virginia and Georgia;

And whereas on the fifteenth day of September last, the President...issued his proclamation, wherein he declared that the privilege of the writ of Habeas Corpus should be suspended...;

And whereas many citizens of the State of Kentucky have joined the forces of the insurgents and such insurgents have on several occasions entered...Kentucky in large force, and not without aid and comfort furnished by disaffected and disloyal citizens...residing therein, have...made flagrant civil war, destroying property and life in various parts of the State; and whereas...combinations have been formed in...Kentucky with a purpose of inciting rebel forces to renew the said operations of civil war within the said State, and thereby to embarrass the United States armies now operating in the...States of Virginia and Georgia and even to endanger their safety:

Now, therefore, I Abraham Lincoln...do, hereby, declare that in my judgment the public safety especially requires that the suspension of the privilege of the writ of Habeas Corpus so proclaimed in the said proclamation of the 15th. of September, 1863, be made effectual and be duly enforced in and throughout... Kentucky, and that martial law be for the present established therein....

The martial law herein proclaimed...will not...interfere with the holding of lawful elections, or with the proceedings of the constitutional legislature of Kentucky or with the administration of justice...in suits or proceedings which do not affect the military operations or the constituted authorities of the Government of the United States.

Once again, the president delegated the drafting of the document to Seward, who sent it to Stanton for revisions. Again the proclamation was legally redundant and as much as said so, merely urging enforcement in Kentucky of the proclamation of September 15, 1863.[31]

If the habeas-corpus proclamations and orders were legally imperfect, some of these shortcomings were attributable to the harried circumstances of their issuance. While the Kentucky proclamation was being drafted, General Jubal A. Early's Confederate troops were nearing Washington, D.C. On the very day of the document's publication, Stanton ordered the commander of the Department of Washington to "take prompt and vigilant measures" to protect the city from attack, and within a week, the president himself would be exposed to enemy fire at Fort Stevens, a fort guarding the capitol. The administration was experiencing a deep political crisis as well. Salmon P. Chase had resigned as secretary of the treasury on June 29, and Lincoln was still forging an agreement with his replacement on July 4. On that same Fourth of July, the president pocket-vetoed the Wade-Davis Bill on Reconstruction of the South and thus more or less threw down the gauntlet before the radical wing of his party.[32]

Yet the latest habeas-corpus proclamation, in its way, maintained a steady adherence to the principles and goals of the Lincoln administration. Above all else, it showed a continuing determination to win the war. It was also consistent with Lincoln's steady desire to avoid political abuse under the habeas-corpus policy. From his first mention of martial law to Winfield Scott less than two weeks after the fall of Fort Sumter to this last proclamation suspending the writ and stating explicitly that martial law should not interfere with normal political life, military goals rather than political ones remained uppermost with Lincoln when restricting civil liberties. Though issued piecemeal and unsystematically, the orders and proclamations were usually provoked by problems of military mobilization—first by obstructions of the routes to the underprotected capital and later by draft resistance.

Suspending the writ of habeas corpus was surely a political liability and was always so regarded by the Lincoln administration. The president and his cabinet spoke of suspension as a strong measure that the American people would tolerate only because they recognized the peril to the Union and shared the government's mission to win the war. In his diary, Gideon Welles dismissed Chase's fears of revolution in the North if the September 1863 order to the provost marshals were issued, by saying: "I have never feared the popular pulse would not beat a healthful response even to a stringent measure in these times, if the public good demanded it."[33] Intimidation at the polls in Maryland, Kentucky, and Missouri, if it gained a handful of votes in such places, surely never equaled the risk of general unpopularity incurred by imposing these odious measures on freedom-loving Americans.[34]

5

The Dark Side of the Civil War

Though military arrests of civilians increasingly brought Southern citizens into the infamous bastilles of the North, throughout the war some of the prisoners, of course, came from the North as well. Among the Northerners were businessmen, entrepreneurs, and peddlers who fell victim to the Lincoln administration's policies, not because they dissented from the war effort but because they appeared to be trying to profit from it by shady exploitation. These Northerners sometimes suffered as harsh a treatment as any civilian, and, indeed, one group of Northern civilian prisoners received the harshest treatment meted out officially to any prisoners, civilian or military, Southern or Northern. To look at these arrests is to see a dark side to the Civil War rarely glimpsed in the numerous military histories of the conflict.

The soldiers charged with overseeing prisoners beheld warfare at its least heroic. Associate Judge Advocate Levi C. Turner, for example, seized a rare moment from routine duties on February 6, 1863, to describe some of the underside of the war effort:

> Claim Agents, who advise and aid in the manufacture of false & fraudulent amounts against the Govt. and present them for payment—and obtain payment—violate no existing Law of Congress, and cannot be punished unless they have violated some state Law to punish forgery & forging.
>
> A military officer makes out a false account for subsistence...and *certifies* to it—if he has no vouchers and it gets the approval of the Adj General, it is paid. The officer takes his money (from $500 to $10,000) and then *resigns.* There is no law of Congress or of any State authorizing the arrest & punishment of such plundering. And *citizens* cannot be punished for presenting fraudulent accounts, unless in so doing they violate some State law against forgery & forging.
>
> A Fraudulent & plundering military or civil officer can be reached and punished, *while in the public service,* by Laws now existing—but having filled their pockets, they *resign,* and are then beyond the reach of military or civil Laws....

Persons, all over the country, are engaged in procuring bogus substitutes and aiding & promoting desertions.

I know of no Law of Congress that provides for the prompt & remedial punishment of such infamous agents—It is true a citizen who encourages and aids desertions can be indicted and tried in criminal courts—a process so slow & inefficient, that prosecutions for aiding desertions seldom occur, and convictions *never.*

War Department solicitor William Whiting, to whom Turner sent his letter, had drafted bills to solve such problems, and they were now before Congress. Had they been law "one year, or even six months ago," said Turner, "millions of dollars would have been saved, and thousands of infamous plunderers would, now, instead of being *at large,* have been serving out righteous sentences of imprisonment."[1] One way to halt such wrongdoing was to have the army arrest such civilians. The role of military arrests of civilians thus expanded further.

Military Arrests for Fraud and Corruption

Examples of such cases abound in the records of civilians arrested by military authorities during the Civil War. A particularly sordid case occurred in February 1864, when War Department agents arrested C. E. Green and J. E. Jenkins and padlocked the offices of Bell & Green, claims agents. These entrepreneurs made their business from the plentiful suffering of widows and orphans in the Civil War. They tracked down claims for deceased husbands' or fathers' back pay and bounties, somehow lost by government departments inadequate to the task of administering hundreds of thousands of pensions. Visiting the auditors' departments in the nation's capital, these claims agents took a percentage of the claim itself, usually 5 percent. Bell & Green, with offices eight blocks from the Treasury Department, were well situated to do this, especially since Victor Bell was, according to President Lincoln himself, "one of my most valued friends" and "one of the cleverest and best business men" in Illinois.[2]

Bell & Green built a thriving claims business and branched out, investing $2,400 of their clients' money in an army sutlership, a business that accompanied the troops and sold them supplies, food, and drink. Other funds went into speculation in Kansas lands. None of the investments prospered. Soon irate widows—and their lawyers—began to hector them for unpaid claims. Bell & Green borrowed time by blaming government red tape for delays: the second auditor of the treasury seemed always to be slow in certifying claims. Using the last of the firm's cash, Bell began speculating in contraband cotton in Memphis, and the situation grew dire. Finally, someone in the office—perhaps the clerk Jenkins, perhaps one of the firm's principals—started forging widows' names to powers of attorney so Bell & Green could collect more from the army paymasters. When they forged a name on Certificate 3,026 for $118.20 in back pay and bounties due the widow of deceased Private John Anderson of Company D, 106th Pennsylvania Volunteer Infantry, Mrs. Anderson's lawyer quickly caught them in their fraud, for Magdalene Anderson could not read or write and signed all her legal documents with an "X." Soon afterward, the War Department shut Bell & Green down and arrested Green and Jenkins. Such military arrests of civilians rarely drew criticism from anyone. Senate secretary C. H. Cobb, for example, applauded the move in the Bell & Green case, noting that the "civil law here is constructed with meshes large enough to admit the escape" of such men but that military law "does not stand upon soulless forms until the soul of justice is eliminated."[3]

The sums of money realized to the government by such arrests grew so great that gradually Judge Advocate Levi C. Turner, who had been going to Old Capitol prison every day but Sunday to interrogate and investigate civilians arrested by military authorities, came to spend much of his time investigating cases of fraud against the government.[4] Back in the desperate period when Seward oversaw the arrests, fraud cases seldom appeared on the lists of civilian prisoners. The administration then had little time for careful accounting and a businesslike system amidst the confusion of hasty mobilization for war. "Honest Abe" Lincoln could hardly afford for his administration to earn a reputation for corruption, and eventually attention was turned to such problems. By the end of the war, government investigators were uncovering large cases of fraud themselves.

On February 18, 1865, for example, military authorities in New York City arrested John N. Eitel, a prosperous clothing merchant and partner in John N. Eitel & Company. They did not tell his family or his business associates the cause of arrest, whisking him out of town quickly to place him in Washington's Old Capitol prison. Complaints and pleas for his release soon poured into the War Department, as was customary when wealthy or influential men were arrested.

Though Eitel had indeed received summary treatment, he was involved in a rather unattractive, if not downright sordid, business that the War Department watched closely. As a sideline to his regular clothing business, Eitel was a recruitment broker. During the Civil War, recruitment for the armed services fell largely into private hands. The government itself at first encouraged private recruiting by offering a two dollar premium to any person who brought in a recruit who was accepted for service. Gradually, this led to private brokers all but taking over the supply side of the recruiting system. And nowhere were they more active than in New York City, where the New York County Board of Supervisors authorized a $300 bounty for volunteers and permitted another committee to use private brokers for distribution of the bounties. When a man volunteered in New York, the broker who brought him in paid the soldier a part of the bounty price agreed on beforehand. Then the soldier would assign the whole bounty to the broker, who would collect $300 from the New York County committee.[5]

Eitel may have gone into this business as a natural outgrowth of his clothing business. In an affidavit written on Eitel's behalf, a clerk in the naval rendezvous at 173 South Street, one of several where recruits were mustered in New York

City, testified that Eitel was in the business of selling clothing to naval recruits. His company sold the goods for an eight or nine dollar profit on each recruit and paid about half the profit to runners who brought the recruits in. This was almost perfectly analogous to the way the bounty brokerage system worked, so it is little wonder that Eitel moved into that as well. The enterprise was made a good deal easier when Eitel's business partner, George Goin, was appointed acting master of the naval rendezvous.[6]

It would be fair to characterize this as "big business" by the standards of Civil War America. Three hundred dollars constituted a substantial sum of money in those days, and there were thousands of recruits for this, the bloodiest war in all of American history. Recruiting in New York City was especially important for the United States Navy. Secretary of the Navy Gideon Welles could grow exercised over any interference with naval recruitment in the city. The navy needed recruits and never really wanted to see any decrease in competition among recruiters to drum up every available man.[7]

Opportunities for fraud were abundant in this system, not only because of middlemen and the vast sums of money involved but also because of the rather primitive record-keeping and accounting practices of the day. Records were all handwritten, of course, by scores of clerks with hands of varying legibility and with limited understanding of the operation. Although the principle of alphabetization was known, apparently the idea of putting alphabetized names on index cards was not, with the result that most Civil War lists appear on long sheets of paper, the names are alphabetized by initial letter only, and there is usually an addendum of names for persons processed after the list was already complete.

One of the clerks in the naval rendezvous witnessed the wild confusion that made record-keeping difficult and fraud simple. For example, the office mustered in more men some days than could be taken aboard the receiving ship on the same day. Overnight, the reluctant recruits might run away, never appearing on the receiving ship but leaving a name as recruited on the office books. Also, for reasons the clerk could never fathom, duplicate records of mustering-in and duplicate receipts given by the recruit for his bounty payments were produced on the receiving ship "with nothing thereon to indicate which was the original." Thus, "a large field for fraud was opened."

Whether Eitel entered that fair field for fraud is unknown, but War Department detective Colonel Lafayette C. Baker thought he did and produced evidence to prove it. Baker drew a rather different picture of the naval rendezvous in New York than that drawn by Eitel's partisans, who attributed its faults to confusion and disorder rather than corruption. Baker recalled:

> It would be impossible to give a correct idea or understanding of the condition in which I found the recruiting business [in New York City]. A large number of persons, of the most desperate and disreputable character, were engaged at the different rendezvous in filling the quotas. The great and urgent demands of the Government to fill up the ranks of our depleted army, were seized upon by these individuals, known as bounty brokers or receiving agents, as a fit time to perpetrate

those forgeries and frauds upon the Government and soldiers, the extent and enormity of which, I believe are unparalleled in the history of the world. These frauds, which robbed the soldier and his family, were but mild offenses compared with the crime of actually aiding the enemies of the Government, by representing a paper enlisted man who never existed.... What was true of the frauds peculiar at the army rendezvous in New York and vicinity, was more than true of the naval rendezvous. Out of seven of these naval recruiting rendezvous, but three could be entered without first passing through a public drinking saloon of the lowest and vilest character, and a substitute or bounty broker's office. In fact, the last two named institutions seemed to be necessary appendages to a recruiting depot.... The high social and official positions of many of the suspected parties, the large pecuniary interests involved in the business, tended to weaken my confidence in my success.

The points of potential fraud brought up in the affidavit of Eitel's clerk were more than corroborated by Colonel Baker:

Another manner of desertion, and by far more generally practiced, was by permitting recruits to desert in transit from the rendezvous in New York to the Island, or receiving ships. For instance, I will refer to the Cedar Street rendezvous. Between the 20th of May, 1864, and the 9th of October, 1864, there were enlisted at this rendezvous, one thousand two hundred and eighty-four men. The books on Governor's and Hart's Islands show but eight hundred and thirteen received from said Cedar Street rendezvous. About a similar deficiency between the actual enlisted and number received, is shown by the examination of the books of the other rendezvous.[8]

If each of those men represented a $300 bounty, this discrepancy alone accounted for a potential $141,300 fraud—substantial money by today's standards and a fortune in the Civil War era.

Baker caused Eitel's arrest, but soon the War Department received a stream of petitions and letters urging his release. His sterling reputation, these pleas stated, made it impossible to believe that the merchant was guilty of any crime. Moreover, anxiety since the arrest had all but driven his family to distraction. His wife was ill, and his business affairs were suffering serious damage in his absence. He never made much off his recruiting sideline, anyhow.

Following receipt of these, the case was referred for review to the Bureau of Military Justice, after which Judge Advocate General Joseph Holt commented on March 15:

The crimes committed by the class of offenders to which the prisoner is alleged to belong are so atrocious & strike so directly at the life of the military service that it is believed as a general rule the parties should not be paroled. The spoils of these crimes are known to have been so enormous, that a monied deposit would afford but an imperfect guaranty for the appearance of the offenders. Exceptions to the rule suggested may be found in cases where the measure of proof so presented is not regarded as full & entirely satisfactory. In the present case, however, Colonel Baker declares that the proofs of the prisoner's guilt are positive. This Bureau cannot therefore recommend that he be paroled.

As he had done many times before, President Lincoln overruled Holt, endorsing the file of Eitel's case on March 17, 1865: "Let this man be bailed, Mr. Dana to fix the amount." Charles A. Dana was an assistant secretary of war and the person who referred the case to Holt for review. He set the bail at $10,000, and Republican placeman Abram Wakeman of New York sent a check for that amount to parole John Eitel.[9]

When he was not setting bail for corrupt businessmen, Dana worked mainly on supply procurements for the army. The War Department oversaw a gigantic enterprise. In the first year Dana worked there, the quartermaster's office paid out nearly $285,000,000 with another $221,000,000 in bills or contracts at year's end awaiting examination before payment. "We bought," Dana recalled "fuel, forage, furniture, coffins, medicine, horses, mules, telegraph wire, sugar, coffee, flour, cloth, caps, guns, powder, and thousands of other things." By the winter of 1863, when he took this job in Washington, the army was already well supplied, the war having been going on for more than two years. Still, Dana purchased "over 3,000,000 pairs of trousers, nearly 5,000,000 flannel shirts and drawers, some 7,000,000 pairs of stockings, 325,000 mess pans, 207,000 camp kettles, over 13,000 drums, and 14,830 fifes."

Dana and his War Department colleagues were well aware that opportunities for fraud abounded. Assistant Secretary of War Peter H. Watson, a former patent attorney, reputedly had a "knack" for detecting frauds. An old law associate and friend of Secretary of War Stanton's, Watson had joined the War Department when Simon Cameron left. Since rumors of graft and corruption, as well as a reputation for inefficiency, had led to Cameron's demise, Stanton and his team surely came to the office with a mission to clean up the operation. Both the secretary and his assistant, Watson, sacrificed substantial income in accepting their poorly paid posts.[10]

Not long after Dana came to the department, Watson uncovered "extensive frauds in forage furnished to the Army of the Potomac." The government was purchasing a mixture of oats and corn for horses and mules. The grains had different weights and sold for different prices, and Watson detected the fraud in the mix and at once arrested those directly involved. He then departed Washington for another project in New York, and while he was gone, the president of the Philadelphia corn exchange and other merchants from the city came to Dana with a proposition. They paid him $33,000 to cover the amount stolen by one of the accused parties, who had confessed, and $32,000 for another. They asked for the release of the prisoners and restoration of the papers confiscated at their arrest. Dana, wisely, did not act on his own but telegraphed Watson.

Meanwhile, the grain merchants' political friends were at work on other fronts. Senator David Wilmot of Pennsylvania approached Lincoln and persuaded the president to go to the War Department to discuss the matter with Watson, who had returned to Washington. Naturally, the assistant secretary defended his action; he wanted the merchants tried by military commission.

The president pointed out that much of the money had been refunded, that grain was necessary to keep the army moving, and that the political unity of Pennsylvania was necessary for the war effort. Watson refused to release them on a verbal order and asked that the president order the release in writing. Lincoln then left the room and told Wilmot that he had accomplished nothing with Watson.[11]

The president's rather mild reactions to frauds by moneyed men contrasted sharply with the intransigence of the army. Men like Stanton, Watson, Turner, Lafayette Baker, and Joseph Holt had less patience than Lincoln with such transgressions. One of the most striking cases began on Monday, October 17, 1864, when federal authorities swept down on some of Baltimore's oldest and largest commercial establishments, scattering horrified female customers and arresting clerks and managers alike. They closed the doors of Weisenfeld & Company, clothiers; Jordan & Rose, dry goods; Isaac R. Coale & Brothers, commission merchants; Charles E. Waters & Co., hardware; A. & L. Friedenreich, gentlemen's furnishings; Simon Frank & Co., jobbers; and Hamilton Easter & Co., dry goods. Some Washington merchants were arrested as well. The prisoners were not at first informed of the cause of their arrest.[12]

The charge proved to be knowingly selling goods that were to be run through the blockade to the Confederacy. All of the arrests apparently followed visits to the establishments by Pardon Worsley and a woman whom he sometimes identified as his wife. Worsley was, according to the editors of *The Collected Works of Abraham Lincoln,* "a notorious blockade-runner," but his notoriety seems to have escaped the notice of the most important retail merchants in Baltimore.[13]

Before the war, Baltimore's businessmen were major conduits for the flow of wholesale dry goods from New York to the South, but by 1864, the pattern of this trade had changed drastically. Hamilton Easter & Co., for example, had specifically instructed its clerks not to sell goods that were likely to go south unless the buyer was a licensed government trader with passes and permits to trade through the lines.

Such precautions, if indeed they were ever taken seriously, failed when Pardon Worsley went shopping in 1864. His first visit to Hamilton Easter & Co., occurred several months before the arrests, when a clerk of some fifteen years' experience waited on Worsley and his female companion, a "Mrs. Chancellor." She lived someplace in Virginia, the clerk recalled later, where, at the time, goods from the North were allowed to pass. When the clerk asked her whether she had a pass, she said that she did not but that the man with her did. Worsley returned in early October to purchase doeskin or cadet-mixed doeskin. The clerk sold him the goods.

Worsley visited the hardware store of Charles E. Waters on October 9, again accompanied at first by a woman. He returned later in the day, asking for percussion caps and gun wads ("which are articles much in demand at this season for Sportsmen," Waters explained to federal investigators later). Then

Worsley asked for military buttons, assuring the clerk that he was a licensed government trader, selling on both sides of the river. The hardware store had none but supplied him with nonmilitary brass buttons.

Such transactions led to the arrests of October 17. Hamilton Easter and his partners, two nephews of his, were arrested along with forty clerks. At Weisenfeld & Company, the lowly cloth-cutters were arrested as well as the managers. A special train carried away ninety-seven astonished merchants, clerks, and tradesmen who were imprisoned in Old Capitol.

The existing records of the cases are fragmentary, but it appears that the principals in the firms were tried by military commission before the end of the month, and Worsley testified against them. Most of the clerks were paroled, but some of the merchants received sentences to be served in the Albany penitentiary in New York.

When civilians of such prominence as these Baltimore merchants were arrested by military authority, there was a good chance that Lincoln might eventually become involved. A man like Hamilton Easter could hardly disappear into prison unnoticed by the larger world. Born in Ireland in 1810, he came to the United States when he was sixteen years of age. He was naturalized and commenced business in Baltimore on his own account in 1830. Since then, as he told the government's investigators, "I have by Integrity, Industry and Honorable dealing entirely unaided by friends, built up one of the largest and most extensive establishments in the country—we are large Importers as well as Jobbers and Retailers." The scale of his enterprise made plausible his defense that doeskin at two dollars a yard was "rather a small . . . profit for a transaction so risky as selling goods knowingly to run the blockade." His business record also made possible the mustering of an impressive list of endorsements of his integrity by prominent Baltimoreans like Enoch Pratt and Johns Hopkins and by the adjutant general of the state of Maryland.

Social prominence provided no real protection from the federal detectives. If anything, the system of military arrests worked on opposite principles and was biased against prominent citizens. The Union authorities hoped to capture influential wrongdoers and, conversely, often showed a willingness to let harmless men of little influence, intelligence, or ability go ("game not worth the candle," Levi Turner called them). Nor was the government system corrupt. When Moses Weisenfeld of Weisenfeld & Company attempted to bribe Colonel J. A. Foster with $7,400 to influence the decision of the military commission, Foster quickly turned the money over to the judge advocate's office.

Those with social prominence, however, could gain the attention of the elected politicians, including the most important one who occupied the White House. For Lincoln to alter the judgment of a military commission, of course, there had to be some evidence of injustice. And by 1864, the advice, if not the consent, of the judge advocate's office in such matters was important as well.

In the case of Abraham Friedenreich, one of the Baltimore merchants arrested on October 17, 1864, influential friends intervened, and Lincoln wrote an endorsement, saying simply: "Hon. Sec. of War, please give this man a hearing." Abraham Friedenreich had been arrested by mistake because his nephew Leon used his uncle's name on the sign at the store. Abraham actually had no connection with the shop. Moreover, his brother was loyal and, as a friend reported, had "wide connections and great influence amongst the Hebrews of Baltimore." Lincoln's secretary, John Hay, reported that the president wanted Friedenreich released "if Judge Turner's report is favorable." Turner agreed that a mistake had been made, and Friedenreich was released unconditionally within a week.

At later dates, the president was several times involved in the Baltimore arrests. On January 29, 1865, he asked Holt for "his opinion whether it would be legally competent for the President to direct a new trial" in the case of Thomas W. Johnson, who had not been allowed to introduce witnesses to refute the testimony of Pardon Worsley. On January 31, Lincoln ordered that Charles E. Waters not be sent to the Albany penitentiary to start serving the sentence of the military commission.

On February 11, 1865, Lincoln handled a long petition written on behalf of three Washington merchants involved in the case, T. W. Johnson, R. M. Sutton, and J. H. Hennage:

> Those men have been tried by military commission, on a charge of selling goods in violation of law, to one Pardon Worsley, an alleged blockade runner, and convicted, they are now in the Penitentiary in execution of sentence, we humbly but earnestly ask their immediate and unconditional pardon for the following reasons, viz:
>
> 1st From our knowledge of them, we cannot believe they have knowingly violated the law—
>
> 2nd As we are informed, and believe, they entered upon their trial with the expectation, that they would have the testimony of their clerks to disprove the allegations made against them—but, their clerks being held at the time, as prisoners, and parties to their alleged guilt, were not permitted to testify. The affidavits of those clerks since released have been shown to us, and in our opinion fully disprove the charges upon which they stand convicted.
>
> They have already suffered severely from confinement in the Old Capitol Prison—have suffered immensely in their business, and their families feel this stroke deeply, and severely—Tis enough, let the hand that saves be stretched forth for their deliverance—let an unblemished mercantile reputation of twenty years standing vindicate them—and shield them from even suspicion, of the guilt that is sought to be fastened upon them by the unsupported testimony of a stranger, of a very doubtful character.

Twenty-five dry-goods firms put their names to the petition, and John W. Garrett, president of the Baltimore and Ohio Railroad, wrote a note on behalf of the signers of the petition.[14]

Finally, on February 17, 1865, President Lincoln wrote Joseph Holt the following letter:

> In regard to the Baltimore and Washington Merchants—clothes dealers—convicted mostly on the testimony of one Worsley (I believe) I have not been quite satisfied.

I can not say that the presumption in favor of their innocence has not been shaken; and yet it is very unsatisfactory to me that so many men of fair character should be convicted principally on the testimony of one single man & he of not quite fair character. It occurs to me that they have suffered enough, even if guilty, and enough for example I propose giving them a jubilee, in which course the Sec. of War inclines to concur; but he tells me you are opposed. I write this to ask your cheerful concurrence.[15]

The letter is one of the many minor masterpieces that lie buried in *The Collected Works of Abraham Lincoln,* wonderfully clear, economical, comprehensive, fair-minded, and, withal, sparkling in tone. Despite the grim subject and the drearily legal and bureaucratic context, Lincoln rose to the occasion with a small gem of a letter. Holt and Turner, although they dealt with similar problems almost daily, produced nothing even remotely like this document.

It was effective too. Holt responded on the same day: "I certainly have no disposition to oppose the impulses of your kind heart, in the matter.... In a conversation with the Secty of War this morning, I said, in allusion to your anticipated action, that I thought the sentence resting in large part on a finding of guilt of attempt to bribe an officer of the government, might, in the exercise of your clemency, be well distinguished from the other cases in which no such criminality was averred." The next day Stanton wrote an order to release Johnson, Sutton, and eight other merchants who had been sentenced to imprisonment for one to five years and had been fined one to fifteen thousand dollars. He excepted Moses Weisenfeld, the man who had attempted to bribe Colonel Foster.[16]

Many years later, Assistant Secretary of War Dana revealed yet another reason for Lincoln's dissatisfaction with the punishment of the Baltimore merchants: Worsley was a double agent. When Worsley traveled in Virginia, he would boast that he had hoodwinked Washington authorities and would bring with him some items of contraband, difficult to obtain, that helped insinuate him into the good graces of the Confederates. Government detectives approved the items he carried, and they amounted to little when weighed against the valuable information about the strength and movements of Confederate armies that he brought back with him. U.S. agents checked his information and it always seemed valid.

In October 1864, however, when Worsley the peddler came back from Richmond to resupply himself, he purchased so many goods that Lafayette Baker would not allow him to go back to the Confederacy. There were, as Dana reported, "uniforms and other military goods, and all this, of course, was altogether too contraband to be passed." Worsley had the receipts for the goods, perhaps $25,000 worth. The authorities arrested him, confiscated the goods, and on October 17 went after the merchants who had sold them to him.

When a deputation from Baltimore came to Washington to protest the arrests, Lincoln took them over to the War Department to confront Stanton. The secretary of war gave an eloquent speech to the delegation, denouncing the rebellion, pointing out its cost in lives, and then showing the nature of the contraband goods Worsley had managed to purchase in Baltimore. According to Dana, the Baltimoreans left quietly.[17] For stern lectures on matters of tainted commerce, one turned to the War Department and not to the president.

The Growth of Awareness of Fraud
Inside the Administration

Until the deep crises following the fall of Fort Sumter and the terrible defeat at Bull Run had passed, the Lincoln administration could hardly afford to concentrate on devising a system of accountability for its procurement processes. The importance of fraud dawned only gradually on the Lincoln administration. As late as December 1862, for example, army disbursing officers were still paying, without audit, accounts submitted by recruiting officers for collecting, organizing, and drilling volunteer soldiers. Finally, General Order No. 298 of December 3, 1862, issued by the adjutant general's office, required audits of recruiting officers' claims before funds were disbursed. If the auditors rejected a claim or detected fraud, they turned the case over to Levi Turner for further examination or, after March 2, 1863, for prosecution. The Provost Marshal General's Bureau was given responsibility in this area and kept some statistics during the war. Of the 4,402 claims acted upon by the bureau before May 1, 1863, 745 were rejected and turned over to Turner. After that date, 4,603 claims were examined and 674 rejected.[18]

The Provost Marshal General's Bureau eventually supervised volunteer recruitment by the states. Recruiting officers had to report to officers of the bureau's recruiting branch to be assigned their duties, and they could not incur expenses without the bureau's authority. No accounts were paid without such approval.[19]

After the desperate early days, when the War Department under Cameron's leadership, along with other parts of the government, had frantically hastened to organize, equip, and arm forces to put down the rebellion, the administration finally had a chance to implement some effective controls. When Cameron was censured by the House of Representatives in 1862 for loose practices in procurements from New York, President Lincoln defended him on May 26 by recalling the desperate situation in Washington after the fall of Fort Sumter. Rather than "let the government fall at once into ruin," the president had availed himself "of the broader powers conferred by the Constitution in cases of insurrection." He initiated military purchases without congressional acts and, fearing the many disloyal persons still holding jobs in the government departments, he relied on unofficial agents "favorably known for their ability, loyalty, and patriotism." Private messengers carried these irregular orders throughout the land. "I believe," Lincoln told Congress, "that by these and other similar measures taken in that crisis, some of which were without any authority of law, the government was saved from overthrow." He had naturally been apprehensive lest the "public funds thus confided without authority of law to unofficial persons" might be "either lost or wasted," but the president could tell Congress in 1862 that not a dollar had been so squandered. Whatever irregularities may have occurred under Cameron, he concluded, "not only the President but all the other heads of departments were at least equally responsible for . . . error, wrong, or fault . . . committed in the premises."[20]

Naturally, "Honest Abe" Lincoln, whose personal rectitude in money matters made his nickname a genuinely apt description, must have looked kindly on efforts to clean up the process of supply procurement, mobilization, and other military activities that invited fraud. The number of arrests for such activities amounted to around 4.7 percent of the prisoners listed, for example, by Levi Turner's Record of Prisoners of State. They constituted a smaller percentage of the arrests noted in other series of records because of Washington's central importance for supply and mobilization.

Of the 1,561 cases in the Record of Prisoners of State where a charge is noted, at least 74 can be classified as fraud cases. Forging discharge papers, selling passes, financial fraud committed while employed in an army or government bureau, buying or selling government property, posing as a government detective in order to accomplish fraud or theft, stealing wood from government piles or reserves, and taking bribes to exempt men from enrollment for conscription were among the charges noted. In addition, some people were held as witnesses in fraud cases or as likely accomplices, like the civilian clerks of defaulting army paymasters.

It is difficult to think of such prisoners as being in the same category as Clement Vallandigham, yet one of the more famous freedom-of-speech cases in the war could be seen almost as a fraud case: the suppression of the New York *World.* On May 18, 1864, the *World* and another metropolitan paper, the *Journal of Commerce,* published a bogus presidential proclamation calling for a day of fasting and prayer and, more important, calling for a new draft of 400,000 men. Later that day, the president ordered the arrest of the editors, proprietors, and publishers of the newspapers and the seizure of their offices by the military. Seward was the cabinet member most eager for the arrests and for the suppression of the troublesome Democratic newspapers, while Gideon Welles, by now a constant and nagging critic of Seward, came to see the papers as the victims of a hoax:

> The seizure of the office of the *World* and *Journal of Commerce* for publishing this forgery was hasty, rash, inconsiderate, and wrong, and cannot be defended. They are mischievous and pernicious papers, working assiduously against the Union and the Government, and giving countenance and encouragement to the Rebellion, but were in this instance the dupes, perhaps the willing dupes, of a knave and a wretch. The act of suspending these journals, and the whole arbitrary and oppressive proceedings, had its origins with the Secretary of State. Stanton, I have no doubt, was willing to act on Seward's promptings, and the President, in deference to Seward, yielded to it.[21]

The knave was Joseph Howard, who schemed with fellow newspaperman Frank A. Mallison to raise gold prices by contriving the ingenious forgery on authentic Associated Press paper and distributing it in the early morning hours when the sharper editors were not at work. Welles seems not to have comprehended the breadth of the government's reaction. Fears that the telegraph had somehow been corrupted seized the War Department, which assumed that the forgery had been carried on the wires of the Independent Telegraph Company, which, unlike its competitor, the American Telegraph Company, did not have wires running directly into the War Department. The Independent's office staff in Washington was arrested and put in Old Capitol prison. Orders were sent to arrest telegraphers in New York, Harrisburg, Philadelphia, and Pittsburgh as well. Howard and Mallison were arrested, too, and put in Fort Lafayette, though Howard's influence, as a parishioner of Republican preacher Henry Ward Beecher, gained him early release on Lincoln's order on August 23, 1864.[22]

Fear that the bogus proclamation was a Confederate plot, willingly abetted by the hated New York Democratic newspapers, lay at the root of the administration's tough response. The sweeping arrests also stemmed from fears of the genuine mischief that would be accomplished if the telegraph system were corrupt. Finally, hatred of fraud, and especially of speculation in gold, played a part in motivating the government as well. Even the skeptical Gideon Welles exploded in indignation at first, calling the proclamation "a cunningly devised scheme—probably by the Rebels and the gold speculators, as they are called, who are in sympathy with them."[23]

The Role of the Loyal Opposition

The president must have smiled on the efforts to ferret out corruption. His wife once described him as a "monomaniac on the subject of honesty," but Lincoln's political heritage actually left him an ambivalent legacy on this score. The Whig party had been at once more stricken with Protestant conscience and more comfortable with business, the financial community, and promoters than the Democratic party. Lincoln was personally honest in money matters, but he was also more sympathetic than Stanton and other army officials to financial wrongdoers. Lincoln himself embodied the split Whig personality on such questions, and he continued to exhibit it as a Republican president.[24]

Probably President Lincoln's political party was also kept on good financial behavior by the constant vigilance and criticism of the Democrats. The sheer enormity of the sums expended on this unprecedentedly large war effort were bound to attract the attention of the opposition political party. When Congress debated a deficiency appropriation for the War Department on March 7, 1864, for example, Democratic Congressman James Brooks noted that it exceeded the cost of any year of the War of 1812 four- or fivefold and that it exceeded the yearly cost of the Mexican War by $63 million. "There is something wrong in all this," he exclaimed, "there must be something wrong." The House could not "abandon its functions, but should look into these expenditures, and see when, how, and where the money has been appropriated, and where these enormous sums of money have gone, who have had them, in what quarters they have been paid, why they are called for, and what they are to effect."[25]

The Democrats were able to circulate a pamphlet in 1864 called *Corruptions and Frauds of Lincoln's Administration,* which focused on the New York Customs House. Yet the information the Democrats exploited came from investigations conducted by a Republican committee in Congress. Chaired by New York Republican Charles H. Van Wyck, the House Select Committee on Government Contracts was established in July 1861. According to historian Fred A. Shannon, the committee "discovered an astounding amount of illegal and fraudulent activities." Stanton's accession to the War Department early in 1862 brought further internal investigation and supervision. And there was obvious administrative muscle behind the clean-up, as is evidenced in the general widening of military solutions to essentially civilian problems in the congressional acts of June 2 and July 17, 1862. These acts, among other things, made some War Department contractors subject to military law and to trial by military courts in cases of fraud.[26]

Even so, Shannon, who wrote in the shadow of the war profiteering scandals of World War I, concluded that, "considering the wealth, population, industrial conditions, and available resources of the two periods even the World War does not present a larger degree of profiteering activities than did the Civil War. As for official corruption," Shannon added, "the utmost efforts of a hostile congress in 1919–1920 failed to discover anything to compare with the conditions of 1861–1865." Nor were accusations of corruption or fraud confined to the halls of Congress. Stereotypes of corrupt government contractors appeared in political cartoons in New York's popular illustrated newspapers as early as the summer of 1861. And the Democratic press everywhere quickly found anecdotes and slogans that criticized government corruption in organizing the war effort at the state and federal levels. "Shoddy" came into use as a term to describe items of poor quality manufactured on government contracts by profiteers. Democrats thus denounced "the shoddyites...who are making fortunes out of the war, by robbing the government and the soldiers." In fact, the term "shoddy" gained its modern meaning in the Civil War. Originally a word that described a kind of second-rate wool made by shredding old woollen goods, "shoddy" came to mean poor material made up to look like superior material because it was said the woolen uniforms sold to the government were made of shoddy. The term "shoddy aristocracy" referred to men who gained their wealth from fraudulent government contracts during the war.[27]

On the whole, most Democratic criticism of fraud and corruption did not much hurt the Republicans, perhaps because the Republicans themselves made some moves to control corruption and fraud. Far-reaching military solutions to this civilian problem were enacted in a bill that sailed through Congress early in 1863. At the instigation of William Whiting, who drafted the bill, Senator Henry Wilson of Massachusetts introduced this new measure to prevent and punish frauds upon the government of the United States. Its most controversial provisions defined defrauding contractors as members of the military forces for punishment of new classes of offenses, and it made discharged soldiers who had committed frauds while in the U.S. forces still liable to military punishment for their crimes. Republican Jacob Howard of Michigan led the floor fight for the bill. Most criticism came from fellow Republicans, especially from Edgar Cowan of Pennsylvania. The only Democratic member to challenge the bill in a speech was a border-state conservative, Garrett Davis of Kentucky, who saw that "the distinctive features of the bill fix its paternity on the Secretary of War. He is prone to bring every person and every act within military law and military courts."[28]

The Democrats did not really press the issue, even though it subjected yet another class of civilians to military law. The Republicans thus took care of fraud and corruption to some degree on their own, the Whiting bill originating entirely in the War Department as a response to a practical problem of enforcement and not, apparently, to Democratic political criticism. The loyal opposition party let the Republicans co-opt this issue, and the Democrats thus lost the bite of their traditional Jacksonian criticism of economic corruption.[29]

Anti-Semitism and Military Arrests

Horace Greeley's New York *Tribune,* a newspaper of Republican political philosophy but as independently cantankerous as its famous editor, criticized administration fraud and corruption early in the war. One article pointed to a sorry lot of cattle sold by a contractor to the government and described them as "fag ends of small lots of scalawags which had been rejected by the speculators and Jew brokers."[30] Like many societies in Europe, the United States was tainted with anti-Semitism at mid-century, and the army proved to be at least as anti-Semitic as society at large and probably more so. The results of prejudice, at any rate, seem apparent in the program of military arrests of civilians.

Such prejudice is evident especially in an area of policy closely associated with fraud and corruption: contraband trading. Trading with the enemy directly strengthened the opposition, whereas fraud and corruption were practiced on the U.S. government itself. But they shared the same basic motivation, greed, and the same image of putting one's narrow personal interests ahead of the interests of the nation. Many in the army deeply resented such crimes.

When, in their attempts to halt trading with the enemy, the army happened to arrest a trader or smuggler who was Jewish, his religion was noted in the record, though no other religion was (skin color was noted in the case of blacks). Thus, Baker's detectives put a store in Fairfax, Virginia, under observation and finally arrested, in June 1862, a man named Rothschild for engaging in contraband trade. Soldiers described Fairfax as the "headquarters" of "these *Jews* who are engaged in the illicit trade."[31] Later that year, when Jacob Sterne was arrested in Baltimore for buying quinine for the rebels, the authorities noted that he was "related to a Jew."[32]

It is not always easy to tell what adverse effect this prejudice had on individual cases. What can best be documented is the army's awareness of the Jewish prisoners' difference. One other thing is clear. Prejudice could easily lead Christians to denounce Jews to the arresting authorities. For example, Emil Lippman was arrested in New York in 1862 for draft fraud. He was a district enrolling officer and was accused of taking money not to enroll someone for the draft. One of the army officers noted, however, that "the affidavits are made by ignorant German Catholics, who hate this German Jew."[33]

The common stereotype that depicted Jews as rootless profiteers who owed allegiance only to mammon and to no country surely must have cast a shadow of dark suspicion on economic activities taking place near the armies—activities that otherwise might have escaped the authorities' attention had they been engaged in by Christians. The most notorious incident of anti-Semitism in the Civil War pitted the army against Jewish merchants who followed in its wake. On December 17, 1862, Ulysses S. Grant, then a major general commanding the Department of the Tennessee, issued General Order No. 11, expelling "The Jews, as a class...from the Department" because they were "violating every regulation of trade established by the Treasury Department, and also Department orders." Historians have described the immediate origins and context of Grant's notorious order: "frequent complaints concerning Jewish involvement in illicit trade" in "the western theatre of war" as well as Grant's personal anti-Semitism and his frustration in trying to control the illegal trade in enemy cotton. Its immediate consequences have been described as well. Jewish citizens and soldiers were outraged by Grant's order, which remained in effect until Caesar J. Kaskel of Paducah, Kentucky, visited Lincoln on January 3, 1863, to protest the ban. The next day, General Halleck revoked Grant's order. One cannot be certain of Lincoln's reasoning, for only Halleck's words survive in the documentary record. Thinking Grant deserved an explanation, Halleck told the general on January 21, "The President has no objection to your expelling traders & Jew pedlers, which I suppose was the object of your order, but as it in terms prescribed [*sic*] an entire religious class, some of whom are fighting in our ranks, the President deemed it necessary to revoke it."[34]

The idea that Jewish peddlers could fairly be excluded remained commonplace in the army, despite the president's revocation of Grant's more sweeping order. Indeed, on January 5, 1863, just after Halleck rebuked Grant, Colonel John C. Kelton informed Grant "unofficially" of the "objection taken to ... Genl Order No. 11. It excluded a whole class, insted of certain obnoxious individuals. Had the word 'pedler' been inserted after Jew I do not suppose any exception would have been taken." No one in the cabinet expressed concern over the issue, and Attorney General Bates, who received a letter protesting the order, nonchalantly passed it along to the president, expressing "no particular interest in the subject" himself. Grant's champion in Congress, Representative Elihu B. Washburne of Illinois, praised the order to the president and then went to see Halleck, who repeated the idea that precision in stating that Jewish "ped-dlers" or "traders" were excluded would have been "all right." The stereotype of the Jewish merchant's willingness to betray his country for gold was perhaps more deeply embedded in the administration than the famed revocation of Grant's order might suggest.[35]

Certainly, many in the army believed fiercely in the stereotype and did not cease to believe in it after Grant's reprimand. This prejudice was, in part, the dark reverse side of the army's rather strict attitude toward financial corruption and fraud. An even larger context for Grant's order, then, should include the deep anti-Semitism of the officer class, as exhibited in military arrests in all theaters of the Civil War. Anti-Semitism in the U.S. Army was a trait shared with other armies in the Western world in the second half of the nineteenth century. Grant's infamous order brings to mind the Dreyfus affair in France later in the century. The British army of Lincoln's era had no Jewish generals. The first Jewish chaplain in the British army was not appointed until 1892.[36]

Yet, as dark as this side of the American Civil War may seem, the picture was brighter perhaps than it might have been. American officers may have shared the prejudices of their class in Europe, but these ideas never resulted in a social upheaval as serious as the Dreyfus affair. They never were rationalized in a widely accepted warrior critique of bourgeois values that might be char-acterized as proto-fascist. Such beliefs could be found in later years in the United States among only a few bizarre militarists like Homer Lea. And the politicians had a better record on the question than the generals. A congres-sional act of July 17, 1862, made it lawful for rabbis to become chaplains in the U.S. Army, and President Lincoln appointed the first rabbi to serve as military chaplain, Jacob Frankel, on September 18, 1862.[37]

Torture

If the frauds and corruption of the Civil War were gradually met to some degree by safeguards and investigations of the administration's own devising, and if military anti-Semitism was somewhat balanced by the president's even-handedness, at least one development in the system of military arrests showed an ominous possibility of degenerating into cruelty and brutality masked by bureaucratic indifference: the rise of torture as a means of extracting confes-sions. The steady decline in the status of Southern noncombatants in the military thinking of the North might lead one to suspect that the likeliest torture victims were the detested Southern guerrillas or scouts. In fact, the victims were not Southerners at all. They were Northerners suspected of deserting from the United States Army.

On the surface, it might appear that the problem of desertion from the U.S.

forces is wholly military and would lie outside the scope of this book. But, as was the case with Confederate deserters, civilians became victims of military arrest because of the practical difficulties involved in capturing deserters, and, more so-called "political prisoners" were created. The military authorities naturally arrested some suspects who proved not to be deserters at all but were, in fact, innocent civilians.

The torture of these civilians by the military would never have been recorded had there not been a further complication in the attempt to arrest deserters: some of the innocent persons arrested on suspicion of desertion from the Union army were British subjects. In 1864, the complaints of these prisoners to British authorities in the United States began to include descriptions of torture. Mathew Murphy, for example, was an Irishman languishing in jail in Alexandria, Virginia, in October 1864. He had been arrested on suspicion of desertion and apparently gave conflicting statements to the authorities. He had been picked up wearing some government-issue clothing and was a "hard-looking" man. Murphy maintained that he had been handcuffed and suspended by the wrists. When British authorities intervened to ask for investigation of the case, prison-keepers in Alexandria explained that they intended that no cruel "punishments" occur, but they admitted that some might.[38]

Handcuffs and hanging by the wrists were rare, but in the summer of 1863, the army had developed a water torture that came to be used routinely. In one case, J. W. Nash, a British subject, was arrested, as most such victims were, when he was about to board a train at a railroad station. He was in the company of two deserters; he was dressed the same as they; and he held the same amount of money. The government detective assumed all three were bounty-jumpers. When Lord Lyons intervened to protect the British subject, he complained about "violent cold water shower baths" applied to Nash. As was customary in State Department enquiries in such matters, Seward asked Levi Turner to investigate. Nash was a prisoner at the Central Guard House in Washington, and the captain commanding there replied to Turner's inquiry thus:

> He was subjected to what is called a shower bath, which consists of a stream of water from a small rubber hose. It is not severe nor at this season of the year very unpleasant, as the prisoners here shower each other for their own comfort, daily.

Nash had been uncooperative, refusing "to give his proper name," and he had no identification papers on him. He was released, said the commander, "after the usual time had expired, viz: eight days." Not being able to find his name missing from any regimental roll, the authorities let Nash go. The captain added "that nearly all the deserters from the army claim to be British subjects."[39]

Seward sent Lyons a copy of Turner's letter about Nash, but the British minister was not satisfied. "This explanation," he protested, "does not show that the cold water was applied in Nash's case, in conformity with any law or regulations as a punishment for a known and proved offense[;] on the contrary it tends to confirm the statement that it is used in the Central Guard House for the purpose of extorting, by the infliction of bodily pain, confessions from persons suspected of being Deserters." In October, Turner replied to the secretary of state, telling him that the captain at the Central Guard House assured him that no one was "'showered,' save as a punishment for violation of rules, falsehood, refusing to give their names & the like."[40]

Another case investigated that summer led to a British protest that the prisoner had been subjected to "a hose of water directed with full and powerful action against his naked person." Again soldiers at the Central Guard House insisted that the "Judge Advocate orders the punishment by shower baths *only* when the evidence against the prisoner is very *clear* and *conclusive* that he is in the U.S. service, or when he does not comply with the rules of the prison. It never was used as a mode of torture as can be testified by all attached to the prison."

Torture is torture, whether it is administered as punishment for a violation of the rules or to extract a confession. *Webster's Ninth New Collegiate Dictionary,* for example, defines torture as "the infliction of intense pain . . . to punish, coerce, or afford sadistic pleasure." By the dictionary's meaning of the word, at least, the "cold-water treatment" was torture, whatever its purpose.[41]

The only possible way to construe the cold-water treatment as something other than torture would be to prove that it did not involve "intense pain." The pain proved adequate to cause several British subjects to mention it in their letters to Lord Lyons; they did not mistake the treatment for a refreshing shower taken for relief in hot weather. James Buckley maintained that he had been subjected to showering for two hours until his skin broke. When asked about Buckley's case, the captain commanding at the Central Guard House said only that Buckley had received the "usual cold-water application." When William Williams, another British subject erroneously arrested as a deserter, was released early in January 1865, after imprisonment since at least the previous October, the captain at the Central Guard House told Turner, "No water has been applied to him since warm weather, and then only what was necessary." But Williams had been in prison a long time, not the "customary" eight days allowed for suspected deserters whose names could not be found on any regimental roles. Such a prisoner might be subjected to many applications of the torture.[42]

Records of these six cold-water torture cases in 1864 are preserved only because the victims were British subjects. There must have been many more cases involving persons born in the United States who had no recourse or protection and thus no way of generating records of their treatment for the archives. The Central Guard House captains, at least three of them, readily admitted using the treatment; one said that the judge advocate ordered it; and one described it as "usual."[43]

The bureaucracy of the Lincoln administration reacted in predictably bureaucratic fashion. Turner, when asked by Seward, in turn asked someone else, the Central Guard House detail, for a report. When it was given him, Turner quoted it accurately for Seward, who relayed it in condensed form to Lyons.

But that was all. No one exploded in indignation or horror. No one issued a special order demanding that such practices cease. No one requested investigation or study. No one asked whether other prisoners than the ones Lyons inquired about received such treatment. No one, except Lord Lyons, asked what law governed such cases. No one expressed any personal outrage or personal feeling at all, including Lincoln's secretary of state.

As usual, the Lincoln administration's internal security system betrayed no signs of corruption in the water-torture affair. No one attempted to cover up the practice, from lowly captain in the Central Guard House to secretary of state. But there was no impulse to correct the abuse; indeed, no one saw it as an abuse. It had become a usual and customary way of handling certain kinds of prisoners. Had the Civil War continued longer—into the summer of 1865 and beyond—such practices might well have increased. As military authority was extended to cover more and more kinds of cases, torture might have spread also. And the history of the war would have become darker yet.

6

Numbers and Definitions

Most military arrests of civilians did not involve torture or prejudice against an ethnic group, but historians nevertheless generally regard them as a dark chapter in the history of the Lincoln administration. How dark a chapter is a question best answered by showing how many victims were arrested. Scholars have differed in their estimates, but no historian or commentator ever maintained that the number was trifling. No one has put the figure below ten thousand. Most have put it higher than that. In the years immediately following the war, the estimate stood at its highest. As it turns out, that high estimate appears as near as any to being correct.

History of Estimates of the Aggregate Number of Arrests

Just after the Civil War, *The American Annual Cyclopaedia and Register of Important Events of the Year 1865* stated that the total number of military arrests in the North had been thirty-eight thousand. This publication had followed the issue closely throughout the war in articles written under the heading "Habeas Corpus." In their final article on the subject, the editors expressed shock and dismay: "The extent to which the arbitrary arrest of citizens without benefit of the writ of *habeas corpus* was carried, is indicated by the records of the Provost Marshal's office of Washington, which shows that from June, 1861, to January 1, 1866, the cases of some thirty-eight thousand prisoners have been reported to that office. Out of this vast number the Old Capitol prison shows upon its record that it has housed for longer or shorter periods sixty-five hundred prisoners of war, forty-five hundred real and fancied offenders against the State, and twenty-five hundred deserters and bounty jumpers."[1]

My search for the exact source of the *Cyclopaedia's* figure proved fruitless. The two-volume *Final Report... to the Secretary of War by the Provost Marshal General,* published in 1866, the year the Provost Marshal General's Bureau was abolished, does not contain the figure. The portion of the report devoted to the Deserters' Branch of the bureau provides figures only for deserters and draft evaders. The other branches of the bureau had little to do with arrests, and most of the first volume of the report is taken up with medical statistics on draftees. Only the Historical Report of the State Acting Assistant Provost Marshal General for Illinois, included as a model of its kind, contains a figure for "persons arrested (not deserters)." The number given is 443, probably too small to constitute Illinois' share of 38,000 arbitrary arrests, if indeed the number of 38,000 was used by the provost marshal general. The historical reports for other states not chosen for inclusion in the *Final Report* exist only in manuscript form in the National Archives and do not contain aggregate figures on arrests. The tone of the Illinois report does make clear the overriding purpose of the provost marshal general's office during the Civil War. The category of "persons arrested (not deserters)" was included as a casual after-thought. Such arrests were considered marginal to the important work of the bureau—arresting deserters and draft dodgers.[2]

The abolition of the bureau and the rearrangement of its records make it unlikely that the source of the thirty-eight thousand figure will ever be found. Even the figure given in the *Cyclopaedia* for Old Capitol prison seems suspect. For one thing, deserters were not kept in Old Capitol. For another, only a handful of "morning reports" from that prison survive which list prisoners resident each day. Arriving at an aggregate figure for the war from "morning reports," even if a more complete run of them existed, would require immense labor, as each prisoner was listed again each morning until he left the prison. Some monthly records survive, too, but they suffer from the same problem of duplication of names. I estimated that it would require ninety working days of data entry to compile the list necessary to establish an aggregate figure for this prison alone, if such records existed. But they were too fragmentary to be useful in any event.

The second person to estimate the number of arrests ignored the figure printed in the *Cyclopaedia*, despite its impressive size and obvious utility for groups hostile to the Lincoln administration. John A. Marshall's *American Bastile*, a bitter history of arbitrary arrests published in 1869 as a sort of book of martyrs, stated that the military prisons "contained, during the short period of four years, as variously estimated, from ten to twenty thousand men, besides women and children—free citizens of free states." When the book was revised in 1885, the estimate rose to "from twenty to thirty thousand men, besides women and children."[3] Marshall was not prone to understatement or much concerned about scholarly precision. He would surely have used the thirty-eight thousand number had he known about it.

The thirty-eight thousand figure came to the attention of American historians later in the century from still another source and as a result of British cultural criticism of the United States. The Englishman James Bryce's influential book, *The American Commonwealth*, appeared in 1888 and contained the statement that "Lincoln exercised more authority than any Englishman since Cromwell."

When James Ford Rhodes, the most important Civil War historian of the turn of the century, read this statement in Bryce's book, he was flabbergasted. Seeking more information, Rhodes turned to the article on "Habeas Corpus" in *Lalor's Cyclopaedia of Political Science,* a forerunner of the *Encyclopedia of the Social Sciences.* There, in an article by the eminent Princeton historian Alexander Johnston, Rhodes for the first time saw the thirty-eight thousand figure. Sources cited for Johnston's figure in *Lalor's Cyclopaedia* included *Reports of the Provost Marshal General* and *Circulars of the Provost Marshal General,* May 15, 1863–March 7, 1865. I was unable to find any *Reports* (Johnston presumably referred to published ones since the title was italicized) except for the previously mentioned publication which contains no aggregate figures. Bound sets of circulars do exist, but they contain only rulings on controversial points, no numbers or charts.[4]

Rather than check the source cited in Johnston's article in *Lalor's,* Rhodes sought a new estimate. The patriotic Rhodes could hardly believe his country's record was worse than England's on this score, even though he was raised during the Civil War in a Democratic family and therefore presumably was disposed to think the worst of the Lincoln administration.[5] He paid a call on the War Department to find out whether thirty-eight thousand civilians really were arrested by the military during the Civil War. Eventually, he asked Colonel F. C. Ainsworth, chief of the Record and Pension Office of the War Department, to search the relevant files. This the obliging colonel did—to earn only a modest footnote in Rhodes's multi-volume *History of the United States from the Compromise of 1850.* Despite "a protracted and exhausting examination," Ainsworth reported that it was impossible to compile a correct figure from the "very incomplete" records. But he could find no authority at all for the figure cited in *Lalor's Cyclopaedia,* which he characterized as "really nothing but a guess." The colonel's clerks did not uncover anything even approaching it; instead, they found records of 13,535 civilian prisoners in the papers of the commissary general of prisoners for the period from February 1862 through the end of the war—the period when the War Department oversaw military arrests of civilians.[6]

Colonel Ainsworth's figure proved influential. The number 13,535 has been accepted by most historians ever since. Despite professionalization in history since Rhodes's and Ainsworth's day and despite the modern movement for quantification as well, most historians of the Civil War have not questioned— have in fact used—the old War Department estimate.[7]

Among those who accepted the new figure was the author of the standard twentieth-century scholarly work on the subject, James G. Randall. A longtime professor of history at the University of Illinois, Randall is now famous for attacking unprofessional standards in Lincoln scholarship. His much-quoted 1936 paper, "Has the Lincoln Theme Been Exhausted?" argued that "the hand of the amateur" had rested too long and too heavily on Lincoln studies.[8] Yet, a decade earlier, when Randall had examined the problem of military arrests of civilians for a chapter in his book, *Constitutional Problems under Lincoln,* his method differed little from that of the amateur or gentleman historian, James Ford Rhodes. Perhaps both scholars had little choice. The relevant War Department records may not have been open for research. Whatever the case, to check the 13,535 figure, Randall merely wrote a letter to Robert C. Davis, adjutant general of the United States, to ask him whether he could now in any way revise Ainsworth's estimate. Major General Davis said that he could not, but he agreed with his predecessor that the real number had to be "much less" than the "exaggerated guess of Alexander Johnston" in *Lalor's Cyclopaedia.*[9]

It is impossible today to verify the thirty-eight thousand figure from its original source or to duplicate the work of Ainsworth's clerks or Davis's. The bureau from which the earliest figure derived was abolished. Likewise, the position of commissary general of prisoners, from whose records Ainsworth said he had derived his figure, had been abolished in 1867 and the records had been turned over to the Prisoner of War Division of the adjutant general's office. They now apparently constitute Record Group 249 in the National Archives. They contain a few fragmentary rolls of civilian prisoners but nothing that could amount to 13,535 names, and, more important, nothing that could possibly be mistaken for a complete listing of civilian prisoners from the Civil War. The largest extant number of lists of civilian prisoners are now located in the Union provost marshal's files in the Confederate Records. These chaotic and fragmentary records terminate in 1867, the same date as the abolition of the position of commissary general of prisoners and that suggests that the Union Provost Marshal's File of Papers Relating to Two or More Civilians may once have been part of the records of the office of the commissary general of prisoners. The War Department transferred them at some unspecified time into Record Group 109, described as Confederate Records, apparently because the prison records contained information on Confederate sympathizers.

The Irrelevance of Mercy

Despite the work of Rhodes and Randall, both of whom set out to prove that Lincoln's record was not as bad as it seemed, there remained room for uneasiness about the extent and nature of military arrests of civilians. Rhodes sought to defend the United States from Bryce's criticism by arguing that "there was in Lincoln's nature so much of kindness and mercy that he mitigated the harshness perpetrated by Seward and Stanton." Randall thought "the greatest factor" in preventing serious abuse of this potentially dangerous power "was the legal-mindedness of the American people." But, like Rhodes, he could not resist emphasizing as "a very great factor . . . Lincoln himself. His humane sympathy, his humor, his lawyerlike caution, his common sense, his fairness toward opponents, his dislike of arbitrary rule, his willingness to take the people into his confidence and to set forth patiently the reasons for unusual measures— all these elements of his character operated to modify and soften the acts of overzealous subordinates and to lessen the effect of harsh measures on individuals." It was a measure of Lincoln's stature that able historians, like Rhodes and Randall, would make such assertions in the very books where they admitted their inability to examine the relevant records. They offered no description of the administrative process by which this alleged tempering occurred.[10]

Emphasis on the personal character of the president continued to occupy a central place in the arguments of Lincoln's defenders. In 1951, Kenneth A. Bernard insisted that "Lincoln used his powers reluctantly and leniently.... While many arrests occurred which were in violation of this attitude of restraint, such mistakes, especially if they came to Lincoln's attention, were usually quickly rectified. Indeed, it is amazing to note the rapidity with which most cases which reached Lincoln were considered and orders for release issued, even though Lincoln was constantly burdened with important problems of the war." Bernard was willing to admit that "Lincoln, under the powers he assumed, might possibly have destroyed civil liberty in these United States." But he added: "That he did not do so is due, in large degree, to his greatness of character." Writing fourteen years after Bernard, Dean Sprague proved to be a more pronounced critic of the system, but in the end, he also defended the man who instituted the arrests. He depicted President Lincoln as "a man caught in an agony.... He was forced to do something that he hated to do, since he felt it was necessary for the preservation of the Union. Nevertheless, his personal intercession again and again was in the direction of leniency." After the first year of the war, with "his slow, sure tread, he moved away from the policy of repression and the North was spared the omnipresent shadow of Fort Lafayette." Sprague made Seward the scapegoat for the program's harsher aspects.[11]

The stubborn persistence of this sentimental argument in historical literature spanning seven decades speaks volumes on the tradition of writing about Abraham Lincoln. But it is of little utility in evaluating the nature of military arrests of civilians during the Civil War. In so vast a system could a president, no matter how kindly disposed, personally mitigate many of the injustices of an otherwise frighteningly dangerous internal security system? It seems doubtful on the surface, and delving into the vast documentary record—itself too voluminous to have been read by a pardon-minded president or any other busy executive—quickly disproves it. For example, a check of the first 309 cases investigated by Judge Advocate Levi C. Turner (most of them involving prisoners in Old Capitol prison in Washington, the one most likely to attract the president's attention) reveals that Lincoln intervened in only one case. And that case involved the mistaken arrest of the brother-in-law of John A. Logan, a brigadier general in the Union army and a figure whose political influence weighed heavily in keeping Democrats in southern Illinois faithful to the war effort.

The president usually reviewed only the cases of men of influence or the cases of men acquainted with men of influence, because these were the only ones likely to be called to his attention. More important, the number, variety, and geographical dispersion of the cases were too great to permit control by one man who was busy with many other pressing problems. The point here is not to question Lincoln's moral qualities. He was by nearly universal testimony a man of compassion, but he was not ubiquitous.

At Old Capitol prison, from which some monthly records survive, the president took action in one civilian case every three months on average from August 1864 to the end of the war. The prison held from 147 to 336 civilian inmates at all times in that period, and some days, as many as fifty new prisoners were brought in.[12] Arrests of civilians occurred frequently enough, involved enough complex and varied issues, and were geographically widespread enough to require systematic bureaucratic treatment. They could not really be shaped by the ad hoc actions of the president in individual cases. He might have prescribed controlling rules to govern the arrests, but, in fact, Lincoln never did. And the War Department did not either. Instead, they accumulated "precedents" case by case, in the way familiar to lawyers raised on Anglo-Saxon jurisprudence.[13]

Thus, the focus in examining military arrests of civilians during the Civil War cannot remain on President Lincoln himself—on the letters he wrote and received or the conferences he held with parties interested in the release of relatives or friends. The focus must shift away from the White House to the War Department and from Washington to the field in order to ascertain who was arrested, what the circumstances of the arrests were, and what government apparatus existed to apprehend suspects and to investigate them once in prison. And the focus must be placed on the voluminously unmanageable records of arrests kept by the government during the war, not on the traditional ones repeatedly examined by historians since Lincoln's day.

The Surviving Records

Many arrest records, of course, no longer survive, but a tremendous number do. For example, I have used the Lafayette C. Baker–Levi C. Turner Papers, 1862–1865 (now in Record Group 94, Adjutant General's Records, War Department Division, National Archives). Though not opened until 1953, these bring together the documents generated by an important government detective (Baker) and the most influential of the associate judge advocates (Turner); many of them are now copied on 137 reels of microfilm.[14] For the most part, they contain the government papers gathered on the cases of persons arrested by military authority. Reels 128–137, mainly Baker's portion of the papers, consist mostly of receipts, financial records, and affidavits gathered in fraud cases. They show the government's zeal to end corruption but are of limited use in finding arrested civilians. In addition, a key portion of this collection, materials that fill an entire library truck, has never been microfilmed and must be consulted in the reading room at the National Archives in Washington. Other relevant records include the Union Provost Marshal's File of Papers

Relating to Individual Civilians, three hundred reels of microfilm of the most various and shapeless materials, including vast numbers of copies of oaths of allegiance, and a more useful series, the Union Provost Marshal's File of Papers Relating to Two or More Civilians, occupying ninety-four reels of microfilm. The latter collection contains on reels 87–94 lists of civilian detainees arranged by prison, perhaps the only such extensive set of lists outside the Turner-Baker Papers. Despite the abundance of records, one could not duplicate Ainsworth's search now even if it were known what he did, because the records have been rearranged since his day. Still, there are hundreds of thousands of pieces of paper that document—candidly if chaotically—the extent and nature of military arrests of civilians during the Civil War.

An examination of these records quickly reveals the unwarranted fame of Colonel Ainsworth's figure of 13,535. I could not find any list of prisoners with exactly 13,535 names on it or any likely combination of lists that would add up to that total. Most important, there is no extensive set of records that might be used to establish the total number of prisoners of state that is not corrupt, spoiled, or tainted. All have prisoners of war indiscriminately mixed in, so that making a meaningful count requires painstaking labor, case by case. It would require months of work merely to check the figures on which this book is based; it required years to collect them in the first place. Colonel Ainsworth's search may have been "protracted," but it did not last long enough to compile a believable total from the likes of these records.

Any count, including those in this book, is also subject to the frustrating vagaries of nineteenth-century record-keeping. There was no systematic record of political prisoners, despite a law passed by Congress requiring it. From beginning to end, when questions arose, the government had trouble locating some prisoners it allegedly held. Records were a hodgepodge of lists, scraps, fragments, letters, and receipts generated by different authorities and agencies as well as by the prisoners themselves and their lawyers, relatives, and other partisans. Among others the judge advocate general's office, the State Department, provost marshals, prisons, and investigative or review commissions produced some records of prisoners of state. Letters by, to, or about a prisoner were sometimes filed with other case records and sometimes copied in letterbooks.

All the documents were handwritten, of course, though some information was placed on printed forms. There do not appear to have been forms standard throughout the army or the War Department, and forms were sometimes used to record information different from the forms' original purpose. All the information required by their printed headings, blanks, or columns was rarely filled in. Professional clerks with neat handwriting completed some of the documents. Others were jotted or scrawled by army officers of low rank who could not spell well or form their letters legibly.

No system was ever imposed for recording names, and the problems that plagued Missouri's records were at work everywhere to some degree. Care was not always taken to get the spelling of last names correct, and the same person's name often appears spelled in different ways. First names and initials appear willy-nilly. Soldiers of modest education spelled names phonetically. Although some long lists were alphabetized, many were not. Even when alphabetized, the lists are rough, merely grouping together all last names beginning with the same letter. Alphabetized lists almost always included addenda stuck on with the names of the latest arrivals in prison.

Nineteenth-century orthography caused problems at the time and not only for historians in later times. For example, John Foy was arrested as a deserter from the draft in Cincinnati in December 1864, but his arrest proved to be a mistake. The drafted man who failed to report was really named John Fay.[15]

The Legal Meaning of "Arbitrary Arrest"

Colonel Ainsworth certainly did not take time to sort out the problems in the documentary records so that he could come up with a reasonably accurate number of civilian arrests. It would not really have borne fruit if he had, for whatever the state of the records, research in them is useless unless one knows what to look for. In his chapter on "arbitrary arrests," James G. Randall quoted Ainsworth's estimate of the number of "citizens arrested and confined in military prisons from February, 1862 to the end of the war." What Rhodes had asked of Ainsworth, was "the number of political prisoners during the late war."[16] But would the number of "citizens arrested and confined in military prisons" equal the number of "political prisoners" that Rhodes sought? Would the number of "citizens arrested and confined in military prisons" equal the number of "arbitrary arrests" that Randall sought?

The official term used by the War Department and the State Department during the Civil War was "prisoners of state," which did not match the terms used by Rhodes and Randall. The term "prisoners of state" differentiated civilians from "prisoners of war" and "United States prisoners." Prisoners of war were captured members of the Confederate armed services. United States prisoners were members of the United States armed services held for crimes committed in their own camps, ranging from murder and theft to drunkeness on duty and conduct unbecoming to an officer. Prisoners of state, then, were civilians held by the military. Nothing more definite can be said about them in the aggregate. Though often called "political prisoners" even by high-ranking members of the Lincoln administration, they were, in fact, civilians of any sort—any person in military prison who was not a Confederate soldier or sailor or a United States soldier or sailor. The civilian prisoners did not have to commit (or be suspected of committing) a "political" crime to be incarcerated in one of the federal forts.

Surely "political prisoner" would be at best a misleading name for most of the civilians arrested by military authority who have been described in this book so far. There is another term that comes to mind, of course, one with broad current usage, and that is "arbitrary arrests." Could all these civilians best be thought of as victims of "arbitrary arrest"? If so, a definition of that term seems crucial to understanding the records. Yet a definition, in the historical literature or in the law, is hard to come by—impossible, in fact.

"Arbitrary arrest" is a term with no precise legal or technical meaning. The term does not appear in the United States Constitution. It was not, in so many words, forbidden by the Bill of Rights. The term cannot be found today in standard legal reference books like *Black's Law Dictionary* or the immense multi-volume *Words and Phrases*. It does not appear with any frequency in American history before the Civil War. One reason for that seems obvious enough: there were no military arrests of civilians before the Civil War—or very few, anyway. As one anti-Lincoln pamphleteer said in 1861:

> There has been no instance of martial law in England for the last hundred and fifty years, and none in this country since the Declaration of Independence, but that given by General Jackson at New Orleans. Washington carried the country successfully through the seven years' war of the Revolution, amidst spies and traitors, without finding a necessity, or feeling himself authorized, to resort to such means. Baltimore, an open, unfortified city, and every way very similarly circumstanced to New Orleans, was successfully defended during the last war by the spontaneous efforts of its citizens, who voluntarily rendered more service in the way of personal exertion, than would have been extorted by any amount of coercion. This, too, amidst daily denunciations from the Federal press against the President for corruption, in getting the country into the war, and for imbecility in carrying it on. ...Indeed the whole idea of extracting, by coercion from freemen, the most effective resistance against foreign invasion is based upon a wrong conception of the genius of a republic. All history proves that the patriotism and public spirit of a republic are more effective in calling forth in the hour of need the utmost energies of a State, than all the coercive powers of the most absolute despotism.[17]

This passage helps explain the infrequent use of the term before Abraham Lincoln became the first president to suspend the writ and declare martial law, but it also directs proper attention to the one small but interesting exception in American history before the Civil War: "that given by General Jackson at New Orleans." In 1815, General Andrew Jackson arrested several citizens under martial law, including a judge who issued a writ of habeas corpus to release the first detainee. These were what, some fifty years later, would be called "arbitrary arrests." Yet that slogan seems not to have attached itself to Jackson's controversial acts—neither at the time of the controversy itself in 1815 nor in the extensive congressional debates over the question in the 1840s.[18]

An indication that the term "arbitrary arrest" gained currency only in the Civil War is provided by an 1864 political broadside from New York City. "Military Despotism! Arbitrary Arrest of a Judge!!" it shouts, but when the reader looks at the fine print below, he reads the story not of some Republican arrest during the Civil War but of Andrew Jackson's dispute with Judge Dominick A. Hall in New Orleans forty-nine years earlier. The New Orleans story was lifted from a standard life of Jackson, and it did not use the term "arbitrary arrest," but the pro-Republican printer added, in language of special significance and irony in the 1860s, the statement that the maker of the "arbitrary arrest" was later elected president.[19]

The use of the term "arbitrary arrest" came late in American history, but it soon took hold of the popular imagination. Lincoln himself once used it in a hotly worded memorandum about arrests in the District of Columbia, telling General Winfield Scott, "Unless the *necessity* for these arbitrary arrests is *manifest,* and *urgent,* I prefer they should cease." But the president wrote that in May 1861, during the brief period when he was still urging extreme caution in the disturbance of civil liberties. The early date does reveal how quickly the term came to mind for some lawyers accustomed to the procedural safeguards required by the Constitution for those accused of capital or otherwise heinous crimes.[20]

The term was soon banished from the legal lexicon of the Lincoln administration. Secretary of War Edwin M. Stanton referred to "extraordinary arrests" when he issued the order of February 14, 1862, extending the War Department's authority over civilian prisoners. The department's shrewd and influential solicitor, William Whiting, entitled his important pamphlet on the subject *Military Arrests in Time of War* and said, "Among the acts of war which have been severely censured is that class of military captures reproachfully styled, 'arbitrary' arrests." These, he argued, were really "discretionary" rather than "arbitrary." They were not made, as the term "arbitrary" implied, "at the mere will and pleasure of the officer, without right, and without lawful authority."[21]

Despite these efforts, use of the term became virtually irresistible. Even Republican apologists sometimes used it. Defending the *War Power of the President* in 1863, J. Heermans, for example, ended his pamphlet with a plea for patriotic support of the war effort. "Then," he concluded, "we shall have no arbitrary arrests in loyal states." Lincoln himself remained liable to slip into usage of the term at any unguarded moment. Even in his famous defense of the policy in the Corning letter, the president at one point promised "that the necessity for arbitrary dealing" with opposition opinion would cease. When he revised the text for publication in the New York *Tribune,* he replaced "arbitrary" with "strong."[22]

"Arbitrary arrest" was a term of "reproach," as William Whiting said it was, more a political pejorative than a technical term of jurisprudence. Looking at what Democrats said during the war confirms this. Since the term lacked deep roots in law or customary legal usage, it was not consistently used in the early period of the war. The first protests of Lincoln's policy, which came only after Taney's *Merryman* decision of June 1, 1861, exploited fears of military despotism without employing a common catchphrase to describe the arrests. The initial uncertainty of the Democratic journalists was evident in an 1861 Pennsylvania newspaper headline: "Freedom of Opinion and the Privilege of Habeas Corpus Inalienable Rights." This embraced a legal contradiction in terms, for a "privilege" cannot be an "inalienable right." It also contradicted the U.S. Constitution, which explicitly allows for the suspension of the writ of habeas corpus under certain conditions, thus making it a privilege and not an inalienable right. Privilege is the term used in the Constitution itself.[23]

In 1862, the Pennsylvania Democratic platform stressed the Constitution as its overall theme, but the party had not yet appropriated the term "arbitrary arrest." The plank protesting that policy instead criticized "suppression of the freedom of speech and of the press, and the unlawful arrest of citizens and the suspension of the writ of *habeas corpus.*" When the editors and proprietors of the Harrisburg *Patriot and Union* were arrested that summer, the Democratic editor in nearby Gettysburg spoke of "Wholesale Military Arrests." When Philadelphia lawyer Horace Binney defended Lincoln's policy, his pamphlet provoked an avalanche of Democratic pamphlets in reply, and this political exchange gave rise to the widespread use of the term among the opposition party. Philadelphia's George M. Wharton, for example, asked in February 1862 whether "the freemen of America...[were] subject to arbitrary arrest?" Even so, other terms lingered. When Baltimore's police commissioners were released from Fort Warren late in 1862, they sued government authorities, and the Democratic newspaper in Gettysburg referred to the victims' "Illegal Arrests." It was not until 1863 that the term "arbitrary arrest" came to be rather consistently used in the Democratic press.[24]

When the war was over, Democratic partisans still were not in complete agreement on the proper term of denunciation. The early editions of John A. Marshall's *American Bastile* carried the subtitle "A History of Illegal Arrests and Imprisonment of American Citizens during the Late Civil War." After 1885, however, the subtitle became "The Arbitrary Arrests and Imprisonment of American Citizens."

Historians of the Civil War, including those who focused on constitutional issues, picked the term up from the Democrats and apparently meant by it the arrests which occurred when the privilege of the writ of habeas corpus was suspended. If that is the proper meaning, then a tally of the number of civilians in Northern military prisons will by no means equal the number of arbitrary arrests. A large number of those civilians would doubtless have been in prison regardless of the legal status of the writ of habeas corpus. They would likely have been in prison no matter what persons or political party were running the Northern war effort. Civilian prisoners included many Confederate citizens—refugees, blockade-runners, and people living essentially in the "no-man's land" of northern Virginia or Tennessee. Such persons would have been held no matter what Lincoln did about the writ of habeas corpus.

President Lincoln had a sharp sense of the profound dislocations caused by any war, and he hit upon a felicitous expression to describe them. When in the summer of 1862 he appealed to border-state representatives to endorse a plan to end slavery by compensated emancipation in their states, he warned them:

> The incidents of the war can not be avoided. If the war continue long...the institution in your states will be extinguished by mere friction and abrasion—by the mere incidents of the war.

He knew the border-state men much preferred "that the constitutional relation of the states to the nation shall be practically restored, without disturbance of the institution." But that could not be done. They had to face "stern facts," discard *"punctillio* ... and maxims adapted to more manageable times."[25] By analogy, the stern facts of war created incidents not only where black men gained freedom but also where white men lost it. The incidental friction and abrasion of war brought these Southern men to the military prisons of the North.

Statistics from Manuscript Sources of Individual Cases

When James G. Randall inquired about government statistics on the number of civilians arrested during the Civil War, he received from Adjutant General Davis a discouraging reply: "I do not believe that it will ever be possible for any one to gather from any source an approximately definite estimate of the total number of such prisoners held by Federal authorities during the Civil War." Davis was correct, but the records, though incomplete and fragmentary, are nevertheless numerous and in some instances richly descriptive. Whatever the limitations of the records, historians must rely on them to answer the long-standing practical questions about the military arrest of civilians under the Lincoln administration—who actually got arrested, by whom, why, and when. Although they may not yield "an approximately definite estimate," these records do tell us what happened and why attempts to fix a definite number are exercises in futility that miss the true nature of the arrests.

It makes a great deal of difference what kind of questions historians ask of these difficult materials. If the questions are not carefully put, the answers will be worse than imperfect: they will be misleading, useless, or meaningless. To ask of these documents how many "arbitrary arrests" there were during the Civil war, for example, will bring no useful answer, for there is no generally agreed-upon definition of the term "arbitrary arrest." To answer the question of how many military arrests of civilians there were under the Lincoln administration is vaguely possible, in the imperfect way that most historical answers must be calculated, but it will not really answer the question most historians of this subject have been seeking. Telling how many military arrests of civilians there were will yield a figure embracing so vast a group of refugees from the Confederacy, unfortunates trapped near the battlefields, and marginal characters incident to all wars everywhere, that the answer would be useful only to place the American Civil War on a continuum that would tell how disruptive of civilian life different kinds of wars in different eras have been. That may be a question worth answering. Indeed, there is growing interest in such questions among historians who eye the specter of atomic warfare and its promise of working vaster destruction on civilian than on military life.

The myriad incidents, large and small, that led civilians to be held for a time by military authorities were too numerous ever to be precisely quantified by historians. The difficulty is illustrated in an unheralded incident, the recon-naissance to Pohick Church in northern Virginia on November 12, 1861. Companies B, C, and G of the First New York Cavalry took part, led by Captain Henry B. Todd. He was on the march, leading forty-odd men, at four in the morning, to join the command of Colonel H. G. Berry of the Fourth Maine Volunteers. The road before them looked heavily traveled, and three local farmers brought in and interrogated reported a Confederate force of five thousand men nearby. Berry sent Todd and his men toward the church to investigate. They saw what they took to be Confederate infantry and cavalry drilling three-fourths of a mile ahead; in fact, what they really saw was probably some other Union soldiers. Berry thought he was outnumbered and ordered the cavalry back in order to withdraw the whole force. When they halted on their retreat, Todd told Berry some of his company were still out and requested permission to bring them back. After an hour, Berry heard firing in the direction Todd had gone. Soon three of the absent men came in; two were wounded and all had plunder strapped to their horses—sidesaddles, bed clothes, and other items. Berry decided to withdraw before Todd returned with the rest of the cavalrymen. A bugler had been killed. Captain Todd, a sergeant, and one private were reported missing.[26]

Todd, it turned out, had been captured by the enemy, and the action he led proved less than glorious. The losses—including seven horses missing—were caused by the officers' allowing their men to straggle and plunder in hostile territory. While sitting in a Richmond prison, Captain Todd read an unflattering report of the action in a Washington newspaper and hastened to write his general in protest. Todd said that the missing men were his advanced vedettes, thrown out ahead of the rest to scout the enemy. A lieutenant had neglected to tell them to come in when Berry and Todd had retreated. "Humanity forbad me to go and leave them," the captain said, and when he and several of his men went out, they were fired on from ambush. Todd said his men had not been straggling and assured the general that Colonel Berry would confirm it. But Berry had seen the men with plunder and got them to admit what they had been up to. The general believed Berry and not Todd.[27]

What became of the Virginia farmers brought in for interrogation? Had they lied about the presence of a sizeable Confederate force to scare the federals off or to confuse them? Did they escape to join their neighbors in attacking the plundering New York cavalrymen? Were they sent to the provost marshal and then back to the Old Capitol prison? Were they actually "arrested" in any official sense beyond being taken for a time from their ordinary peaceful pursuits by military authorities? Were these men "prisoners"? No one knows, and probably no one will ever know. There are myriads of such incidents described in the *Official Records*, and heaven knows how many others were never documented or published.

Incidentally, in 1862 Todd began working to arrange his exchange for some Confederate officer in Union hands. Once free, he returned to duty in Washington, but his combat career was over. Doubtless his superiors, remembering the Pohick Church affair, decided Captain Todd had best take a desk job. Ironically, he became provost marshal in Washington and spent much of his

time committing to the Old Capitol prison captured Virginia farmers, like the
ones who scared off Colonel Berry in 1861 with their tales of five thousand
Confederate soldiers.[28]

One example, on a far grander scale than the Pohick Church affair, will
illustrate fully the necessarily inadequate documentation of civilians made
prisoner by the friction and abrasion of war. On July 23, 1862, General John
A. Pope, commanding the Army of Virginia, wrote the president:

> Have you yet considered the order I proposed to issue yesterday which directs
> all male citizens living within the lines of the army under my command and in
> the rear of it to be arrested—such as take the oath of allegiance and give sufficient
> security for its observance to be allowed to remain at home and pursue their
> accustomed avocations; such as do not to be conducted South and put within the
> lines of the enemy, with a notification that if hereafter found within the lines or
> in the rear of the U.S. forces they will be considered and treated as spies? I find it
> impossible to make any movement, however insignificant the force, without having
> it immediately communicated to the enemy. Constant correspondence verbally
> and by letter between the enemy's forces and the so-called peaceful citizens in
> the rear of this army is carried on which can in no other way be interrupted.

Lincoln apparently approved the idea, for Pope's General Order No. 11 stipulated precisely what he described in this letter. How many arrests ensued is
unknown and probably unknowable. The number of potential victims was
surely greater than Pope's available manpower could have arrested.[29]

The effect of warfare on men like the Virginians mentioned in Pope's order
has been too little studied, but, interesting as such incidents and the questions
raised by them are, they are not relevant to the questions that have piqued
the interest of historians and political scientists and constitutional lawyers
heretofore. What these scholars have wanted to know all along is the fate of
civil liberties in the Civil War—did Americans on the home front enjoy traditional freedoms of speech, press, and assembly, as well as political liberty,
or were these civil liberties suffocatingly curtailed or dangerously threatened?
Most, even Lincoln's defenders, have answered that they were dangerously
threatened. The president's defenders have relied on the plea that the mag-
nanimous character of the president himself prevented a serious curtailment
of liberty. Except by pointing to liberties that did survive—quoting from the
widespread criticism of the administration from newspapers that went unscathed—historians have heretofore had no effective way of answering the
question.

By abandoning the empty concept of "arbitrary arrest" and refining their
questions, historians can attempt more satisfying answers based on more systematically gathered evidence. How many civilians were arrested in the North
by the military? How many of these civilians were Northerners (and not Southern refugees or even spies)? Of the arrests in the North, how many took place
in the border slave states and how many north of the border states? When did
they occur, and why were the prisoners detained? How many of them involved

what would normally be considered political oppression or partisan exploitation?

In a sense, all Northerners admitted the necessity of detaining Confederate citizens found near or behind Union lines—by never uttering a word of protest against the practice. And even Democrats during the Civil War often conceded the necessity of military arrests of civilians in the border states, though they hardly endorsed *all* the arrests made in those areas. On the other hand, most Democrats and a few influential Republicans criticized most of the arrests made north of the border states. Above the militarily contested and politically uncertain slave states of Maryland, Missouri, and Kentucky, most of the North was militarily secure during the greater part of the war. Pennsylvania suffered brief invasions that did not penetrate deeply into the state. John Hunt Morgan's Confederate cavalrymen made raids along the Ohio River. Even New England witnessed a tiny raid. Such incidents caused temporary panic. But as soon as the Civil War was over, some people could see that there really had been little need for numerous arrests of civilians north of the border states or, more precisely, north of actual battlefield areas. Yet no one then knew how extensive the practice was. No one knows now.

One way to refine the examination of these arrests is to approach the question historically, as this book has done thus far. The Civil War was, after all, an event of four years' duration. Most historians would say that it was militarily and politically quite different at the end from the way it began. The sort of warfare that was waged, it is generally agreed, became less and less restrained; some argue that it prefigured twentieth-century warfare in its indiscriminate destructiveness. At the firing on Fort Sumter, neither side contemplated any change in the status of slaves. By 1863, ex-slaves were toting muskets and shooting their former masters, legally (if they were in uniform).

Heretofore, incidents behind the lines have not benefited from such discriminating study as the battles have enjoyed, and the mass of military arrests of civilians has hardly been looked at. Surely a war of such dynamic character effected changes behind the lines as well. Examining the arrests historically, as they developed over time, and not as one rather abstract sociological group of unfortunate people, yields a new and more complex portrait of the Lincoln administration, its differing policies administered by different people from various government departments.

Though we will never be able to ascertain exactly how many people were arrested, analysis of the available statistics can settle, once and for all, the larger questions about civil liberties under Lincoln. One need no longer rely on subjective assessments of the justice of the handful of court cases well known to Civil War historians: *Merryman, Vallandigham, Milligan,* and perhaps a few others. The internal security system was vast, and historians must look at large numbers and a wide variety of cases to make their assessments.

To obtain as comprehensive a view as possible, several sets of records require examination. The best available set, apparently gathered and put in systematic order at the time of the Civil War, comes from the State Department while

Seward was in charge of military arrests of civilians. Chapter 1 of this book provided a count of them and a statistical analysis, but these arrests preceded the period for which Ainsworth's figure of 13,535 is supposed to account.[30]

After the State Department files, the best available records are located in the Turner-Baker Papers at the National Archives. Unlike the Seward records, these were never published and appear not to have been systematically filed in the Civil War era. Rather, they were found in Levi C. Turner's office, apparently in rather chaotic order, at his death in 1867. Clerks were assigned thereafter to put them in order. The documents were kept separate from the rest of the records to which they logically belonged (the records of the judge advocate general's office). They were maintained by the judge advocate general's office until 1894 when they were turned over to the Record and Pension Office. The War Department allowed some to be printed in the *Official Records* but kept the rest confidential, to be consulted only for official purposes.[31] Eventually, they became part of the records of the adjutant general's office, as that was the record-keeping branch of the War Department. They were closed to the public until the middle of the twentieth century but then opened and partially microfilmed. Among the unmicrofilmed portions of the collection is a neatly written manuscript entitled Record of Prisoners of State, which, though not otherwise identified, appears to list civilian prisoners committed to Washington military prisons and lists them in order of commitment. Most were arrested in Virginia, Maryland, or the District of Columbia, which comprised the region for which Levi Turner was responsible as an assistant judge advocate. The Record of Prisoners of State contains the names of 2,187 persons.

The most useful parts of the rest of Turner's files appear on microfilm. Each case was assigned an arbitrary number, which usually appears on every sheet in the person's file. The cases were microfilmed in numerical order, but otherwise, no systematic order was maintained—by name alphabetically, by type of offense, or by chronology. Some cases were assigned two different numbers, in error, and the relevant documents were thus separated. A handwritten alphabetical name index appears in the microfilm. Some individual cases require more than a reel of microfilm for all the documents. Most cases include several documents. Every case that appears on the 137 reels of microfilm was examined for this book. After eliminating cases duplicated in the Record of Prisoners of State as well as military cases, one finds 3,255 persons' files in this collection. Thus, 5,442 citizen prisoners' records can be found in the Turner-Baker Papers.

There is almost no overlap between these records and the ones from the West. Chapter 2's exceedingly conservative estimate of the number of citizen prisoners that passed through St. Louis's Gratiot Street and Myrtle Street prisons, based on the Provost Marshal General's File of Two or More Civilians, equalled 5,014 cases. No accounting could be made for the many Missouri prisoners who never made it to St. Louis, but the records of army courts-martial, an altogether different set of records which contains the transcripts of trials of civilians by military commissions, make clear that many such prisoners spent their time of incarceration in camps and forts in Springfield, Cape Girardeau, Rolla, Palmyra, or some other military stockade, for which no systematic records appear to be extant.

The existence of careful records of trials by military commission is an anomaly when considered in the context of the other hopelessly incomplete records of civilians held by military authority. They are an anomaly in nineteenth-century legal record-keeping as well. In Lincoln's era, court recorders were not used, at least below the appellate level, and there are no transcripts of ordinary civil or criminal trials. The only paper records generated below the appellate level—unless it was a sensational case covered daily by newspaper reporters—were the official documents filed with the court: pleadings, receipts for witnesses' fees, affidavits, verdicts, and court orders. Courts-martial were a different matter, and during the Civil War they all had an official recorder. Since the military applied most of the rules of courts-martial to trials by military commission, these civilian trials, uniquely, generated transcripts. Though filed with the multitudinous records of courts-martial and therefore difficult to retrieve, systematic records of trials by military commission exist in one central location: the Records of the Proceedings of the United States Army General Courts-Martial, 1809–1890 (Records of the Office of the Judge Advocate General [Army], RG 153, in the National Archives). Apparently, when it was determined that a civilian prisoner would be tried by military commission, the available papers in the case were forwarded to the commission. Therefore, very few persons whose cases appear in the records of trials by military commission also left records in the Turner-Baker Papers.

The total number of military trials of civilians, 4,271, could more or less be added to the aggregate, if it were not for the terrible state of the Missouri records. The Missouri records are so chaotic, incomplete, and uninformative that they merit only sampling; therefore, the statistics for St. Louis cannot be compared exhaustively, case by case, against the records of trials by military commission. Considerable duplication is probable. Eliminating all the 1,940 Missouri trials (for fear that they might duplicate persons included in the overall estimate for St. Louis prisons), one still finds 2,331 trials for persons unaccounted for in other sets of records under discussion here.

Taking only these collections of arrest records—the manuscript Record of Prisoners of State, the microfilmed cases in the Turner-Baker Papers, the provost marshal general's records for St. Louis, and the records of trials by military commission (excluding those from Missouri)—one quickly arrives at a total of at least 12,787 citizens arrested. This number nearly matches Ainsworth's estimate without even attempting a count of thousands and thousands of other documents. It does not attempt an estimate for records that are clearly lost; for example, the 12,787 persons include only a handful of Kentuckians. Political prisoners from that bitterly divided border state were usually sent to Camp Chase in Ohio, and there are almost no extant records for Camp Chase. Levi C. Turner reported to the secretary of war that he had reviewed reports of 2,020 cases from Camp Chase.[32] The reports are now missing, but these were almost certainly civilians. Louisville was only a transfer prison where persons were held briefly before being dispatched elsewhere; such places, at best, generated chaotic records. The manuscript Record of Prisoners of State, though it is a 550-page ledger, lists prisoners brought into Washington only from December 1862 to October 1864. Many of those incarcerated in Washington's prisons from February–December 1862 and from October 1864 to the end of the war are not accounted for. In other words, the Record covers only twenty-three of the thirty-eight months accounted for by Ainsworth in coming up with the 13,535 figure. If prisoners were held in proportion to the Record for the other relevant months, one could quickly add another 1,340 to the aggregate. Likewise, the number of trials by military commission held in Tennessee suggests that a substantial number of political prisoners were held there, as does the well-known history of bitter internecine strife within that state during the war. But there is no systematic record of Tennessee cases, except for those that resulted in trials, surely a small portion of the total number arrested.

From the 12,787, one must subtract the 122 duplicate records found by individual search in the registers of trials by military commission. That still leaves 12,665.

One could add many more arrests to the 12,665. For example, fragmentary records from New Orleans in the Provost Marshal General's file on two or more civilians list 633 civilian prisoners (after removing those tried by military commission—only 4 persons—from the total). All of these prisoners appear to have been arrested between the spring of 1863 and the spring of 1864. Only blockade-runners, hostages, and refugees are specifically noted. Thus, one can easily bring the total to 13,298. And there must have been many other prisoners taken in New Orleans between the time of Union occupation in 1862 and the end of the war, a period more than twice as long as that represented in these fragmentary records. Many must have been arrested for reasons other than those noted in this small batch of records.

One could add hundreds and hundreds more from similar fragmentary records. Associate Judge Advocate Turner, for example, arranged two exchanges of women and children with Confederate authorities at City Point. Altogether, 1,222 women and children were sent south in these two trips, yet the number of females and children among the arrest records I have read numbers only in the dozens.[33] There is no point in going further with this exercise. It is clear that far more than 13,535 civilians were arrested. And it is clear that, wherever the prison was located, most of its civilian prisoners came in as victims of the war's incidents and friction.

On one point, anyone familiar with the records must agree with Colonel Ainsworth: no one will ever know exactly how many citizens were arrested by military authority during the Civil War. But to answer the question most historians have been asking does not require knowledge of that total. Since even Democrats in the Civil War sometimes conceded the necessity of the arrest of civilians in the border states, the crucial test of the "arbitrary arrest" question lies north of the border and the District of Columbia. How many Northerners were arrested there?

Among the 5,442 civilians listed in the Turner-Baker Papers, only 624 can be positively identified as having been arrested above the border states and the District of Columbia. The Turner-Baker Papers constitute the best overall set of records of civilians arrested in the eastern United States after the War Department assumed control of these matters. The next summer, the War Department issued the August 8 orders, when all arrests throughout the nation were supposed to be reported to Levi C. Turner. Among the 624 arrests in the North were at least 287 caused by the August 8, 1862, orders and made before those orders were relaxed in September (after most recruitment quotas were filled). In fact, 434 of the 624 arrests occurred in 1862, most of them after August 8. If one leaves out the disastrous War Department experience with these ill-conceived orders, then the number of arrests north of the border states falls dramatically. In the microfilmed portion of the Turner-Baker Papers, thirty arrests are of unspecified date, and there are eighty-six in 1863, sixty-two in 1864, and twelve in 1865. The Record of Prisoners of State, from the unmicrofilmed portion of these papers, contains only thirty-six arrests in the North (twelve in 1863; twenty-four in 1864). Fifteen of these were telegraphers picked up after the bogus draft proclamation appeared in the New York *World.* Most of the rest were held for perpetrating frauds on the government. If many Northerners were being arrested, they certainly were not being held in Old Capitol prison. Of the 154 arrests in the North of known charge in the microfilmed Turner-Baker Papers, few involved questions of freedom of speech, press, or assembly. Most prisoners were rather ordinary characters arrested on suspicion of contraband-trading, desertion, or draft-dodging.

In fact, it stands to reason that, after the disastrous August 8 orders were modified, most of the civilians arrested in the North must have been arrested as suspected deserters or draft evaders. The final report of the Provost Marshal General's Bureau, issued in 1866, contained astonishing statistics on desertions. The bureau arrested and returned to the army 76,526 deserters, well over 2,000 a month from the birth of the bureau in 1863.[34] Moreover, the bureau never stated how many were arrested among the 161,286 citizens who failed to report, but it must have been a sizeable number. Though never mustered in, they were deserters under law.[35] Levi Turner alone investigated the cases of 929 alleged deserters who were held in Washington, D.C.'s Forest Hall prison.[36] Since arresting deserters was hardly an exact science, thousands of mistakes must have been made. Many were forgivable, perhaps—arrests of drunks or troublemakers who boasted aloud that they had deserted, for example, or men found dressed in military clothing or in the company of deserters. But it seems clear from the testimony in the water-torture cases that government detectives or provost marshals were eager to arrest as bounty-jumpers any man of draft age holding substantial cash and boarding a train.

The likelihood is that most such arrests resulted in confinement for eight days in a post guard house, and few such cases generated prison records available to historians today.[37]

Of the 154 persons arrested north of the border states and the District of Columbia whose cases, complete with charge, found their way into the microfilmed Turner-Baker Papers, at least 36 were suspected of involvement in draft evasion, that is, an action that amounted to more than merely speaking ill of the draft. People suspected of involvement in contraband trade or blockade-running—persons caught with suspiciously large quantities of items useful to the Southern war effort or persons arrested immediately after debarking from a ship in New York or some other harbor—numbered twenty-five, and twenty-nine persons were suspected of defrauding the War Department or swindling recruits. Thirteen persons were arrested for words they were alleged to have spoken or written about the war. Four of these cases are undated and probably stem from the period after the August 8, 1862, orders. Only two cases were "politically" significant. The rest involved drunken boasts in barrooms or expressions of delight at the assassination of President Lincoln. Some of the latter showed genuine humor and creativity. John Bailey of New Jersey, for example, draped his hog pen in black for Lincoln's funeral.[38] The Union authorities were not amused.

More were arrested for being ex-Confederate soldiers or citizens of Confederate states—fifteen—than were arrested for criticizing the war effort or the government in Washington. Against these must be weighed the two politically significant cases found in these records: those of Albert D. Boileau, a Philadelphia newspaper editor, and John U. Andrews, a minor Democratic politician arrested in 1863 as a ringleader of the draft riots in New York City.[39]

Looking at these cases in detail is instructive. After Boileau took over the Philadelphia *Evening Journal,* the paper began criticizing the Lincoln administration sharply. One editorial early in 1863 described the war as having entered a phase with "no other purpose than revenge, and thirst for blood and plunder of private property" and declared the Lincoln administration "incapable of . . . winning victory in the field." The order for Boileau's arrest, issued by General Robert Schenck, commander of the Middle Department (which included Philadelphia), specifically mentioned an editorial that praised Jefferson Davis's annual message at the expense of Lincoln's annual message. Late at night on January 27, Boileau was arrested and taken to Fort McHenry. Democrats protested vigorously, though the reaction fell short of anything legally or physically troublesome to the Lincoln administration. Boileau recanted, disclaiming responsibility for inspiring the articles in question and promising not to permit others like them. He was then released, and his newspaper, temporarily suppressed after having criticized the government for arresting its proprietor, began publishing again.[40]

In the midst of the draft riots in New York City, John U. Andrews, Virginia-born but a resident of the city for four years, had spoken to the rioters from the roof of a shanty across the street from the burning draft office. Two newspaper accounts of his speech, including the one in the *Caucasian,* were in substantial agreement that Andrews had told the crowd that conscription, as constituted under the Lincoln administration, should be resisted and that he was willing to be their leader in doing so. Though arrested by the New York police, Andrews became a political prisoner in Fort Lafayette. Eventually, he was tried in federal court in 1864 for conspiracy to levy war against the U.S. and to resist U.S. laws. The jury found him guilty and sentenced him to three years at hard labor.[41] Perhaps the most remarkable aspect of the Andrews case lies in its being the only one of the New York City draft-riot cases resolved by federal rather than local or state authority. Moreover, the case was tried in a federal court, with a jury, not by a military commission. Faced with the greatest act of resistance to conscription in all of American history—the greatest civil disturbance in all of American history—the Lincoln administration left trials to the local New York courts except for the case of John U. Andrews.

At least two Northern states attempted to determine the number of civilians arrested within their borders. A select committee on military arrests in the Ohio general assembly found only eleven by 1863, but this was a Republican committee and lacked genuine investigative zeal. A similar committee from the Indiana House of Representatives, however, was controlled by Democrats and could find only forty-three cases by the beginning of 1863. Of these, over a third (fifteen) were arrested as a result of a riot that ensued when citizens were first drafted in Hartford City.[42]

It seems unlikely that any significant number of arrests of politically important individuals have been overlooked. Such cases provoked speeches in legislatures, pamphlet protests, and newspaper copy. They were few in number. Most prisoners of state came from incidents of war.

Aggregate Prison Statistics

Although records of civilian prisoners far exceeding what has survived to this day must have existed at one time, the best continuous record available is the abstract of monthly returns from the principal United States military prisons printed in the *Official Records.* These records do not tell exactly how many civilians were arrested by military authority in the Civil War because they provide only monthly totals, and those totals, therefore, contain many of the same persons from month to month. Moreover, not all prisons reported regularly at all periods. Nevertheless, they can still provide a check of the scale and nature of military arrests of civilians as described herein.

These prison records probably exist less because the government desired an accounting of the extent of the internal security system than because the government feared overcrowding in the prisons. Overcrowding led immediately to problems of sanitation or disease as well as military security—all troubles that prison officials wanted to avoid. Besides humanitarian considerations, legal concerns, fears of providing grist for enemy propaganda mills as well as fears of losing control of masses of poorly treated captured soldiers guaranteed that the government would monitor the numbers of prisoners, especially as they became numerous with the progress of the war.

Looking, then, at these figures, as they were accumulated gradually, can one say that the total number of arrests suggested in this book could be wildly wrong? Let us examine as a test the high point of the war for numbers of civilians in Northern military prisons, November 1864, when 2,006 were in prison, up from 1,727 in October. Noting the figures for individual prisons, one can gain reassurance that the nature of the arrests was indeed as has been described.[43]

	October 1864	November 1864
Alton, Ill.	235	246
Camp Chase, Ohio	153	138
Camp Douglas, Ill.	—	4
Camp Morton, Ind.	31	29
Elmira, N.Y.	46	78
Ft. Columbus, N.Y.	—	1
Ft. Delaware, Del.	92	421
Ft. Lafayette, N.Y.	43	39
Ft. McHenry, Md.	158	198
Ft. Pickens, Fla.	22	5
Ft. Warren, Mass.	91	96
Ft. Wood, N.Y.	—	—
Johnson's Island, Ohio	30	29
Little Rock, Ark.	61	48
Louisville, Ky.	16	5
McLean Barracks, Ohio	18	25
Memphis, Tenn.	9	8
Nashville, Tenn.	139	136
New Orleans, La.	—	—
Old Capitol, D.C.	336	219
Point Lookout, Md.	201	240
Rock Island, Ill.	23	26
St. Louis, Mo.	—	—
Ship Island, Miss.	4	—
Wheeling, W.Va.	19	15
	1,727	2,006

Who were the 2,000-odd citizens in military prisons in Union hands in November 1864? Too few lists of individuals from each fort survive to be certain, but applying what can be learned from other sources about the general nature of the prisons suggests an answer. No military prison was reserved exclusively for civilian prisoners, and most held only small numbers of civilians compared to prisoners of war. The two prisons that held mostly civilians were

Forts Lafayette and Warren, and they were small. These two prisons held 135 citizen prisoners in November 1864, most of them surely blockade-runners. On November 6, 1864, Lieutenant Colonel Martin Burke, commanding at Fort Lafayette, reported to army inspectors that the "number of prisoners confined in Fort Lafayette is 116, composed of prisoners of war, blockade-runners, and a few political prisoners." Secretary of War Stanton informed the U.S. Senate on February 18, 1864, that a special commission under General John A. Dix would "investigate the cases of persons arrested and detained in Fort Lafayette and other military prisons of the Eastern Department, which have been used by direction of the President for the custody of persons seized by naval officers while engaged in blockade running or illicit trade." Forts Lafayette and Warren had long since become essentially repositories for blockade-runners.[44]

Fort Delaware held the largest numbers of citizen prisoners in November (421), many more than it usually held (the October figure was only 92). All Northern military prisons were subject to similar sudden fluctuations in numbers of inmates because of transfers to or from other military prisons or because of influxes of prisoners from the military front. Reasons for transfers varied: inspectors sometimes required them to relieve crowding and improve sanitary conditions, the prisoner-exchange officer sometimes needed them to be concentrated together for mass exchanges with the South by carefully computed arithmetic formulas, and security may have dictated occasional moves to foil escape plots. The reason for the sudden fluctuation at Fort Delaware seems apparent in the report of September 30, 1864, of the special commissioners sent to examine the cases of prisoners of state there and in Fort McHenry. The commissioners reported that "the papers in these cases were in many instances necessarily meager, many of the parties having been arrested in haste by officers on the march and upon representations made by the loyal people in disturbed districts, sufficiently justifying the officers who arrested them, but inaccessible to us." The commissioners reported further that "no case was referred to them in which it did not appear from the prisoner's own statement that he was properly arrested, and none were confined who were willing to disavow sympathy with the rebellion or who were not the implacable enemies of the Government, whom it was dangerous to release." In other words, Fort Delaware was generally the recipient of civilian prisoners sent in from the Middle Department—the same sorts of Virginians and Marylanders who populated the cells of the Old Capitol prison.[45]

Alton, the Old Capitol, and Point Lookout jointly held 705 of the 2,006 citizen prisoners recorded for November 1864—over a third of the total number. There is no reason to think that their prisoners were different in general composition in 1864 from what they had been earlier in the war. Alton held mostly the Missourians sentenced to imprisonment by military commissions in that state. There were some 2,000 such trials during the war, and the guilty often earned sentences of one to ten years' duration. Therefore, a Missourian once in Alton was there for a long time—unlike the ordinary prisoners of state who did not face trial and who often gained release in a matter of weeks.

Old Capitol's prisoners are the best known. Figures for November 1864 are not available, but there are good records for Washington's prisons for almost a two-year period preceding that month in the Record of Prisoners of State. These have been discussed at length elsewhere in this book and show the classic pattern—few Northerners and many citizens of Confederate states.

The other prisons on the list left fewer documents for historians. Point Lookout, like most of the prisons under discussion here, was primarily for Confederate prisoners of war, over ten thousand of them in November 1864, who surely made the 240 civilians seem insignificant, even difficult to locate. Situated in Maryland on a promontory where the Potomac River flows into Chesapeake Bay, it was close to the Confederacy but protected by Union naval supremacy. The civilian prisoners at Point Lookout were most likely Southern civilians who came to the fort mixed in with large groups of prisoners of war. Point Lookout received Confederate enlisted men (rather than officers) who were sent in as prisoners by the Army of the Potomac. A few civilians often became intermixed with the thousands of prisoners of war.[46]

Moreover, civilians, like prisoners of war, were exchanged. And like prisoners of war, they had to be gathered in staging areas. Prisons like the Old Capitol were used as staging areas for citizen prisoner exchanges. Altogether, according to the commissary general of prisoners's consolidated report of exchanged and paroled prisoners of war issued in December 1865, at least 2,156 Confederate citizens were exchanged during the Civil War. This was a minuscule part of the 329,363 Confederate prisoners of war paroled and exchanged during the war, and the civilians' relative insignificance helps account for the prison officials' haphazard record-keeping for their cases.[47]

The other two prisons on the list showing sizeable numbers of civilian prisoners are Camp Chase and Nashville. Despite the survival of almost no records at all from these two posts, it seems evident that Kentuckians and West Virginians were held in the first and Tennesseeans in the latter. William B. Hesseltine's study of Civil War prisons identified Chase as "a depot for the confinement of political and military prisoners from Kentucky and western Virginia." Louisville served only as a transfer prison, sending prisoners from Kentucky and farther south to Chase, Johnson's Island, and other midwestern prisons located father north. Nashville Military Prison held Tennesseeans detained by the area's Union military commander.[48]

Thus, most of the 2,006 prisoners noted in November 1864 can be accounted for on the same general principles already discussed in this book. In other prisons, much the same pattern applies. The seventy-eight scattered civilians among the 7,800 Confederate enlisted men in Elmira, New York, probably came mixed in with the prisoners of war on transfer from Point Lookout, just as they had come into Point Lookout from the Army of the Potomac in Virginia.

One could find more names than are available in the record series examined for this book. But it seems quite clear that such cases of arrests from the North above the border states would involve more of what can be seen in these records—more draft evaders, suspected deserters, defrauders of the government, swindlers of recruits, ex-Confederate soldiers, and smugglers. There would be more of the same, surely, and in roughly the same proportions.

Historians like James G. Randall have distorted the picture of state prisoners as much as critics of Lincoln have. This is Randall's characterization of the victim of military arrest:

> But as a rule the men confined in the Old Capitol Prison, Fort Lafayette, Fort McHenry, or the other state prisons were there for good reason. They had been acting as Confederate agents, furnishing supplies to the enemy, encouraging desertion from the service of the United States, committing outrages upon Unionists, stealing military supplies, destroying bridges, engaging in bushwhacking, making drawings of fortifications, carrying "treasonable" correspondence, intimidating loyal voters, or otherwise materially assisting the enemy. Many of the prisoners were actual spies.[49]

Although not wholly inaccurate, this description does not, in fact, inform the reader of the most striking qualities of "the men confined in the Old Capitol Prison, Fort Lafayette, Fort McHenry, or the other state prisons": *they were mostly citizens of the Confederacy.* In other prisons, like St. Louis or Alton, they were mostly citizens of the border states.

On the other hand, the traditional concern of Lincoln's critics has been with "the limits of dissent," as Frank Klement put it in his biography of Clement Vallandigham. Critics of the Lincoln administration have seen in the internal security system a potential for partisan abuse and political exploitation. Again, whatever the merits of the arrests described in the foregoing paragraphs, most had nothing to do with dissent or political opposition in the loyal states above the border states.

The cases involving political freedom, freedom of speech, press, and assembly—the few cases reported in the newspapers of the day and written about by historians since the Civil War—show up in the important series of records from the highest authorities in Washington. They show up in newspapers. They appear in memoirs and in history books. Such cases, if they did not reach the president's or the cabinet's attention—as did the Vallandigham case and the case involving the New York *World*—were likely at least to reach the attention of Turner or Holt. Thus, besides the obscure men caught in Pennsylvania as suspected deserters and other such historically anonymous characters, one also finds in the series of the Turner-Baker Papers under discussion here Albert D. Boileau, editor of the Philadelphia *Evening Journal.* Although the records of many civilians who were incarcerated in Northern military prisons have been lost, it is not likely that the records of any large number of those arrested who were politicians or newspaper editors have been. They are disproportionately memorable. It is the arrests of poor refugees and suspected bushwhackers that have been lost to history.

Precise figures are not available, but historians can nevertheless be precise about what the available figures mean. They indicate that there were few arrests above the border states. They show that after 1862 a majority of the citizens arrested were citizens of the Confederacy. They suggest a variety of causes for arrest of Northerners, among which, speaking, writing, and gathering in political groups were rare. They also show that more citizens were arrested than Ainsworth, Rhodes, Davis, Randall, or any historian writing since the turn of the century had thought. There were more arrests, but they had less significance for traditional civil liberty than anyone has realized.

7

The Revival of International Law

After the war, Major George B. Davis reflected, "The Federal Government...
succeeded in placing in the field armies of unexampled size, composed in great
part of men taken from civil pursuits, most of whom were unfamiliar with
military affairs and so unacquainted with the usages of war. These armies were
carrying on hostile operations of every kind, over a wide area, and questions
of considerable intricacy and difficulty were constantly arising which required
for their decision a knowledge of international law which was not always
possessed by those to whom these questions were submitted for decision.
Conflicting decisions and rulings were of frequent occurrence...and great
harm not infrequently resulted before decisions could be reversed by competent authority."[1] Davis's observation held true for sea as well as land, and
in fact, the best example came in the Lincoln administration's "conflicting
decisions and rulings" on the treatment of blockade-runners. Through it all,
however, an attempt to adhere to international law, albeit fumbling and at
times half-hearted, proved to be a hallmark of this aspect of the Civil War.

The Laws of Blockade

The percentage of prisoners of state who were blockade-runners rose as the
war continued. This stands to reason, as the Union blockade grew increasingly
effective after its paper beginnings. Yet the number of such prisoners by no
means kept pace with the growing effectiveness of the blockade. Modern
scholarship shows that the naval blockade increased fivefold in effectiveness,
as federal captures of ships rose from one in ten in 1861 to one in two by
1865.[2] Yet the number of men held in Northern military prisons for running
the blockade by no means quintupled.

139

A look at any nineteenth-century work on the law of blockade explains the difficulty under which the administration labored. For example, *Lalor's Cyclopaedia of Political Science* (1881) contained an entry on "Blockade" written by Theodore S. Woolsey, an American authority on international law, which states: "The penalty for breach of blockade is confined to the ship and its cargo; no punishment can be visited upon the crew." This long-standing principle of American law reached back to an unambiguous opinion of the attorney general in 1796: "A citizen of a neutral state who, for hire, serves on a neutral ship employed in contraband commerce with a belligerent power, is not punishable personally, according to the law of nations, though taken in the act by that belligerent nation to whose detriment the trade would operate." The principle received special emphasis in the standard antebellum work on the subject, Henry Wheaton's *Elements of International Law,* first published in 1836. This treatise embalmed the anti-British, freedom-of-the-seas doctrines of Wheaton's Jeffersonian youth.[3]

On the eve of the Civil War, the principle was reiterated. In a text published in 1860, Theodore S. Woolsey's father, Yale president and early political scientist Theodore Dwight Woolsey, all but characterized the punishment of blockade-runners' crews as barbarism:

> The penalty is confiscation, and it falls first on the ship as the immediate agent in the crime. The cargo shares the guilt, unless the owners can remove it by direct evidence.... Besides this penalty on cargo and vessel, the older text-writers teach that punishment may be visited upon the direct authors of a breach of blockade. Even de Martens... declares that corporal pains, by the positive law of nations and by natural justice, may be meted out to those who are guilty of such a breach. But the custom of nations, if it ever allowed of such severities, has long ceased to sanction them.

The idea was certain to penetrate the highest levels of policy-making in the Lincoln administration, for Henry W. Halleck, who would become general-in-chief for a time, had also written a text on international law published in 1860. In it, Halleck said: "The penalty for breach of blockade is the confiscation of the ship, and as a general rule, of the cargo."[4]

The same rules applied by analogy to the contraband traders and smugglers who worked the bays, inlets, and rivers along the boundaries between Union and Confederate territory. Thus, when a man named Henry Snowden was arrested in Baltimore for blockade-running and trading with the South, Levi Turner had to inform an overeager provost marshal that "The Acts of Congress prohibiting trade with insurrectionary districts forfeits [*sic*] on condemnation in the United States Courts, all goods found proceeding to, or coming from said districts, and in no other way punishes the parties interested therein." Turner knew "of no law, or pretence of law, by which a Provost Marshall is authorized to seize goods and persons in the loyal states, on suspicion of being designed for, or engaged in, prohibited commerce, *and to hold, or dispose of the same, at his pleasure.*"[5]

On the open seas, most blockade-runners fit the traditional definitions of international law—they were neutrals. The greatest number were British because the Old South had a weak seafaring tradition. Moreover, Great Britain was still the world's premier sea power, and she was accustomed to plying the sea lanes to carry Southern cotton for her mammoth textile industry. To make matters more difficult for the U.S. Navy, Southern men captured running the blockade might try to escape punishment by claiming to be foreign neutrals anyway. And accurately identifying citizenship was made more difficult by the substantial numbers of resident aliens in America and the traditional reliance of the American merchant marine and the navy on seafaring men of foreign birth.

International law might dictate a policy that was ignored in practice, of course. Throughout the Civil War, most blockade-runners, even after capture by the Union blockading squadrons, went scot-free, as international law prescribed. But from the start, some temptations to ignore international law and consistent American precedent proved irresistible. For example, Commander Maxwell Woodhull of the USS *Connecticut*, after capturing the British schooner *Adeline* running the blockade in 1861, then exacted as a condition of release of the crew (who were British neutrals) "that they should enter into an engagement not to be employed in a similar proceeding, in the future." Secretary of State William H. Seward soon told Secretary of the Navy Gideon Welles that requiring such agreements was illegal:

> It occurs to this department that, as the requirement referred to is not warranted by public law, the commanders of blockading vessels should be instructed not to exact any similar condition for the release of persons found on board of vessels charged with a breach of blockade. It may be lawful to detain such persons as witnesses, when their testimony may be indispensable to the administration of justice, but, when captured in a neutral ship, they can not be considered and ought not to be treated as prisoners of war.

State Department concern over possible violations became so great that on October 8, 1861, Seward ordered the release of all sailors in Fort Lafayette who were being held as blockade-runners.[6]

Both short-range interests and long-standing tradition dictated Seward's position. The prime goal of the Lincoln administration's foreign policy was to prevent British recognition of the Confederacy, and it would have been ill-advised to provoke irritation over violators of the blockade. Moreover, America had in the past almost always been the neutral shipper buffeted by conflicts between France and Great Britain. The State Department, therefore, found it easy to invoke the principles of international law that had traditionally protected neutrals. As a nation, the United States was historically soft on blockade-runners.

The tables were turned in the Civil War, however, and the United States found herself in Great Britain's usual position. As British historian Peter J. Parish puts it, the gamekeeper had turned poacher, and American courts began to invoke rules long despised by American international lawyers. The embrace and even expansion of the judicial doctrine of "continuous voyage" is a routinely cited example, but other measures taken by the executive branch, heretofore overlooked by historians, tended in the same direction.[7] Consider the case of J. M. Vernon, a passenger of uncertain nationality who was captured on board the *Huntress* running the blockade. Joseph Holt ruled: "He should be treated as other persons captured under like circumstances have been— as a quasi prisoner of War & as such exchanged. He certainly should not be allowed to remain within our military lines."[8] Even the judge advocate general did not really comprehend the true status in law of some neutrals running the blockade, and surely no treatise on international law defined "quasi" prisoners of war.

In 1863, the USS *Fulton* captured the British ship *Banshee* trying to run the blockade off Wilmington, North Carolina. Of the *Banshee*'s crew of forty men, thirty-nine claimed foreign citizenship and thereby a right to immediate release unless required as witnesses in prize court. But the captain had run the blockade eight times, and this was his third capture. Several of the crew had run the blockade from three to seven times. A majority of them had run it more than once. It was galling to contemplate the release of such men in the knowledge that it assured resumption of their activities. In December, Brigadier General E. R. S. Canby, giving an opinion for the War Department, noted that the Confederate government and some Southern state governments now owned many blockade-running vessels in whole or in part. This new development, Canby argued, made blockade-runners "no longer neutrals, engaged in traffic, on their own account, but employees of the Rebel government, who have violated the neutrality of the Foreign Powers, whose subjects they claim to be, and have subjected themselves, when captured, to the same treatment as any other prisoners of war."

Canby recommended their detention at least until they could prove their neutrality by something other than their own statements. The "practical effect of their discharge," he warned, would be "to throw upon the cities, to which these prizes are taken, a large number of restless and desperate men, eager to avail themselves of any enterprize, which may inure to the disadvantage of the Government—with the fullest opportunities for collecting and communicating information, they are spies, of the most dangerous character, without any of the responsibilities attached to that employment."[9]

In December 1863, the U.S. attorney for the District of Massachusetts, Richard H. Dana, Jr., wrote Gideon Welles with a similar message:

> We have found in the prize steamer *Cornubia* letters which prove that steamer, the *R.E. Lee,* and *Ella and Annie* and others of their class are the property of the Confederate Government and that their commanders are in the service of the Confederate Navy Department. This raises the question whether, in like cases, the Government will detain foreign seamen found on board as prisoners of war. The letters also show that they are under orders to conceal these facts while in neutral ports, in order to escape the rules applicable to public vessels of belligerents.

Welles ruled laconically, "The persons captured on the boats mentioned and others in like cases to be detained as prisoners."[10]

The State Department would doubtless have taken a dim view of this recommendation had some incriminating dispatches not fallen into Seward's hands at almost the same time. On January 11, 1864, Seward wrote Secretary of War Edwin M. Stanton, calling this recently intercepted correspondence to his attention. He sent along copies of a letter from Alex. Collee & Bros., of Manchester, England, to Captain John Wilkinson of the Confederate navy, expressing the company's willingness to dispose of cargo brought to England on the blockade-runner *Giraffe*, a vessel owned by the Confederate government. Another letter, from Confederate Ordnance Bureau head Josiah Gorgas to the captain commanding the *Cornubia*, instructed the officer not to carry any "insignia of vessels of war," not to reveal that the Confederate Ordnance Bureau owned the ship, and instead to make it seem as though the Confederate government merely placed their cargo on the ship. "With the knowledge obtained through these papers," Seward concluded, "it will be proper for you to direct, that henceforth, British blockade violators, be detained in custody and not released as heretofore."

On January 30, 1864, however, Welles asked Seward why the secretary of state still forwarded to the Navy Department inquiries from the British minister about "British blockade-runners who were retained in custody on his [Seward's own] suggestion." Seward "said he wished that course pursued, but the change of policy required time to effect the change."[11]

British violators of the blockade, or at least some of them, were subsequently imprisoned, until May 16, 1864, when Welles revised the order. But even with this change, release was not necessarily as rapid as it had been: "When the neutrality of a vessel is doubtful or when a vessel claiming to be neutral is believed to be engaged in transporting supplies and munitions of war for the insurgent government, foreign subjects captured in such vessels may be detained until the neutrality of the vessel is satisfactorily established. It is not advisable to detain such persons under this instruction, unless there is good ground for doubting the neutrality of the vessel." Normally, British blockade-running sailors were released. So routinely was this done that the famous Confederate blockade-runner, Captain John Wilkinson, for example, referred without any further explanation to "a blockade-runner who had once been captured and released, being a British subject."[12]

Violators of the blockade who were not foreign subjects had always been liable to be detained. At times, the Navy Department referred to them as prisoners of war, and some may have been so designated, but others appeared among the prisoners of state and "quasi prisoners of war" at Forts Lafayette and Warren and in the Old Capitol prison—until a surprising thing occurred.

In January 1864, General John A. Dix, commander of the Department of the East, received orders to appoint a military commission to review the cases of men confined in Forts Lafayette and Warren. On February 11, the commission ruled that blockade-runners should be discharged from custody, "on the ground that no violator of blockade is, by the laws of war, personally liable to punishment for his act. In all such cases the proceedings are *in rem;* ... forfeiture of ship and cargo and loss of wages are the only penalty imposed by the law of nations for breach of blockade; ... blockade-running is an offense against the municipal law of the State or district in which it was committed, and the U.S. courts cannot punish the blockade-runner whose acts have not been done within such limits; ... blockade-runners may be lawfully detained until they are brought ashore, and if needed as witnesses before the prize courts can be held to give evidence, but other and further detention is an act of power, not of right." Thus, the military commission decided that all blockade-runners, Confederate or foreign, were entitled to immediate release after any necessary testimony before prize courts.

There was no matter of enlightened self-interest here, as in the case of British blockade-runners, where obeying the law also served the purposes of U.S. foreign policy. The federal authorities did not have to worry about outraging Confederate opinion, as they did British opinion; indeed, some of the blockade violators seem to have thought themselves likely prospects for prison. The remarkable military commission further ruled, against its own interests, "that no persons except such as are in the military or naval service of the United States are subject to trial by military courts, spies only excepted; and that except in districts under martial law, a military commission cannot try any person whatsoever not in the U.S. military or naval service for any offense whatever."[13] The commission thus ruled itself without jurisdiction over the cases and, with this sweeping dictum, anticipated the famous *Milligan* case by two years.

Thereafter, the commission investigated the cases in the forts and, early in June, recommended the release of the blockade-runners. Dix and his fellow commissioners did not choose words that would make their decisions more palatable to the War Department's judge advocates. For example, in the case of E. O. Murden, captain of the blockade-running steamer *Victory* captured on June 21, 1863, the commission ruled: "Were there a federal court in North Carolina the Commission would recommend that he be turned over to the civil authorities." When such infuriatingly impractical rulings and recommendations reached the men in Washington who for years had borne the responsibility of apprehending such persons, there was a predictable outburst. Judge Advocate Levi C. Turner complained bitterly to the War Department in June:

I respectfully submit that the prisoners in question are blockade violators, radically variant in character (type) and status from those to whom the laws of war have heretofore applied. Indeed, they are not blockade-runners in the acceptation of the term to which the laws of war apply. They are, generally, whether citizens or foreign subjects, employes of the rebels. A large portion of the vessels engaged in blockade-running are owned in whole or in part by the rebel Government or by the rebel States. Vessels engaged in blockade-running must be thus owned, as required by the rebel authorities. The men engaged in this blockade-running are not neutrals, engaged in traffic on their own account, and only liable to forfeiture of vessels and cargo if captured, but they are either employes or aiders and abettors of the rebels. Not only this, but these blockade-runners are generally spies, in rebel employment, of the most repulsive and dangerous character, enjoying the largest facilities for collecting and communicating information fatally injurious to the Government, if permitted (in virtue of the commission's interpretation and application of the laws of war) to disguise their real character in the habiliments of blockade-runners as defined by the laws of war. Those blockade-runners are usually desperate, reckless men, and to discharge them is simply releasing unprincipled, traitorous men, whom the public safety requires should be prohibited from going at large.[14]

Turner adapted Canby's old arguments and stretched them beyond the kernel of truth they originally contained.

Turner and Canby did have a point. The Confederate government and the state governments in the Confederacy did own some of the blockade-runners. Moreover, the Confederate Congress enacted legislation requiring that all ships running into Confederate ports after February 6, 1864, reserve half their cargo space for the needs of the government. Blockade-running captains on the Texas Gulf late in the war all held government commissions.[15]

Southerners who ran the blockade were not neutrals in the sense that international law meant, but America had so long been the neutral victim of blockaders that she was punctilious in protecting their rights. For persons like those who sat on the Dix commission, it was difficult to cease thinking like neutrals and to begin acting like blockaders. Turner, exasperated, said that if the ruling of the Dix commission were upheld, all efforts to arrest blockade-runners would be "labor in vain."[16]

Most of the prisoners were released as recommended, but in August the Navy Department ruled that "Pilots and seafaring men, excepting bona fide foreign subjects, captured in neutral vessels[,] are always to be detained. These are the principal instruments in maintaining the system of violating the blockade and it is important to hold them. Persons habitually engaged in violating the blockade, although they may not be serving on board the vessels, are of this class and are likewise to be detained." Thus, on March 18, 1865, over a year after the Dix commission's ruling, Welles sent Stanton a letter explaining that a number of naval prisoners then at Fort Warren had "been confined upward of two years, as they were experienced seamen and it was deemed inexpedient to release them." Since the last great blockade-running port, Wilmington, had fallen on January 15, 1865, Welles noted that the "closing of the principal ports used by blockade-runners had deprived these parties of their power to inflict serious injury upon the Government or to aid the insurgents." Indeed, Welles regarded it as "a matter of no great importance to detain them longer" and turned them over to the War Department with the recommendation they be exchanged. The list included forty-two blockade-runners.[17]

The Lincoln administration's record on the question of blockade-running was not spotlessly correct. Declaring a blockade of ports in rebellion rather than those of another country was legally problematic, as even Lincoln's own

secretary of the navy believed.[18] That question aside—and after all most Northerners probably did not question the nature of the blockade—there were still
many internal disagreements and changes in policy. The administration was
as often guided by immediate interest and expediency as by timeless principles.
Occasionally, even temporary anger dictated policy, as in the case of the
Banshee. Different departments of the government interpreted the law differently, with the result that some kept prisoners that others would have released,
and vice versa. Practical measures taken to win the war sometimes played
havoc with American legal precedents and with international law. Moreover,
the interests of the Navy Department and the State Department diverged considerably on the question of blockade, and the division was widened by Welles's
personal dislike of Seward.

Not many officers knew the law. Henry W. Halleck noted in 1866 that the
"experience of our officers, both volunteers and regulars, in the great civil war
just terminated, has proved that this subject [international law] has been too
much neglected, not only in our colleges, but also in our two great national
schools—the Military and Naval Academies."[19] Indeed, the laws accepted in
American practice were not officially codified or collected until over twenty
years after the Civil War, in Francis Wharton's three-volume *Digest of the
International Law of the United States,* published by the Government Printing
Office in 1886. The slender texts of Halleck and Woolsey did not approximate
its bulk, and Henry Wheaton's *Elements of International Law* had too much
natural-law emphasis and paid too little attention to the accumulated opinions
of secretaries of state and of attorneys general to solve the problems of the
Lincoln administration.

As a result, naval officers apparently had no comprehensive instructions until
near the end of the war. After the blockade had been in effect for well over
a year, on August 16, 1862, Welles reminded himself "to get up a complete
set of instructions, defining the points of international and statute law which
are disputed." The memorandum was prompted by Seward, who had asked
Welles to restrain naval officers in the blockading squadrons in some ways.
Welles realized that "a set of instructions for our naval officers in matters
relating to prize captures and enforcing the blockade" was needed, but he
apparently never found the time to write it.[20]

A year later, Welles, Lincoln, and Seward were still struggling over the
proposed instructions. Welles thought Seward's policy toward the British on
the blockade was "subservient," constituting a "want of policy." "He has no
convictions, no fixed principles, no rule of action, but is governed and moved
by impulse, fancied expediency, and temporary circumstances," Welles complained privately. In a letter to Lincoln, the navy secretary objected in particular
to Seward's and the president's proposed instructions, which would have
stated: "You will not in any case detain the crew of a captured neutral vessel,
or any other subject of a neutral power on board such vessel as prisoners of
war or otherwise, except the small number necessary as witnesses in the prize
court."[21]

Welles's view was that all the persons captured on neutral vessels should be sent in as witnesses—a subterfuge for temporary imprisonment of neutral citizens perhaps but also a policy dictated by a practical problem. "In the war in which we are now engaged," Welles thought, "it must be remembered that no inconsiderable portion of the persons captured on some of the vessels, claiming to be neutral, are rebels. It is impossible for the captor to decide who, or how many are rebels. It certainly is not advisable to go counter to the rule so framed by all the Courts, nor to release captured rebel prisoners." The editors of *The Collected Works of Abraham Lincoln*, the only historians who have investigated this problem, could find no further reference to the proposed set of instructions nor any "final disposition" of the matter. The record remains silent, and the resulting confusion evident in the lists of prisoners seems proof that blockading officers failed to receive clear instructions.[22]

Blockade-Runners As Prisoners of State

In the early months of the war, when Seward oversaw prisoners of state, civilian blockade-runners constituted 12.9 percent of the persons arrested by military authority. In the period after February 15, 1862, blockade-runners constituted 14.1 percent of the prisoners recorded in the Turner-Baker Papers. In cases where the charge is noted, 459 of 2,810, or 16.3 percent, were blockade-runners. These records are biased toward inflated figures for arrests that took place on water. For much of the period, Levi C. Turner reviewed arrests in Washington, Maryland, and Virginia—coastal regions honeycombed with rivers and trade routes as old as the earliest settlements on the North American continent. Moreover, he was at one time sent on a special assignment to investigate the prisoners kept in Forts Lafayette and Warren, traditionally the military prisons in which blockade-runners were held.[23] The records of these investigations remained in his papers at his death and consequently found their way into the eclectic Turner-Baker Papers (microfilmed portion).

Even so, the number of blockade-runners did not markedly exceed those imprisoned in Seward's days as keeper of the prisoners of state. Certainly, they did not increase fivefold, as did captures of vessels on the blockade. All in all, the most notable feature of the statistics on prisoners held for running the blockade is their relatively low number. The larger steamers had crews of forty or more men, and the 459 prisoners who were charged as blockade-runners in the Turner-Baker Papers could hardly have accounted for all of the crews of the many ships captured during the Civil War. The administration let most blockade-running men go free. Those they retained appear to have been older and more skilled and therefore of more immediate value to the Confederacy.[24]

Talk of blockade-running men naturally conjures up images of swashbuckling sons of the sea running valuable cargoes from Wilmington or Mobile to Nassau and Bermuda to be transshipped to England. In fact, prison records show them to have been a motley and often rather unheroic crew. Under the rubric of "blockade-runners" fell forty-nine prisoners who were not captured at sea. Rather, they were apprehended by the Potomac Flotilla and may never have been to sea in their lives.

This unheralded and unheroic little fleet made, as its commander reported, "almost daily expeditions up the rivers, in the creeks, and through the marshes of the northern neck of Virginia." They aimed at stopping what the secretary of the navy described as the work of "petty contrabandists," different from "the more open blockade-runners." Only the early Civil War naval historian James Russell Soley recognized the important but inglorious "work of the flotilla in the Potomac," which "was chiefly confined to the suppression of the small attempts at illicit traffic which are always found along a frontier of belligerent operations."[25]

When one person charged with "running the blockade" was brought to Fort McHenry, a federal official complained that the "phrase" was "so vague."[26] He was right, but history has little appreciated the difficulties, semantic or otherwise, in characterizing the variety of offenses connected with illicit trade in time of war.

Arresting blockade-runners, whether on inlets or at sea, posed the usual legal and practical problems involved in military arrests of civilians. Consider, for example, the case of the *Ella and Annie*. This Charleston blockade-running ship of 905 tons burden, carrying 480 sacks of salt, 500 sacks of saltpeter, 281 cases of Austrian rifles, 500 barrels of beef, and 42 cases of paper, encountered the *Niphon,* a ship of the North Atlantic blockading squadron fresh from an unsuccessful chase off Wilmington. Acting Rear-Admiral Stephen P. Lee reported that the captain of the *Ella and Annie,*

> finding himself intercepted, put his helm up and endeavored to run down the *Niphon.* This attempt was partly avoided, though the *Niphon* was struck about the fore rigging, and her bowsprit, stern, and starboard boats carried away. At the moment of collision Acting Master Breck reports he opened upon the enemy with shell and canister and carried the prize by boarding.

Three men on the *Niphon* were slightly injured, and the 38-man crew of the *Ella and Annie* was sent to New York. There some of them, at least, were imprisoned in Fort Lafayette.[27]

The case seemed on its way to the customary resolution. The prize master would bring the *Ella and Annie* into a port where a prize court would adjudicate the claims to the vessel and its cargo. The sailors on the Southern vessel, or at least some of them, would be held as witnesses and then, probably, released. Soon, however, complicating facts began to surface. Breck discovered that the captain of the *Ella and Annie,* F. N. Bonneau, held a commission as lieutenant in the Confederate navy, and one of the crew members held appointment as master in the Confederate navy.[28] Lieutenant Bonneau, worried about his status as a captive in enemy hands, volunteered this information because he wanted to be treated as a prisoner of war. A special investigative commission maintained, on the other hand, that Bonneau had commited a "piratical act" by using force to try to resist capture and should be tried in a federal court. Levi C. Turner decided to await a decision from the attorney general; the case was too complex even for the experienced judge advocate.[29]

The common sailors on board the *Ella and Annie* presented yet another problem. They were not, apparently, enlisted in the Confederate navy, but like most such blockade-runners, they were civilians. Thus, twenty-one-year-old South Carolinian Edward Hessell, on his first voyage to challenge the Union blockade, found himself in Fort Lafayette threatened with treatment as a prisoner of war because the *Ella and Annie*, by trying to run down the Union gunboat rather than simply to elude her pursuer, had committed an "act of war."[30] C. P. Jervey, the first mate, was in an equally ticklish situation. The son of British subjects, he called Scotland home but had always "followed the sea" (as the contemporary expression went). He told his interrogators in New York that he did not want to run the *Niphon* down but was obliged to follow the captain's orders.[31] Eli G. Whitney, a twenty-year-old serving as third mate, had followed the sea since he was eleven years old. His attempt to get home to Charleston on the *Ella and Annie* constituted his first attempt to run the blockade, and furthermore, he maintained that he had been stationed below decks and knew nothing of the aggressive action against the *Niphon*.[32]

As was often true of complicated cases, the Union authorities decided to hold these men not as prisoners of war but as prisoners of state. Several were held until 1865. In a way, this was a reasonable compromise. The men were not hanged as pirates, but they did not gain the usual speedy release.

Only a few of the prisoners held for blockade-running had not actually been picked up at sea or in the act of carrying illicit goods. Among those few were unfortunates like William E. Chenoweth of Washington, D.C., a barber and embalmer's agent, who was arrested on May 21, 1864, for having boasted to a man on the street that he had run the blockade several times. Like many such loudmouths, Chenoweth had apparently been drinking. He paid for his folly by spending over six months in the Old Capitol prison.[33]

Men apprehended at sea often proved perplexing as well. Oystermen, for example, caused many problems. Blockaders could hardly distinguish them from smugglers, whether spotted from afar or examined up close. Further complications arose from the conflicting interests of the Treasury and Navy departments. Salmon P. Chase, who issued permits for loyal fishermen to go oystering in patrolled waters, sometimes furnished a chart to the commander of the North Atlantic blockading squadron marking the limits within which the individual oystermen promised to work.[34] Army commanders of the Department of Virginia and North Carolina might issue such permits as well, thinking that within their powers.[35] Whatever the charts and paperwork said, a blockader often could not see a runner more than a hundred yards away and needed to react quickly to perform its duty. Nothing kept blockade-running ships out of the charted waters, and oystermen were often picked up by blockaders by mistake.

Another practical difficulty of blockading was handling men after capture.

As soon as the government let some men go, others began to demand that they had a right to be released as well. International law stipulated holding only the number necessary as witnesses for prize court; the rest were supposed to be released. Later, investigating commissioners reviewing the plight of civilians languishing in unpleasant fortress prisons noted, with some sympathy, pleas that everyone else aboard the vessel had already been set free. William G. Thompson, captured on the *Minna* in 1863, wrote from Fort Warren to the secretary of war early in 1864 to point out that "the whole crew officers and all were discharged because[,] I suppose[,] they claimed English Protection but I have not got that low yet."[36] Benjamin Shinft, formerly of the Confederate navy (by his own admission), captured on the sloop *Ann Eliza* in 1864, complained to authorities that he was still in prison, though the owner and the captain of the vessel had already been released.[37] When the navy vowed to hold pilots, whatever their status as civilians or foreigners or sailors, the problem of identifying a vessel's pilot arose. John Neill, from the *W. H. Gladding,* surmised that he remained a prisoner in Fort Lafayette because he was being held as a pilot, but he insisted he was not.[38]

The other large series of records of arrests made mostly in the eastern United States, the Record of Prisoners of State, in the unmicrofilmed portion of the Turner-Baker Papers in the National Archives, likewise supports the claim that the work of the Potomac Flotilla was diligent. Of the 2,187 prisoners recorded in these files, 228, or 10.4 percent, were arrested for violating the blockade or were picked up by gunboat crews on the waters around Washington, D.C. Of these, at least ninety-five were arrested by the Potomac Flotilla, which thus accounted for at least 4.3 percent of all civilians in Washington military prisons. Actually, the percentage must have been higher. In cases where the charge is noted or ascertainable, blockade-runners constitute 14.6 percent (228 of 1,561), and the Potomac Flotilla accounted for 6.1 percent of the total arrests (95 of 1,561).

The Lincoln administration adhered so strictly to international law and American traditions regarding freedom of the seas that most blockade-runners went free. In marveling at the government's leniency toward these damaging troublemakers, one should not lose sight of the nature of these prisoners. They were "prisoners of state." Though civilians, they certainly had little or nothing to do with what are customarily thought of as political freedoms endangered by mobilization for war. Their exact status was difficult to determine at the time and has been ever since, but they are best thought of as prisoners created by the incidents and frictions of war.

Hostages

From the earliest years of American history, Europeans had been expecting warfare in the New World to degenerate to the level of savagery associated with its native inhabitants, the American Indians. By the middle of the nineteenth century, only traces of this assumption remained—seen, for example, in the feathered savages commonly used in French political cartoons to represent the combatants in the American Civil War. The Europeans proved wrong, and war continued to be waged on as civilized a plane in the New World as in the Old. Even the more savage-appearing practices, like the taking of civilian hostages, were in fact routine measures in European wars as well as American.[39]

"Taking hostages," in the words of a recent article, "is something associated with the excesses of modern terrorism."[40] But it was a commonplace occurrence in nineteenth-century warfare, and taking civilian hostages as prisoners of state became a routine practice in the American Civil War. International law condoned it, and the American laws of war codified it. Abraham Lincoln, Ulysses S. Grant, and Robert E. Lee all dealt directly with situations involving hostages without condemning the practice. The documentation for several hostage incidents during the Civil War is solid, yet the cases have not been studied before. Mark M. Boatner's *Civil War Dictionary,* for example, has no entry for "hostages." *The Collected Works of Abraham Lincoln* and other standard references for the period do not index the word. Hostages simply do not constitute a familiar category of thinking about the American Civil War.

Hostages were mentioned in James Kent's *Commentaries on American Law,* a standard source for Civil War lawyers and statesmen. Before the war, interest in the subject focused on maritime cases in which a hostage was given to an enemy to guarantee that a prize crew would carry a captured ship to port and pay a ransom for that safe conduct. Henry W. Halleck had described maritime hostages in his *Elements of International Law and Laws of War.* Theodore Dwight Woolsey, whose *Introduction to the Study of International Law* appeared on the brink of war in 1860, referred to hostages as a "mode of securing the faith of treaties, formerly much in use but now almost obsolete." The most recent example he could find was the Treaty of Aix-la-Chapelle in 1748. Woolsey also mentioned the use of hostages for ships' ransoms.[41]

Statements like Woolsey's—"now obsolete"—provide an index of American innocence about warfare on the eve of 1861. After the firing on Fort Sumter, the United States descended into the hellish experience of modern war on a grand scale, and the taking of hostages soon became part of that experience. Tennessee was to see some of the worst civil strife of the war, and one of the first hostage incidents occurred in that state. In May 1862, U.S. military authorities held twelve citizens of Rutherford County, Tennessee, as hostages "to secure the safety of Murfreesboro, and ... to guard against the repetition of unlawful acts." Military governor Andrew Johnson released them upon their subscribing $10,000 bond, promising loyalty, and disavowing "all irregular warfare."[42]

The fiery Johnson quickly adapted himself to this sort of warfare, and on June 5, 1862, he wrote President Lincoln as follows:

There are seventy east Tennesseeans now lying in prison at Mobile [many] of them the most respectable & valuable citizens of this section[.] They are there simply for being union men[.] They are treated with more cruelty than wild beast[s] of the forest[.] I have taken this day steps to arrest seventy 70 vile secessionists in this vicinity & offer them in exchange & if they refuse to exchange I will at once send them south at their own expense & leave them beyond our lines with the distinct understanding that if they recross or come again within said lines during the existing rebellion they shall be treated as spies and with death accordingly[.] Does this meet your approval[?] It is no punishment now to send secessionist[s] north[.] in most instances they would rather go to the Infernal regions than to be sent south at this time[.]

Although Governor Johnson did not actually use the term "hostage," he was embracing the concept in fact. So did Lincoln, in his somewhat gingerly worded reply: "Your despatch about seizing seventy rebels to exchange for a like number of Union men, was duly received. I certainly do not disapprove the proposition."[43] Lincoln distanced himself from Johnson's action by use of a double negative, but the import was clear: he approved.

Available documents reveal few hostage incidents in the eastern theater in the first twenty-one months of war. Perhaps the first involved some Pennsylvanians captured by the cavalryman J. E. B. Stuart in 1862. The Confederate agent for prisoner-of-war exchanges, Robert Ould, explained that they were "held only in retaliation for captures of non-combatant citizens of the Confederate States." In February 1863, two citizens of Loudoun County, Virginia, Stephen R. Mount and Volney Purcell, were held as hostages for a Mrs. Davis of Snickersville. They were released later in the spring, but nothing else is known about the case.[44]

The "White Flag Affair" and the Fredericksburg incident, both of which occurred in the spring of 1864, revealed more vividly the stresses that could bring Christian warriors to take hostages. Around the first of June, three men appeared on the Virginia side of the Potomac at Edwards Ferry and hailed the federal pickets on the other side. The men on the Confederate side apparently waved a white handkerchief and shouted that they were refugees who wanted to cross over. Two soldiers from the Second Massachusetts Cavalry climbed into a boat and rowed out to fetch them. As they neared the other shore, the white flag was dropped and the three so-called refugees, along with five other concealed men, began firing on the boat. One of the Union soldiers was wounded in the thigh, and both were taken prisoner. The eight captors then robbed the two soldiers of their pistols, their watches, and some of their clothing. The captives were taken to nearby Leesburg, Virginia, and were, according to Union sources, "paraded through the streets to the gratification of the robbers and citizens."

Union general Christopher C. Augur acted quickly. He arrested two men, John L. Rinker and George C. Ryan, who were the fathers of two of the alleged ambushers. By June 4, with the approval of Henry W. Halleck, he had also arrested "eight of the principal secessionists of Leesburg" to be "held as hostages" for the return of the two Union soldiers and the surrender of the eight White Flag ambushers. When the two Union soldiers were sent back to Union lines, Judge Advocate General Joseph Holt determined on June 9 that "the two oldest and least noted for disloyalty" among the hostages would be released in exchange. Holt added a sharply worded endorsement of Augur's action: "The crime having been com[m]itted under the shelter of a flag of truce was one of unspeakable baseness & atrocity, & was in every way in keeping with the spirit of the rebellion. The government may well resort to all means known to civilized warfare to compel a surrender of the criminals."[45]

Though it may sound like an oxymoron now, Holt was correct in saying that hostages were among the "means known to civilized warfare." About a year before the White Flag Affair, the United States War Department had published General Order No. 100, "Instructions for the Government of the Armies of the United States in the Field." Drafted by the pioneering political scientist Francis Lieber, the instructions constituted the legal code for the United States forces during the last two years of the Civil War (and indeed provided the basis for America's laws of land warfare until 1956). General Order No. 100 devoted 2 of its 157 articles to hostages:

> 54. A hostage is a person accepted as a pledge for the fulfillment of an agreement concluded between belligerents during the war, or in consequence of a war. Hostages are rare in the present age.
> 55. If a hostage is accepted, he is treated like a prisoner of war, according to rank and condition, as circumstances may admit.

Lieber realized that hostage-taking could not be ignored as a part of modern warfare.[46]

Ironically, hostages appear to have become less and less rare after Lieber's code was published in April 1863. Moreover, the practice of taking hostages never seems to have lived up even to the rough standard of justice in General Order No. 100. The White Flag hostages present perfect examples of the savage gulf between the law and the practice of war.

Lieber, like the authorities before and after him, described the taking of hostages as a passive sort of affair in which the enemy would be pressing its citizens on reluctant federal soldiers, begging "acceptance" of them as pledges for future acts. "Acceptance" certainly implies a previous "offer." The articles have a drily contractual tone about them, as though reasonable Christian gentlemen would reach some formal agreement, and one party would throw some hostages into the bargain as guarantors of good faith.

Such was not the case in any Civil War hostage incident. In the White Flag Affair, Augur acted fast without any previous agreement with Confederate authorities. The Virginians did not offer and Augur did not "accept" the eight Leesburg citizens. General Augur seized them in order to force the Virginians to agree to hand over the perpetrators of the crime. Hostages were a recognized means of civilized warfare, but the persons taken from Leesburg did not fit the legal definition of "hostages." They were called hostages, but they might have been described more accurately as prisoners taken in reprisal for an atrocity.

Even as retaliation, taking hostages may not have been entirely illegal. Lie-

ber's code stated that "All prisoners of war are liable to the infliction of re-
taliatory measures." Johnson's "vile secessionist" prisoners were civilians, and
not prisoners of war, but then he was seizing the civilian secessionists in
retaliation for the holding of other civilians by the Confederates. In the end,
Lieber had to say that the "law of war can no more wholly dispense with
retaliation than can the law of nations, of which it is a branch."[47]

To be sure, some sort of atrocity had been committed in the White Flag
Affair. Henry Clay Ryan, one of the men who ambushed the pickets, denied
that a white flag was used to decoy the federal soldiers over the river and
maintained further that the two men "were treated kindly" while they were
prisoners. But he admitted taking three pistols and "one or two watches."
Stealing the personal property of prisoners of war violated the standard rules
governing prisoners of war everywhere. Article 72 of General Order No. 100
stated: "Money and other valuables on the person of a prisoner, such as watches
or jewelry, as well as extra clothing, are regarded by the American Army as
the private property of the prisoner, and the appropriation of such valuables
or money is considered dishonorable, and is prohibited."[48] If Ryan was willing
to admit to this notoriously "dishonorable" practice, he may well have been
capable of others. Some citizens of Leesburg expressed their disgust at the
whole ambush and did not deny that a white flag had been used.

Loudoun County, where the incident occurred, lay in the heart of "Mosby's
Confederacy," an area always ripe for atrocity. Indeed, one of the participants
in the ambush was a member of White's cavalry, a partisan unit often associated
by Union authorities with Mosby's rangers. The citizens of Leesburg did not
blame Mosby, but in a petition asking for release of the hostages they did
depict their county as a theater of anarchic guerrilla violence: "Ever since the
advance of the Federal Army through this country we have been without any
civil government, and without any regular protection from the army. The
consequence has been, a spirit of lawlessness. . . . Armed men claiming to be
soldiers are continually passing thro' the county committing all kinds of
depredations."

The Confederates' version of the incident, although it admitted some irregularities, differed considerably from the Union version. Ryan maintained that
he and the other seven men had been ordered by Colonel White to find a
deserter named Mansfield. Hearing that he had become a ferryman, they came
to Edwards Ferry and called to him. When the two men in the boat proved
to be federal soldiers, they had no choice but to start shooting.

After the smoke had cleared and federal authorities had resolved on a course
of action, how did they pick the eight leading secessionists of Leesburg? No
one knows, but records of interrogations of the hostages show that none of
them could take the oath of allegiance to the United States in good conscience.
Five admitted voting for Virginia's secession ordinance. Two maintained they
did not, and the views of one man are unknown. Hostage William S. Pickett,
a 41-year-old hotelier had served fifteen months in the commissary of the
Seventeenth Virginia Infantry and had a brother in Confederate service as well.

Dr. Armstead Mott, a physician aged forty-two, had served two years as a surgeon in the Confederate army. Charles F. Fadeley, a 47-year-old farmer, had a son in the Eighth Virginia Infantry. Thomas Edwards (47), Dr. William Cross (60), and John P. Smart (69) had among them five nephews in the Confederate army. Edgar L. Bentley, a 46-year-old farmer, had no relations in Confederate service, and little is known about W. H. Gray, who described the hostages as a group of "elderly and peacible citizens." Drs. Cross and Mott had allegedly tended both Confederate and Union sick and wounded, as was common practice among conscientious physicians in this war-torn region.

Smart and Dr. Mott were the first to be discharged (on June 13, 1864), following Holt's guidelines. The rest were transferred from Old Capitol prison to Fort Delaware, where they remained until August. By that time three of the perpetrators of the ambush had been captured (though one subsequently escaped by jumping from a train en route to Washington from the prisoner-of-war camp in Elmira, New York). After that, the hostages, except Ryan and Rinker, were discharged (on August 25). Once Henry Clay Ryan and Noble Rinker were in custody, their fathers, who had been held in their stead, were released also (in December). And somehow, a mere boy of sixteen, George W. Ryan, arrested with his father, was lost in the shuffle, shipped off to Fort Warren in Boston harbor, and not released until Christmas Eve in 1864, his arrest apparently forgotten until that date.[49]

The Fredericksburg incident involved larger numbers and, if anything, more wholesale justice than the White Flag Affair. Sitting behind the Rappahannock River, astride the halfway point on the 100-mile route from Washington to Richmond, Fredericksburg, Virginia, bore the brunt of several military campaigns. On May 8, 1864, in the aftermath of the Battle of the Wilderness, about sixty federal soldiers, "most of them slightly wounded" according to Fredericksburg historian S. J. Quinn, straggled into town. They were armed, and the local citizens, fearing mischief and pillage, demanded their surrender as prisoners of war to go to Richmond, or their departure across the river to an area infested with Confederate guerrillas. Most of the dispirited soldiers chose surrender and were promptly delivered to the nearest Confederate post. From there they went to Richmond and military prison.[50]

Federal authorities regarded such action by alleged noncombatants as a violation of law, and the secretary of war ordered the provost marshal to arrest about sixty citizens to be held hostage for the return of the soldiers. Colonel Edmund Shriver responded to the order with a letter that suggests a chilling picture of Civil War Fredericksburg, a bustling commercial town of five thousand people before the war:

> In relation to the apprehension of 60 prominent citizens of this city, as ordered by you, I have to report that on account of the very few males who are now present it has been impossible to get that number ready to dispatch to-day, but I hope to send them to Washington tomorrow. There are 30 in custody, 9 of whom are suspected of having been engaged in conveying our wounded to Richmond.

Lieutenant George Mitchell, a member of the United States unit that captured Fredericksburg just after the incident, gave an equally telling description of the ravages of war. The town began to fill up with Union wounded, and Mitchell was appointed assistant provost marshal and assigned the task of requisitioning food and shelter for them. He quartered wounded men "in every house in the city," but he "found very little food in the homes" for the "starving" wounded.[51]

The War Department in Washington had probably been moved to drastic action by the vision of wounded soldiers captured by Virginia citizens and sent to Confederate prisoner-of-war camps. Lieutenant Mitchell, however, described a different scene:

> I found out that our soldiers had straggled in without any escort whatever and commenced to break into the stores which were closed [it was Sunday]. A great many of these men, were nothing but stragglers, and were not wounded. There is sufficient proof when they arrived into the city a day and a half before the escort, and ambulances.

Mitchell thought all the Fredericksburg hostages but three should be released, the exceptions being two scouts and a man who was armed with a saber and firearm when captured.

Judge Advocate General Holt took a sterner view of the affair: "the fact of their being stragglers does not change the character of the conduct of the citizens, whose action was not based upon any such distinction, and who have no authority to administer discipline to United States troops." The tone of this note was more measured than his note about the White Flag Affair. He said nothing in this instance about atrocities that justified the use of any means known to civilized warfare.[52]

Nevertheless, the hostages remained in prison. The sixty-four Fredericksburg citizens taken by the provost marshal included nine merchants, six laborers, four shoemakers, three shopkeepers, three tailors, three carpenters, two brick-layers, two farmers, two blacksmiths, two bootmakers, a gunsmith, an artist, a pump-maker, a tinner, a carriage-maker, a landlord, a miller, an architect, and nine men of unknown occupation. They ranged in age from sixteen to seventy but thirty-one of the hostages were fifty years old or older, including six men who were sixty years old or older—a special cruelty dictated by the scarcity of young men not in uniform in the heavily mobilized Confederacy of 1864.[53]

On May 31, the Fredericksburg common council met to consider the problem and wrote the Confederate secretary of war, James Seddon. George H. C. Rowe, a local citizen who had arranged a prisoner exchange back in 1862, another time of trouble for Fredericksburg, went to Washington in June and forged an agreement with Edwin M. Stanton for an exchange. Rowe himself took responsibility for the safety of the federal soldiers released by Confederate authorities in Richmond. After difficulties encountered crossing the lines and delays in meeting with the United States authorities in Washington, Rowe managed the exchange. Most of the citizen hostages marched the last twelve miles home in July.

Before the release of the Fredericksburg hostages, President Lincoln had written a letter to Secretary of War Stanton about one of the prisoners:

> Understanding that Mr. John J. Chew, of Fredericksburg Va.—is now in arrest as a hostage for our wounded soldiers, carried by citizens from Fredericksburg into the rebel hands at Richmond, and understanding that Mr. Chew, so far from doing anything to make him responsible for that act, or which would induce the rebels to give one of our men for him, he actually ministered, to the extent of his ability, to the relief of our wounded in Fredericksburg, it is directed that said John J. Chew be discharged and allowed to return to his home.[54]

Lincoln was not, apparently, aware of the War Department's discovery that most of the federal soldiers had been stragglers rather than seriously wounded men. But he otherwise clearly understood the principles involved in the taking of the Fredericksburg hostages: some of the citizen prisoners bore some responsibility, direct or indirect, for the act, but others were held merely to "induce" the Confederate authorities to give up their Union prisoners in exchange. Lincoln seems not to have desired to make a direct statement endorsing the practice of hostage-taking, but he endorsed it indirectly more than once.

In a less-well-documented incident, Dr. Samuel K. Jackson and Joseph Mead, already prisoners in Old Capitol in Washington, were selected by the War Department as special hostages for James Hamilton and J. P. Culbertson, who were themselves imprisoned as hostages by the Confederates in Salisbury, North Carolina. After thirteen months in prison, Culbertson and Hamilton were released because they promised Confederate authorities they would work for the release of two civilian prisoners held in the North. Accordingly Hamilton wrote Secretary of War Stanton on September 2, 1864:

> Mr. Culbertson and I were prisoners for thirteen months, held as hostages. [Robert] Ould [Confederate exchange agent] alleges that you hold citizens on insufficient or no charges who are not connected with military organizations. He released us because we promised to try to effect the release of Smithson and Reverend Doctor Handy.... Ould proposed to release all civilians and capture no more. He proposes to exchange the soldiers, man for man, and hold the excess, and says you might hold hostages for the negro soldiers if they refuse to exchange them. This much I promised to say. Could you not capture and hold as hostages, say, two or three for one, some prominent citizens of Virginia to procure the release of the seven citizens who are remaining in prison at Salisbury, N.C.? West Virginia did so for some of her citizens, and they were sent North. I received some intimations that such a course would prove successful.

Having thus technically upheld his end of the bargain, Hamilton went on to describe the "severe" treatment of prisoners in the South, "food deficient in quantity and quality" and "no letters." Part of his letter, at least, yielded results. Colonel William Hoffman, the United States commissary general of prisoners, wrote General Albin Schoepf, who was in command at Fort Delaware:

> The Secretary of War directs that the twenty-six citizen-prisoners, recently sent from this city [Washington] to Fort Delaware as hostages for a like number of

citizens of Pennsylvania now in confinement in Salisbury, N.C., shall be treated and fed as far as practicable in the same manner that the prisoners are for whom they are hostages.

I enclose herewith a letter received from Mr. James Hamilton, late a prisoner at Salisbury, giving an account of the food and treatment he received while there, and I respectfully request you will make the treatment of the hostages referred to correspond with this in all particulars as far as practicable.[55]

General Ulysses S. Grant was an old hand at the taking of hostages, and practices did not change therefore when he became general-in-chief. On February 16, 1865, Grant wrote General Robert E. Lee about some alleged hostages known to Grant only from a clipping from a Richmond newspaper:

Inclosed I send you communication ... relating to James Monnehause ... and an extract from the Richmond Examiner, dated December 8, 1864, containing statement of the capture of thirty-seven Union citizens and their commitment to Castle Thunder, to be held as hostages for the good treatment and return of Confederate citizens alleged to have been captured by us. Previous to the receipt of the enclosed communication and before any attention was called to the extract from the Richmond Examiner, I directed the release of all persons held by military authority within the Department of Virginia and North Carolina against whom sufficient evidence could not be found to convict them of the offense with which they stand charged, and also such as were imprisoned without proper charges, if any such there were. Similar orders were intended to be given throughout the entire military command of the United States, but before such orders are now given I desire information as to the truth of the statement of the Richmond Examiner ..., and, if true, the names of the persons held by us for whom they were seized and held as hostages, and when and where captured, that their cases may be inquired into and the proper action had as to each. I would respectfully propose the release and exchange of all citizen prisoners now held by military authority, except those under charges of being spies or under conviction for offenses under the laws of war on both sides.

It is, perhaps, a little more surprising to discover Lee's ready acceptance of the practice, but he did respond thus:

I have ... submitted your proposition to release citizens held as prisoners by either party to the Secretary of War. I shall be glad if some arrangement can be made to relieve such persons from unnecessary suffering. I have no knowledge of the facts mentioned in the extract from the newspaper, but will direct inquiry to be made. I gave no order for the arrest of any citizen, and if it be true that those mentioned were taken by any of our forces, I presume they are held as hostages generally for persons of the same class in the custody of the Federal Authorities, and not for particular individuals.

The issue was resolved, apparently, by an agreement already put in place by the commissioners of exchange, but the correspondence reveals Lee's quiet acceptance of the notion of taking hostages "for persons of the same class" held by the North.[56]

Christian soldiers of all nations routinely took hostages. Lee, Grant, President

Lincoln, and many others accepted the practice. Whether the writ of habeas corpus was suspended or not, such civilian prisoners—a by-product of some of the most abrasive friction of war—were going to swell the numbers of "political prisoners" in a war fought on this continent. Though war may have been at least as savage in the New World as in the Old, American generals and statesmen usually sought to live up to the standards of international law.

8

The Irrelevance of the Milligan *Decision*

After General John A. Dix produced his controversial ruling on military commissions while reviewing the cases of blockade-runners, he told General Halleck that he would like to hear Francis Lieber's answer to the key question:
"Can any military court or commission, in a department not under martial law,
take cognizance of, and try a citizen for, any violation of the law of war, such
citizen not being connected in any wise with the military service of the United
States?" The political-science professor's answer differed from the negative
opinion of the Dix commission:

> ... undoubtedly a citizen under these conditions can, or rather must, be tried by
> military courts, because there is no other way to try him and repress the crime
> which may endanger the whole country; it is very difficult to say how far martial
> law extends, in cases of great danger arising out of war; and ... it must never be
> forgotten that the *whole country* is always at war with the enemy; that is to say,
> every citizen is an enemy to the opposing belligerent, and that there is in case of
> war—especially in a free country where no "cabinet wars" are carried on—by no
> means that distinction between soldier and citizen which many people either
> believe to exist or desire—as though the citizen could quietly carry on all possible
> mischief with reference to the army, which is in fact his own army, and with
> reference to the war, which is as much his war as that of the army.

In stressing the difficulty in measuring "how far martial law extends," Lieber
approached the formal constitutional doctrine of the Lincoln administration,
to the degree it had one. The president, when confronted by the argument
that military arrests should not be made "outside of the lines of necessary
military occupation, and the scenes of insurrection," replied in the Corning
letter that they were "constitutional *wherever* the public safety does require
them ... as well where they may restrain mischievous interference with the

raising and supplying of armies, to suppress the rebellion, as where the rebellion may actually be."[1]

This is not to say that Lincoln—or Lieber, for that matter—embraced the idea of warring on whole populations with whole populations. In fact, Lincoln's doctrine applied (as stated in the Corning letter) only to arrests. He did not actually mention trials by military commission. But after Lincoln's habeas-corpus proclamation of September 24, 1862, military arrests and liability to trials by military commission went hand in hand in administraion policy. The president had there proclaimed "that during the existing insurrection and as a necessary measure for suppressing the same, all Rebels and Insurgents, their aiders and abettors within the United States, and all persons discouraging volunteer enlistments, resisting militia drafts, or guilty of any disloyal practice, affording aid and comfort to Rebels against the authority of the United States, shall be subject to martial law and liable to trial and punishment by Courts Martial or Military Commission." Like other early administration documents restricting civil liberties, this one contained dubious legal terminology. Few people would have agreed that a civilian could be tried by a "court martial."[2]

Plenty of people also thought the use of military commissions outside insurrectionary districts or zones of occupation was of dubious legality. In fact, the use of military commissions to try civilians proved in the end to be the part of the internal security system that most damaged the reputation of the Lincoln administration. It supplied the Democrats an issue after Congress's Habeas Corpus Act of March 3, 1863, had answered their early objection that the president had usurped a power of Congress in suspending the writ. The infamous Vallandigham and Milligan cases (the names most often brought up by historians in discussions of civil liberties in the Civil War) involved trials by military commission "outside the lines of necessary military occupation"—in Cincinnati and Indianapolis, to be exact. Moreover, what the U.S. Supreme Court condemned in the policies of the Lincoln administration with the famous *Milligan* decision of 1866 was the use of trials by military commission in such circumstances. The court did not directly challenge the suspension of the writ of habeas corpus or the resulting military arrests of civilians.

Despite their crucial importance to the reputation of the Lincoln administration, trials by military commission have been little studied and less understood. Though extensive treatments of a few individual trials exist, most notably those of Lincoln's assassins and of Milligan and Vallandigham, no overall study of military commissions exists that might provide an institutional and a historical context for these more sensational trials.[3] Heretofore, no one has shown how many trials there were or what kinds of persons were usually the defendants. No one has revealed where most of the trials occurred or when. No one knows whether the famous—or rather, infamous—trials mentioned above were typical or atypical. No one has described courtroom procedures or precedents that dictated the conduct of these commissions.

Evidence on trials by military commission is available and far more manageable, comprehensive, and exhaustive than the fragmentary records of military arrests of civilians. The army required that the transcripts of all general courts-martial be sent to the judge advocate general. His bureau maintained complete registers of all cases of general courts-martial and, because military commissions amounted to little more than courts-martial for civilians, the clerks recorded the trials of civilians in their registers along with the far more numerous courts-martial. They are buried among the records, but the treasure is there if one digs.[4]

Martial Law and Trials by Military Commission

Democrats commonly characterized martial law as no law at all, and their contention was buttressed by impressive historical precedent. The early interpreters of English law had denounced martial law, and even the Duke of Wellington had dismissed it as the mere will of the commander.[5]

Despite such allegations, trials by military commission in the Civil War were marked by procedural regularity. The best proof of this is the existence of the records themselves. In one way, these records were more complete than those for ordinary civil and criminal trials. In Lincoln's day, courts below appellate level did not regularly employ recorders, and trials generated written records consisting only of pleadings, affidavits, formal orders, and the like. Trials by military commission were all recorded. Further proof lies in the rulings of higher authorities disapproving some commissions' verdicts. More than one officer usually reviewed a case before the president, including the judge advocate general, Joseph Holt. Holt was a loyal Kentuckian who retained an unflinching hatred for rebels and traitors, but he was also a punctilious lawyer and military administrator who repeatedly overturned the decisions of trials by military commission (as well as courts-martial) for what can only be called legal technicalities. In 1865, military lawyer William W. Winthrop assembled some of Holt's decisions in a digest which would influence military justice for a long time. In 1866 and 1868, Winthrop produced enlarged editions and in 1880 a lengthy annotated one.

In one of his opinions, Holt showed that the army wished to provide superior safeguards for the liberties of civilians as opposed to soldiers tried by its courts. "As an exception . . . to the rule that military commissions are to be constituted in all respects like courts-martial," the army ruled, "the minimum number of members . . . has been fixed by usage at *three*. To establish a military commission with but two members would be contrary to precedent."[6] Other revealing opinions can be found in the transcripts of individual trials. For example, when Holt reviewed the sentence of Mary Clemmens, ordered beyond the Union lines for attempting to smuggle envelopes, combs, and needles (contraband of war) through Union lines to the Confederacy, he complained:

The Commission is constituted with three members:—but none of these is designated as Judge Advocate, recorder, or otherwise—nor is any officer, not included in the detail, appointed by the order as Judge Advocate.

There is no mention or appearance of a Judge Advocate in any part of the record—but the sentence is subscribed "John J. Jones—Lieut. Col. & recorder."

Further it is stated that the Commission was duly sworn—but does not add "in the presence of the accused,"—Nor does the Record show that the accused had any opportunity of challenge afforded her.—These are particulars, in which it has always been held that the proceedings of a Military Commission should be assimilated to those of a Court Martial. And as these defects would be fatal in the latter case, they must be held to be so in the present instance.

Holt declared Clemmens's sentence "inoperative."[7]

In a decision that affected several men tried by a military commission in New Mexico, Holt held all the sentences inoperative because "Two surgeons were placed upon the original detail of the commission—and both are found to have sat, as members of the Commission, upon the trial of all the cases." The problem with surgeons, he explained later, was that they were not as familiar with military law as other officers, and they might be called away suddenly to perform their medical functions. For those reasons, they had rarely been used on courts-martial, and he reminded the officers that the "rules of law applicable to ... Military commissions should always, as far as possible, be assimilated to those which prevail in the case of Courts Martial." Detailing surgeons to a military commission was a practice best avoided.[8]

Before Holt's appointment as judge advocate general on September 3, 1862, standards may have been lax, particularly since most such trials before that date occurred in loosely controlled Missouri. Even so, departmental commanders who first reviewed the transcripts and approved sentences also insisted on some procedural regularity. Especially when military commissions were held in remote camps, like that of a cavalry brigade in Shannon County, Missouri, the informal circumstances invited procedural transgressions. Robert Ward, the defendant in one such case, was convicted of being a guerrilla in arms and sentenced by a colonel, a lieutenant colonel, and a lieutenant to confinement at hard labor for the duration of the war. General Samuel R. Curtis, then commanding the Department of Missouri, disapproved the findings "on account of informality." The record, Curtis complained, did not show that court and judge advocate had been sworn or that the accused had been asked whether he objected to any members of the court. General Curtis added that the commission could not ignore the defendant's right to have counsel present, and he particularly chided the Missouri cavalrymen for the loose narrative way in which they had drawn up the charge against Ward: "A Military Charge should always contain two distinct parts. The charge declaratory of the offence and the specification which should allege or specify the act, with time, place and circumstance."

In the case of William H. Walthal, tried in Springfield, Missouri, in August

1862 for violation of the laws and customs of war, the guilty finding was disapproved because no evidence "was adduced supporting those portions of the specifications which alone support the charge." Occasionally, commanders chided the courts for giving defendants too much latitude. This was rare, but in the case of Herndon H. Lee, tried in Springfield, Missouri, in August 1862, General Curtis "observed that there has been a wide departure in these proceedings from the rule that requires the prisoner to speak for himself, by the advice of counsel if he desires, but not through his counsel. A disregard of this rule has caused the record to be encumbered by a number of frivolous motions and pleas."[9]

Just as the departmental commander might be overruled by Holt, the judge advocate general could be overruled by the general-in-chief. When Henry W. Halleck occupied that post, he did not hesitate to correct Holt. It is surprising, in light of Halleck's reputation as a stickler for legalisms and a bureaucrat mired in detail, that he sometimes faulted Holt for ignoring substantial justice in the pursuit of procedural correctness. Thus, when in June 1863 General Jeremiah T. Boyle, commanding the District of Kentucky, court-martialed Michael King, a civilian, for violation of Burnside's notorious General Order No. 38 (the same one under which Vallandigham had been arrested), Holt peremptorily invalidated the proceeding. He pointed out to the thickheaded Boyle that a civilian could not be tried by court-martial for using disloyal language in public in violation of that order, though he could still be tried by a military commission. Halleck then advised that the sentence be remitted but the trial itself not be questioned. "To formally set aside the proceedings," he told Holt, "would tend to involve General B. in a civil prosecution." In another case illustrative of Halleck's work with Holt, the general-in-chief let a sentence in a court-martial stand, because the "ends of justice" were "attained," even though Holt thought the specifications in the charges failed to provide venues and were therefore defective.[10]

Department commander, judge advocate general, and general-in-chief could all be overruled by President Lincoln—if he reviewed the case. Death sentences were automatically referred to him, but the corollary to that rule was that other cases were not routinely referred to the president. Moreover, *all* military death sentences were sent to Lincoln, and the numbers of courts-martial greatly outstripped the number of civilians tried by military commission. The papers must have accumulated in great piles in the office. Nevertheless, Lincoln gave the civilian cases attention when he could, and even civilian cases involving lesser sentences reached him by the unsystematic route of personal appeals by wives or mothers or by special pleading from politicians or other persons of influence. When he reviewed a case, the president was usually inclined to mercy.

Such cases as came to him haphazardly, from appeals, Lincoln customarily referred to the judge advocate general for review. Thus, when John F. McCarthy was sentenced to three years' hard labor by a military commission in Baltimore in 1864, Lincoln received an appeal and wrote on the file: "Judge Advocate

General please report on this case." Holt then decided that the charge of assisting desertion had not been sustained by proof and that this five-year veteran of the regular army should be pardoned.[11]

A persistent mother from Lincoln County, Tennessee, saved her son from execution even after Lincoln had once approved his death sentence. Jesse A. Broadway had served as a Confederate soldier in Braxton Bragg's army. He left it and returned home, only, he maintained, to be forced back into arms by Confederates who said they would take him back to the army as a deserter if he did not cooperate. He was thereafter allegedly with a group called Blackwell's guerrillas, who committed various outrages. A military commission in Tullahoma, Tennessee, sentenced Broadway to death in April 1864, and Lincoln approved the sentence on July 8, a day on which Lincoln apparently reviewed thirty-five cases. Then Broadway's mother made a second appeal. She said her son was ignorant and easily influenced and that he had no counsel at his trial. Holt told the president that she was wrong about her son's lack of counsel, but the judge advocate general noted that a witness who originally testified at the trial now maintained that Broadway could not have been with the guerrillas at the time of the outrages. It seemed strange that the witness had not bothered to say so in court, but Holt was willing to leave the decision up to the president without any recommendation of his own. Lincoln commuted the sentence to three years' hard labor.[12]

The president inclined toward mercy but he was not a pushover, as the case of Jourdan Moseley shows. Moseley, accused of being a guerrilla in Lincoln County, Tennessee, and of robbing and murdering a citizen, was sentenced to hang by a military commission in Tullahoma, Tennessee, in 1864. On July 8, Lincoln approved the sentence. Moseley's mother mustered an appeal, forwarded by Andrew Johnson (without comment), but it amounted to little more than the claim that Moseley was young and thought he was under orders. Holt reviewed the case again but recommended execution, citing an anecdote (not used at the trial) that depicted Moseley robbing the victim while he was dying in relatives' arms. Lincoln's secretary, John G. Nicolay, then wrote on September 7, 1864, "Upon the within report of the Judge Advocate General the President declines to interfere with the due execution of the sentence in this case."[13]

In partially completed ledgers found in the Executive Mansion labeled "U.S. Army Court Martial Cases" and listing documents sent to Lincoln from 1863 to 1865, there are records of 210 civilians sentenced by military courts.[14] One hundred eighty-four of them note the action taken by Lincoln, and they break down this way:

	Approves Punishment	Mitigates Punishment
On recommendation of judge advocate general	39	33
Despite recommendation of judge advocate general	5	14
On general's recommendation		14

	Approves Punishment	Mitigates Punishment
Despite general's recommendation	3	2
On strong judge advocate general recommendation	20	3
Despite strong judge advocate general recommendation		8
No recommendation	23	20

More often than not (95 of 184 cases), Lincoln followed the recommendation of the judge advocate general, and this statistic reveals Joseph Holt's power and Lincoln's trust in his ability. But when Lincoln defied Holt's advice, it was most often to indulge mercy. Lincoln followed a general's recommendation in fourteen cases to mitigate punishment in defiance of the judge advocate general's view that the general should be ignored.

The system of trials by military commission definitely stood a step above no law at all, and it embodied mercy as well as military justice. The president did have some systematic effect here, because certain kinds of cases were always referred to him, because good records were kept of these trials, and because their quantity was limited enough to allow Lincoln to examine a substantial percentage of the cases. This is unlike the situation for the masses of prisoners of state who were not tried by military commission.

Although they provided a brand of justice removed some degrees from no law at all, trials by military commission were, nevertheless, an embodiment of an extremely tough policy, suitable, if at all, only for wartime. Occasionally, they furthered some policy goals of the Lincoln administration other than subduing rebellious Southern whites.

Consider the case of West Bogan, a black man tried for murder by a military commission in Arkansas early in 1864. On December 15, 1863, he had killed Monroe Bogan, a citizen of Phillips County, Arkansas, with an axe. West Bogan did not deny he had killed the man, and the military commission sentenced him to be hanged. The mitigating circumstances were such, however, that General Frederick Steele, commanding Union forces in Arkansas, recommended a lighter penalty. West Bogan had killed his master, not in the big house but near the slave quarters, not by night but in broad daylight, and not by calculated stealth but immediately after unanticipated provocation. Monroe Bogan had a reputation for cruelty and for overworking his slaves, and he had blocked West Bogan's path while the black man was on an errand that required the use of an axe. When he physically assaulted the black man for refusing to obey his orders, West fought back and killed Monroe.

Joseph Holt also recommended mitigation:

The Administration of the government must and does recognize the colored population of the rebellious states, as occupying the *status* of freedmen. This office, in considering the present and kindred cases necessarily accepts this recognition with all its legitimate consequences.

It is therefore held that Monroe Bogan, when he met his death, was in violation of law and right holding the prisoner in absolute slavery—not only holding in slavery, but also imposing upon him ceaseless toil and cruel punishments.

Holt told the president that this was grounds for mitigation, but he left the degree up to Lincoln. Where mercy and antislavery conviction coincided, there could be no doubting Lincoln's course: "Sentence disapproved," he wrote cryptically on July 8, and West Bogan was freed.[15]

Such an episode reveals how revolutionary Lincoln's legal revolution really was. Despite the numbingly legalistic style of the Emancipation Proclamation, its practical workings, when enforced by the Republicans in Washington who formulated the policy, were revolutionary. Holt's interpretation, if applied to "kindred cases," put the administration perilously near the invitation to slave insurrection that English conservatives and alarmed Confederates had seen in the Emancipation Proclamation when it was first announced in 1862. But such incidents were hardly common and, even when they did occur, might be dealt with by Union soldiers less well informed or motivated than Holt and the commander-in-chief. The intention of the policy is nevertheless clear.

A Brief History of Trials by Military Commission

Some civilians had been tried by military courts in the United States long before the Civil War, and the practice would continue long afterward. Some civilian employees of the army were held to be triable by military law. These were typically in the Quarter Masters Department and many were citizen teamsters. Other persons were specially identified in the registers of courts-martial who may have held a more complex relation with the military service than ordinary enlisted men or officers—hospital stewards, farriers, artificers, contract surgeons, wagon masters, laborers, telegraph operators, detectives, boatmen, contract nurses, inspectors, blacksmiths, and paymasters' clerks. Hundreds of citizens who were employed by the military and hundreds of persons in the borderline occupations were tried by courts-martial or military commissions during the Civil War. American Indians were occasionally tried by military commissions during and after the war.[16]

A broadening of the categories of persons subject to military justice occurred during the Civil War, though the principal difference in that period as opposed to others lay simply in the far greater numbers of such cases that came up in the enormous armies of this conflict. For example, the sixtieth article of war stated that "All sutlers or retainers to the camp, and all persons whatsoever, serving with the armies of the United States in the field, though not enlisted soldiers, are to be subject to orders according to the rules and discipline of war." In March 1865, the judge advocate general held that

> To restrict the term—*"serving with the armies of the United States in the field"*—to those persons only who may be employed with an army when immediately operating against the enemy, would be a construction not in accordance with the

spirit of our military law, and not in keeping with the necessities of our military establishment.... it is deemed not too much to hold that *the entire army, as at present mobilized and actively employed for the prosecution of a civil war and for the suppression of a vast intestine rebellion, is an army in the field*; and that all persons engaged with it, whether in the camp or at a station, upon services made necessary or desirable by the wants and circumstances of the military body, are triable by a court-martial within the provisions of this article.[17]

In this instance, Holt's ruling upheld the trial of an acting assistant surgeon assigned to duty in the prisoner-of-war camp at Elmira, New York.

The most notable expansion of categories of civilians triable by military law—unrelated to the administration's declarations of martial law or suspensions of the writ—came in the category of military contractors. The congressional act of July 17, 1862, made some army contractors triable by courts-martial. Congress went further in 1863 and made defaulting contractors a part of the army, subject to the articles of war. For its part, however, the United States Congress proved unable to stomach such a development for long. Congress's act of July 28, 1866, fixing the peace establishment after the end of the Civil War, effectively repealed the portion of the 1862 law which made contractors a part of the army.[18]

For a four-year period, then, Congress indulged the military establishment's view that it must be able to deal with its direct suppliers by the methods of military discipline and justice. Thus the trials of contractors listed in the judge advocate general's registers of courts-martial were technically courts-martial and not trials by military commission. Yet this redefinition of the contractors' status was brief, and they maintained civilian status in some ways throughout the period. Therefore, the statistics here count the trials of contractors as trials by military commission rather than courts-martial. Their number was not large, anyhow, and they were specially identified in the registers as contractors.

During the Civil War, the army conducted at least 4,271 trials by military commission.[19] Their geographical distribution was reminiscent of that of the more abundant military arrests of civilians. More than half (55.5 percent) of the trials by military commission occurred in the strife-torn border states of Missouri, Kentucky, and Maryland. By far the largest number from any single state occurred in Missouri, 1,940 of the 4,203 trials for which location is known, or a staggering 46.2 percent.

As Chapter 2 revealed, Missouri was the unhappy birthplace of trials by military commission in the United States. The size of the problem posed by Missouri is perhaps suggested by the astonishing fact that trials by military commission in Missouri exceeded in number the total number of trials in all the occupied Confederacy during the Civil War (1,940 in Missouri; only 1,339 in the eleven Confederate states). The trial records, which are more systematic, clinch what the fragmentary arrest records for Missouri suggested: an appalling amount of civil disorder met with appallingly Draconian government force. When compared with the records of other slaveholding border states, the contrast is sharper yet. There were 193 trials by military commission in Maryland during the Civil War and 200 in Kentucky. Missouri saw over nine times as many trials as either of those border states. And Delaware, the only border state to escape guerrilla warfare and military invasion, was the site of no such trials at all. The incidence of military commissions was a function of civil disorder.

Outside the sometimes intractable border states, Confederate territory occupied by Union forces during the Civil War accounted for another 31.9 percent of the trials by military commission. The greatest number of these occurred in Tennessee, 692 (over half the total number from the occupied Confederacy and 16.5 percent of the total number of trials by military commission during the Civil War).

The geographical location of the trials provides a clue to the nature of the defendants. In Missouri and Tennessee, the military commissions mostly tried men accused of guerrilla activities, horse stealing, and bridge-burning. Often, the only real question was not whether the defendant engaged in such actions but whether he was under orders in a regularly organized military unit when he did so. In this regard, not much had changed from the earliest such trials in 1861 (discussed in Chapter 2).

A typical case was that of Samuel Bryant, of Newton County, Missouri, tried in Springfield, Missouri, in the autumn of 1862, two weeks after his capture by federal troops. With a lieutenant colonel acting as judge advocate to conduct the trial, two majors and a captain heard testimony that Bryant had violated the laws of war by belonging to a guerrilla band in Newton County, that he had aided the band to rob a man named Samuel Long and other loyal citizens in March 1862, and that he had shot at a loyal citizen. When the commission asked for a statement from the defendant after the testimony, he said only, "I am a soldier in the Southern army." Connections with the Confederacy were always tenuous at best, and often such avowals were little more than fantasy on the partisan's part. The military commission found him guilty and sentenced him to hard labor for the duration of the war.[20]

Despite the way in which Bryant's and many similar guerrilla cases appear to hinge on the military nature of the groups to which the defendants belonged, testimony at the trials almost never focused on this. Accusations of belonging to guerrilla groups and denials that such groups were actually regularly organized Confederate military units were rarely supported by witnesses; witnesses usually testified only about specific acts of violence or theft.

Perhaps because of the often unverifiable nature of charges involving guerrilla activities, military commissions sometimes relied on charges that the defendant had violated a previously taken oath of allegiance by aiding a guerrilla band or joining the Confederate army for a time. Despite the impressive legal precision at the top in Washington, out in the field in Missouri, justice could prove rough, and cases involving oaths of allegiance reveal this glaringly. These oaths were usually administered by someone from the provost marshal's office, and one copy went to the oath-taker, one to the department headquarters, and one copy was supposed to be filed with the provost marshal or commanding officer. Yet they were almost never produced as evidence at the trials, and the reason is clear if one reads the charges, specifications, and testimony: the army obviously could not find them. Therefore, specifications often proved vague about the date when or the place where the oath was taken. Sometimes, the army specified only the month in which it was allegedly taken. This was again a function of the army's sloppy record-keeping. The main reason to have the oaths taken was to be able to punish the oath-takers if they later engaged in disloyal activities, but this would be legally compelling only if the army produced the oaths they had earlier affirmed. More often than not, the U.S. Army in Missouri could not produce the documents.

In the cases sampled from Missouri in which violation of the oath of allegiance was at issue, only one had a copy of the oath attached to the trial record.[21] Most military commissions could state only the month, year, and county in which the oath was taken. A minority cited a specific day and place. In the trial of Saunders M. Utterbach in St. Louis in May 1863, the army could charge only that he took the oath "about the month of March 1862."[22] Some cases were complicated. Aaron Dean, for example, was tried in Rolla in the spring of 1863 for violation of his oath of allegiance, which he took on March 26, 1862. On August 20, he joined "Pickett's Rebel Regiment." The defendant's explanation was that the oath was prescribed by general order in Phelps County, and he took it, but only orally. When an order required Missouri men to enroll in the (loyal) militia, Dean left Phelps County to avoid service. Like many Missourians, he preferred not to fight at all but, if forced, proved especially reluctant to fight *against* the South. In Shannon County, he was induced to join a Confederate regiment. This was against his convictions, he maintained, but the Confederates would have conscripted him anyway had he not joined. Despite Dean's coming voluntarily into federal lines after his discharge from Confederate service by reason of illness, the military commission sentenced him to a year's hard labor at Alton.[23]

In this, as in many other instances in the Civil War, one can see the way conscription worked to make war harder on noncombatants. Many Americans would not have chosen any action, for or against the Union, had it not been for the threat of enrollment and eventual conscription.

The decisive measure in Missouri came on July 22, 1862, when General John M. Schofield, commander of the Department of Missouri, ordered a complete enrollment of the Missouri State Militia. Thus, every white male in Missouri between the ages of eighteen and forty-five had to take the oath of allegiance and enroll in the militia or refuse the oath and be listed as disloyal for purposes of assessment. The latter would mean being assessed, along with all other "rebels and rebel sympathizers" in his locality, to pay for any damages caused by guerrillas.[24]

Another difficult case involving the oath of allegiance was that of Francis M. Armstrong, tried in Jefferson City, Missouri, in the summer of 1863. The government could say only that he took the oath in 1861 or 1862 before serving, after August 10, 1862, in a "rebel regiment" called "Caldwell's Missouri Infantry." The army also charged that Armstrong was a "military insurgent" because he came within Union lines as a Confederate soldier to recruit for the Confederate army. Armstrong pleaded guilty to all but coming within the lines to recruit. He was sentenced to be shot. On November 20, 1863, Joseph Holt urged that the sentence be carried out:

> The offenses committed by the accused are such as have been frequently committed by perfidious men in the border states, and such as have in many cases been adjudged by Military Commissions as fully meriting the enforcement of the death penalty. There are no reasons shown in this case why the judgment... sanctioned as it is by usage, which has obtained, where men have deliberately violated their solemn oath of fidelity to their Government... should not be approved.
>
> There is no doubt that in many instances the oath of allegiance is taken with a premeditated purpose to make it a cover....
>
> He may be considered a Messenger or agent, who attempted to steal into territory occupied by our Army, with a purpose to further the interests of the enemy. Professor Lieber in his code... expressly lays it down, that such offenders, if captured are not entitled to the privileges of prisoners of war, but may be dealt with according to the circumstances of the case.

Moreover, Holt added, General Order No. 30 of the Department of Missouri prescribed the death penalty for such an offense. On February 9, 1864, President Lincoln commuted the sentence to imprisonment during the war.[25]

Although the justice dispensed by military commissions in Missouri may have been rough, so were the defendants. In a sample of fifty Missouri trials in which the defendants were charged with serious crimes of guerrilla activity or joining the Confederate army, nineteen (or 38 percent) admitted their guilt in some way. Four simply entered guilty pleas in the trial.[26] Three others pleaded guilty to part of the accusations against them but not to all.[27] Thomas G. Tuttle, accused of violating his oath of allegiance by harboring a Confederate captain, pleaded guilty before a military commission in St. Louis, but added that he did not know that the man was a Confederate officer. The court found Tuttle guilty and ordered him into the Confederate lines. Other defendants in the fifty sample cases admitted the truth of some of the charges against them in their final statements or in previous statements given in interrogation by provost marshals before the trials.[28]

Of the remaining thirty-one Missouri cases sampled, two resulted in acquittals, a mere 4 percent rate and down from that found earlier in the war.[29] The principal problem in Missouri remained always the widespread guerrilla warfare. This forced awful choices on Union authorities as well as on the individual citizens of the state. For example, Robert T. Johns of Greene County was tried by a military commission in the county seat, Springfield, in the fall of 1862 for violating the laws of war by letting rebels lurk in his neighborhood without reporting them to the U.S. military authorities. He pleaded guilty but explained that one of the men lurking in the high brush was his brother-in-law, on whom he did not want to inform. He had discovered the men hiding in the brush only by accident and did not wish either to harbor them or to inform the authorities about them. But neutrality in areas involved in guerrilla warfare often proved impossible. The commission sentenced Johns to six months at hard labor.[30]

There were lesser cases, to be sure. Although the occupying army seems generally to have learned the lesson that it was impractical to try Missourians merely for uttering sentiments against the war effort, still, once in a while military commissions had such cases come before them. Among the seventy-six Missouri case records examined from trials occurring after January 1862, only two involved free speech. On June 24, 1864, James Sullivan said in the presence of two witnesses in Rolla: "I am a Jeff Davis man." He was tried in Rolla and convicted of "disloyal conduct." He was sentenced to pay a $50 fine and to be confined in a government fort for a month, but the army remitted the prison sentence. Elbert B. Rankin was tried in Rolla in April 1863 for saying, "I am a Rebel and am one of General Price's . . . Men," and for taking Confederate money in payment for a mule. His guilty plea earned him one year at hard labor.[31]

Among the difficult cases involving less-serious offenses was that of James Jackson of Audrain County, Missouri, tried in May 1863 for violating his oath of allegiance. He had taken the oath several times, but then he sent a letter to his son, T. R. Jackson, a soldier in the Confederate army. The elder Jackson entered a guilty plea but said he did not know he was violating the law when he wrote his son. Because of his "advanced age, ignorance of Military Law in such cases, and his previous good character," the major general who reviewed the case required Jackson only to post $2000 bond for good behavior.[32]

In trials in Missouri involving less serious charges than engaging in guerrilla activities, acquittals may have been more frequent. Eight are found in this small sample, more than for the more numerous cases of serious crimes.[33]

The only other area outside the Confederacy and the border states with a sizeable frequency of military commissions was Washington, D.C., where 271 trials (or 6.4 percent of the Civil War total) were held. These were a function both of the District of Columbia's peculiar situation as a border area itself (accounting for the number of Virginians and Marylanders tried in the District for aiding the Confederacy) and also of the heavy concentration of military attention and personnel in the District throughout the war. Washington became the focal point for the use of military trials to ferret out corruption in the businesses which supplied the military, the result of extending military law to cover government contractors and other perpetrators of fraud on the swollen military establishment under Lincoln.

Another revealing feature of the distribution of trials by military commission was the early and persistent occurrence of these trials in the western territories. Despite sparse population and an oblique relationship to questions of slavery, states rights, and national unity, territories like New Mexico and Arizona saw the steady application of trials by military commission from late 1861 on. Though their number is not statistically significant, the fact that this region had any such trials at all makes it even more obvious that traditionally lawless and violent areas far from federal control were likely to see the application of military justice and trials by military commission. Altogether, seventy-two trials by military commission took place in the territories (thirty-seven in New Mexico, twenty-one in Arizona, and fourteen in the others). They began long before the writ of habeas corpus was suspended in these areas and seem rarely to have been used for crimes related to the war. Rather, the soldiers used them to police their posts, prosecuting thieves, receivers of stolen government property, or persons who sold liquor to soldiers.[34]

Such was not the use Lincoln or anyone else in Washington had in mind when authorizing suspensions of ordinary civil liberties. Associate Justice Joseph G. Knapp of New Mexico made a point of exposing the abuse of this system in the western territories. He protested so vigorously to Attorney General Bates that General Halleck rebuked the military in New Mexico:

> Judge Knapp...in a communication to the Attorney General, has complained among other things that under your authority military commissions in your department have taken cognizance of and adjudicated upon actions of debt, trespass, &c., between persons not in the military service. I am directed by the Secretary of War to say that military commissions and military courts in your department have no jurisdiction of such cases, and that their decisions are entirely null and void. Moreover, the individual members may thus render themselves liable to punishment and damages. The practice, if it exists, should be immediately discontinued.

The conflict between Knapp and Brigadier General James H. Carleton, commanding in Santa Fe, led to a minor constitutional crisis when Knapp refused to hear cases at a session of the supreme court in protest of the army's preventing him from traveling out of the territory without a military pass. Halleck's letter settled the crisis, but Knapp's justifiable agitation apparently provoked Attorney General Bates to say: "There seems to be a general and growing disposition of the military, wherever stationed, to engross all power, and to treat the civil government with contumely, as if the object were to bring it into contempt."[35]

Five percent of the trials by military commission occurred in states of the North above the border states and the District of Columbia. There were 212 such trials in the forty-three months from their first institution in September 1861 through April 1865—less than five trials per month on average north of the border states and Washington.

Some Northern states never saw a single such trial. Of the New England states, only Massachusetts was the site of any military commissions, and that was because Fort Warren was located in Boston harbor. None was held in Michigan or the border state of Delaware. Only one occurred in Wisconsin.

Many of the trials on Northern soil, in fact, had border-state citizens or Confederates as defendants. This was true of most of the military commissions held in Ohio. They usually tried Kentuckians caught rendezvousing to head south to join the Confederate army. A majority of the trials by military commission in Pennsylvania were caused by two outbreaks of violent resistance to conscription, one among farmers in the Democratic stronghold of Columbia County and the other in the coal-mining regions. Of the sixty-four civilians tried by military commission in Pennsylvania, twenty-three were tried in Harrisburg in October 1864 as a result of the alleged "Fishingcreek Confederacy" conspiracy. This wartime myth had its origins in the murder of an assistant provost marshal who was hunting for draft dodgers in notoriously slack Columbia County. Afterward, stories circulated that disloyal farmers and deserters in the hundreds had constructed a fort in a remote area of the Fishingcreek Valley. Nervous and gullible Union authorities believed them and sent for help. Eventually, federal soldiers arrested forty-four citizens and sent them to Fort Mifflin; twenty-three of these were tried for entering combinations against the execution of government acts, urging others to evade the draft, joining armed squads, and being involved with the Knights of the Golden Circle. Though resistance to the draft was high in the area, there was no such well-organized and militarily potent conspiracy. Fifteen men were acquitted altogether. Only three of the guilty men received prison sentences in addition to fines; the other five were assessed fines only.[36]

Another thirteen or more trials in Pennsylvania stemmed from draft resistance, mainly by Irish-Americans, in the coal-mining regions in Luzerne County. They were accused of interfering with the execution of congressional acts, of cooperation with or membership in the Knights of the Golden Circle, draft resistance, telling coal companies to cease operations until the draft was suspended, threatening provost marshals, calling the conflict a "Nigger War," and inciting others to resist the draft in October 1863 by calling conscription a law for the rich against the poor.[37]

President Lincoln's ultimate defense of such trials, as for all the other restrictions of accustomed civil liberties during the Civil War, was that they were temporary and would end when the war did. In truth, he did not readily admit that such trials even existed. Defending the Vallandigham arrest from protests by the Ohio State Democratic convention in June 1863, President Lincoln said:

> The military arrests and detentions, which have been made, including those of Mr. V. which are not different in principle from the others, have been for *prevention,* and not for *punishment*—as injunctions to stay injury, as proceedings to keep the peace—and hence, like proceedings in such cases, and for like reasons, they have not been accompanied with indictments, or trials by juries, nor, in a single case by any punishment whatever, beyond what is purely incidental to the prevention. The original sentence of imprisonment in Mr. V.'s case, was to prevent injury to the Military service only, and the modification of it was made as a less disagreeable mode to him, of securing the same prevention.

Such a statement at least stretched the truth in the case of Vallandigham and was altogether untrue in respect to many others. Since the president repeatedly reviewed the results of trials by military commission in his White House office, the statement did not stem from ignorance either. Sentences to hard labor or prison terms fixed by years (and not the duration of the conflict) were punishments, pure and simple. Lincoln did not want to admit that the alternative military-justice system for some civilians had been set up. He must have hoped its disappearance at war's end would erase the military trials of civilians from national memory.[38] But trials by military commission did not end with the war, and Americans never forgot them. The trials continued after 1865 only in the occupied South under Reconstruction. Lincoln was at least half right in his faith: the American people in the North did not lose the right to trial by jury throughout the indefinite peaceful future. But what happened in the South? What was the postwar history of trials by military commission?

The Irrelevance of the *Milligan* Decision

The history of trials by military commission during Reconstruction featured a landmark Supreme Court decision, *Ex parte Milligan*. No treatment of the subject of trials by military commission during or after the Civil War is possible without consideration of this decision, the most famous Supreme Court decision of the era. The standard history of the U.S. Supreme Court, written by Charles Warren and published in three volumes in 1923, devoted a whole chapter to "The *Milligan* Case," describing it as "one of the bulwarks of American liberty" and "the palladium of the rights of the individual."[39]

Thus launched to fame by Charles Warren in the 1920s, *Ex parte Milligan* has become one of the most widely anthologized decisions of the United States Supreme Court. John Garraty's *Quarrels That Have Shaped the Constitution* included an essay by Allan Nevins describing the *Milligan* decision as "a great triumph for the civil liberties of Americans in time of war" and concluding with more extravagant praise even than Warren's:

> The line was now emphatically delineated. The Supreme Court established the rule that, no matter how grave the emergency, and no matter how public the excitement, the civil authority is supreme over military authority; that wherever such civil authority is established and its ordinary judicial procedures are operating, its protections of the citizen shall remain absolute and unquestionable. The heart of this decision is the heart of the difference between the United States of America and Nazi Germany or the Soviet Union.[40]

In Michael Belknap's *American Political Trials*, published in 1981, Frank Klement contributed an essay on *Ex parte Milligan* describing the decision as "a notable victory for civil rights" that "has stood the test of time."[41] More recently, historian Emma Lou Thornbrough, in yet another group of papers on important decisions of the U.S. Supreme Court, stated that *Ex parte Milligan* has been "long regarded as a landmark in the history of civil liberties."[42]

The decision was not always so esteemed. At the time of its issuance, it was regarded as a blow to Republican attempts to reconstruct the South, and was therefore denounced by Republicans and cheered only by Democrats and white Southerners. The Republican *Harper's Weekly*, for example, likened it to the *Dred Scott* decision: "It is not a judicial opinion; it is a political act.... The *Dred Scott* decision was meant to deprive slaves taken into a Territory of the chances of liberty under the United States Constitution. The Indiana decision operates to deprive the freedman, in the late rebel States whose laws grievously outrage them, of the protection of the freedmen's Courts." In Virginia, on the other hand, the pleasantly surprised Richmond *Enquirer* said, "It has inspired us with new hope for the future of our institutions.... If the authority of the Constitution shall be vindicated, the South is safe and the end of her troubles approaches."[43] In short, *Ex parte Milligan* at first had an entirely partisan reputation.

This constitutes a cruel irony, for the decision was anything but partisan. In fact, its high status in modern times is, in some considerable degree, a function of the nonpartisan origins of the decision. This, after all, was a Supreme Court to which Abraham Lincoln had appointed five members. The majority opinion itself was written by David Davis, Lincoln's old Eighth Judicial Circuit friend and presidential campaign manager in 1860.

Despite its origins, the *Milligan* decision had little practical effect. It was written in thunderously quotable language (one key to its modern prominence as a landmark case, perhaps), and the ruling was clear enough as far as it went, though it by no means cleared up all the complex civil liberties issues caused by the Civil War. The Supreme Court declared martial law and military trials of civilians illegal in the United States wherever local courts were open and functioning. "Martial Law," the court ruled, "cannot arise from a *threatened* invasion. The necessity must be actual and present; the invasion real, such as effectually closes the courts and deposes the civil administration.... Martial rule can never exist where courts are open, and in the proper and unobstructed exercise of their jurisdiction. It is ... confined to the locality of actual war." With this, the court freed Lambdin P. Milligan, a Democratic politician. He had been arrested in 1864 in sickbed in his home in Huntington, Indiana, for involvement with the Sons of Liberty, a disloyal organization. He was tried in Indianapolis by a military commission consisting of eleven army colonels and one general and was sentenced to hang. By 1866, Milligan's case reached the U.S. Supreme Court, which scolded the Lincoln administration in no uncertain terms when it declared Milligan's military trial illegal:

> The Constitution of the United States is a law for rulers and people, equally in war and in peace, and covers with the shield of its protection all classes of men, at all times, and under all circumstances. No doctrine involving more pernicious consequences, was ever invented by the wit of man, than that any of its provisions can be suspended during any of the great exigencies of government. Such doctrine leads directly to ... despotism.[44]

Despite unmistakable condemnation, trials by military commission continued. From the end of April 1865 to January 1, 1869, another 1,435 such trials occurred—and still more in 1869 and 1870. The use of military commissions, naturally, was declining throughout the postwar period. From May 1, 1865, to the end of the year, 921 trials occurred. The next year, the number fell to 229. There were 181 in 1867 and merely 104 in 1868. In 1869 and 1870, they occurred only in Texas and Mississippi.

The many trials that occurred after April 1865 dealt with old issues of war as well as new ones of Reconstruction. The trial of Lincoln's assassins was typical of these. It dealt with an illegal killing of a Union leader which occurred before all the Confederate armies had surrendered. Many of the other trials dealt in peacetime with events of war.[45]

A series of trials by military commission in Louisville provide vivid examples of such retrospective trials. Robert P. Johnson, W. K. Bruce, Ira Stewart, and William Bruce were tried for being guerrillas in Kentucky's Webster and Union counties in April 1865. William Bruce was acquitted but the rest were sentenced to three years' hard labor in prison in Frankfort. These activities had occurred at what had turned out to be the last moments of the Civil War. Other military commissions tried defendants accused of crimes that occurred well before war's end; presumably, the defendants were men only recently apprehended by federal authorities. Thus, Claude V. Higgins drew a sentence of ten years at hard labor from a military commission in Louisville after the war. The commission heard testimony on his guerrilla activities in Jefferson County, Kentucky, which allegedly occurred in August 1864.[46]

Other cases were bound to arise as long as the army attempted to maintain order in the conquered South. These trials often did not involve any sharply defined political or racial issues that would now be associated by historians with Reconstruction. A postwar military commission in Louisville sentenced George Albert to a year's confinement at hard labor for receiving stolen United States property—medical supplies such as quinine and opium. He had received them in May 1865 before being mustered out as a private from the 77th Volunteer Reserve Corps, but he was tried after having reentered civilian life.[47] There was also the case of Thomas D. Barney, tried in May 1865 in Martinsburg, West Virginia, for having aided soldiers to desert on March 12.[48] The perennial problem of liquor sales to soldiers also continued after the war. In Richmond, Virginia, in 1865, for example, a local saloon-keeper was convicted of perjury for having lied in his testimony before a military commission that was trying his boss for selling liquor to soldiers. For his perjury, William Heitmuller was given a sentence of one year in jail, a stiff $500 fine, and the further penalty of serving time in prison until the fine was paid.[49] As late as 1867, people were being tried in the South for selling liquor to the occupying troops.[50] As in wartime, military commissions after April 1865 were also used to halt corruption, now mainly involving treasury agents defrauding the government, which employed them to acquire cotton that was formerly the property of the Confederate government.[51]

Occasionally the trials by military commission in occupied territory after the Civil War exemplified the original intent of Winfield Scott in establishing such institutions in military law: they restrained an undisciplined soldiery occupying pacified territory. For example, Private Michael Shehan, of the 8th Maine Volunteer Infantry, was tried in Richmond, Virginia, for assault with intent to kill and rape. All that was established to the satisfaction of the military commission was that on August 3, 1865, Private Shehan threw Maria Wade of Richmond into the Kanawha Canal. He was subsequently sentenced to nine months' confinement at hard labor.[52] Sometimes the military commissions tried Union soldiers for crimes committed against other Union soldiers, as in the case of Private Thomas Jones of the 111th Pennsylvania Volunteer Infantry, convicted in Louisville of having robbed a hospital steward of $60 on July 17, 1865.[53]

Strife between the races in the occupied South occasionally amounted to nothing politically more interesting than an ordinary brawl. Two privates of a Massachusetts infantry regiment were tried by a military commission in Richmond for an August 10, 1865, assault on Christopher Blunt, a black man. The disagreement occurred in Blunt's house, where the two soldiers rented rooms from Blunt's wife. The soldiers were acquitted.[54]

Such trials by military commission, and others aimed at bringing better justice to freedmen than they would get at the hands of Southern whites, continued apace, until April 2, 1866, when President Andrew Johnson proclaimed war at an end. This was followed closely by the preliminary announcement of the *Milligan* decision. The Supreme Court chose to announce its decision at the end of its term in April but to leave the publication of the reasoning behind the decision until December. This must also have been unsettling to the army, even before the publication of the full opinion in December made it clear that the army was employing illegal trials. It was not clear whether Johnson's April proclamation restored the writ of habeas corpus and ended trials by military commission in the South or whether it did not do so. Eventually, the army, with Johnson's approval, said that such trials should not be used "where justice can be attained through the medium of civil authority"—not the same thing as saying, as the Supreme Court had, that trials by military commission could not be held where civil courts were functioning again.[55]

There were twenty-four military commission trials in May 1866; then the numbers fell precipitously. In the period from June 1866 through April 1867, only twenty-eight trials by military commission occurred, less than three per month. On July 17, 1866, the War Department had issued orders to use the trials for certain kinds of cases alone.[56]

On March 2, 1867, Congress passed the first Reconstruction Act, which among other things empowered commanders of Southern military districts to use military commissions to prosecute offenders instead of local civil courts if necessary to protect persons and property, suppress insurrection, and punish criminals. Military commissions increased in frequency in May, the number that month (thirty-five) exceeding the total for the previous eleven months added together (twenty-eight). They continued, at a slower rate, until the Southern states were all readmitted to the Union.[57] Then they ceased.

The *Milligan* Decision from Reconstruction to Modern Times

The *Milligan* decision fell so quickly from prominence in the nineteenth century that the entry on "military commissions" in the prestigious *Cyclopaedia of Political Science,* published in 1881, gave the case little mention. Instead, John W. Clampitt, who wrote the article, justified the use of military commissions in many situations seemingly denied by the Supreme Court seventeen years earlier:

> When war prevails in a portion of country occupied or threatened by an enemy, whether within or without the territory of the United States, crimes and military offenses are often committed which can not by the rules of war be tried or punished by courts martial, and which at the same time are not within the jurisdiction of any existing civil court. The good of society demands that such cases be tried and punished by the military power, by referring them to a duly constituted military tribunal composed of reliable officers, who, acting under the solemnity of an oath and the responsibility attached to a court of record, examine witnesses, pass upon the guilt or innocence of the arraigned parties, and determine the degree of punishment to be inflicted for the violation of law.

Thus, Clampitt justified the use of military commissions in areas "occupied or threatened by an enemy" in direct defiance of the explicit language of *Ex parte Milligan,* which decreed that martial law could not arise from a merely "threatened" invasion. He went on to endorse their use in other ways forbidden by the Supreme Court: "It is . . . held, that many offenses which in time of peace are civil offenses become in time of war military offenses, and must be tried by a military tribunal even in places where civil tribunals exist." This included some capital crimes and thus conflicted with the Constitution's requirement of jury trial for all capital crimes. Relying on the standard texts employed by the judge advocate general of the army since the war, Clampitt said:

> . . . while the letter of the article [in the U.S. Constitution] would give force to such a declaration, yet in construing the different parts of the constitution together, such interpretation must give way before the necessity for an efficient exercise of the *war power* which is vested in congress by that instrument. . . . this principle has been recognized by the legislation of the country since an early period in its history, by the adoption of the fifty-seventh article of war, in the fact that it has from the beginning rendered amenable to trial by courts martial, for certain offenses, not only military persons, but all persons whatsoever. This article was first adopted by the congress of the confederation, and remained unchanged at the formation of the constitution.

Clampitt felt it necessary to devote plenty of room in his article to the army's persistent view that some offenses were triable by military commissions "even in places where civil tribunals exist" despite "the letter" of the Constitution—and again in defiance of the language of *Ex parte Milligan.*[58]

Clampitt enumerated the uses of military commissions in the Civil War accurately, pointing to cases of "unauthorized correspondence with the enemy; blockade running; mail carrying across the lines; drawing a bill of exchange upon an enemy; dealing in confederate securities or money; manufacturing arms, etc., for the enemy; furnishing articles contraband of war to the enemy; publicly expressing hostility to the government of the United States or sympathy with the enemy; entering the federal lines from the enemy without authority; violating a flag of truce; violating an oath of amnesty or of allegiance to the government; aiding prisoners of war to escape; unwarranted treatment of federal prisoners of war; burning and destroying bridges, railroads, steamboats, and cutting telegraph wires used in military operations; recruiting for the enemy within the federal lines; engaging in *guerrilla* warfare; assisting federal soldiers to desert; resisting or obstructing an enrolment or draft; impeding enlistments; [and] conspiracy by two or more to violate the laws of war by destroying life or property in aid of the enemy." He noted as well certain "ordinary crimes" assumed to be within the jurisdiction of military commissions during the Civil War: "attempts to defraud the United States, misappropriations of public money and property and embezzlement of the same, bribery of and attempts to bribe United States officers, breach of the peace, rape, arson, receiving stolen property, burglary, riot, larceny, assault and battery with intent to kill, robbery, homicide, and the crime known as 'murder in violation of the laws of war.'" In fact, this old encyclopedia article provides a more comprehensive and representative list of crimes tried by military commissions than any historian was heretofore able to provide.

Clampitt's defense of trials by military commission was significant, for when he spoke of their use in capital cases, he definitely knew what he was talking about. He had served as defense counsel for Mary Surratt in the most sensational trial by military commission in all of American history, the trial of the persons accused in the plot to assassinate Abraham Lincoln. In the *Cyclopaedia,* Clampitt merely stated: "From such jurisdiction, however, are very properly excepted such offenses as are clearly within the legal cognizance of the criminal courts of the country, when such courts have been left in the full operation of their usual powers, upon the establishment of a military government, or the *status* of martial law." In other words, he saw trials by military commission as plugging loopholes in the law in wartime, and with such use he seems to have been reasonably content.[59]

But the trial of Mary Surratt was another matter. Clampitt was on record as saying that that particular military commission had been "organized to convict" and was "illegal."[60]

Clampitt's article on military commissions appeared in America's first political-science encyclopedia. Political science was born as a self-conscious dis-

cipline toward the end of the nineteenth century, and a characteristic of the
first generation of American political scientists—a certain professional cynicism
about paper constitutions—led them not to be especially admiring of such
Supreme Court decisions as *Ex parte Milligan.* John W. Burgess, who held the
prestigious Lieber Chair of Political Science at Columbia University, thus predicted in 1890: "It is devoutly to be hoped that the decision of the Court [in
the *Milligan* case] may never be subjected to the strain of actual war. If,
however, it should be, we may safely predict that it will necessarily be
disregarded."[61]

Ex parte Milligan, as it turned out, did not have to be disregarded in World
War I, a distant foreign war of brief duration for the United States and a contest
that hardly taxed the population or resources of the vast United States. Nevertheless, the internal security measures taken by the Woodrow Wilson administration in World War I appeared to threaten civil liberties on the home front
in serious fashion. Wilson did not act in the same way Lincoln had. Wilson did
not make arrests by executive authority using his "war powers" and then await
endorsement by Congress long after the fact. He employed the Justice Department and eschewed trials by military commission.

With no enemy invasion or rebellion, suspending the writ of habeas corpus
in World War I was out of the question. Yet civil liberties nevertheless suffered
under extraordinary measures taken by the Democratic administration. Congress provided President Wilson with the Espionage Act of 1917, under which
2,168 trials were conducted, resulting in 1,055 convictions. Among those
convicted were over 150 leaders of the Industrial Workers of the World, a
nominee for the United States Senate (J. A. Peterson, a Republican from Min-
nesota), and the famous socialist leader and presidential candidate, Eugene
V. Debs.[62]

Wilson did not need such new legislation to accomplish some of his goals.
He was not following in Lincoln's footsteps in any specific legal way, but the
arrests of 6,300 enemy aliens were accomplished by means of a proclamation
implementing the Alien Enemies Act of 1798, part of the old Alien and Sedition
Acts that made John Adams so odious to Jeffersonian Republicans. Without
using military authority directly, the Wilson administration met the ubiquitous
problem of draft resistance in part by allowing the "arrest" of some forty
thousand Americans by the American Protective League's quasi-vigilante
"Slacker Raids."[63]

In reaction to the government assault on civil liberties in World War I and
in the "Red Scare" that followed the war, the reputation of the *Milligan*
decision suddenly grew. Thus, Charles Warren's history of the Supreme Court,
with its fulsome praise of the decision, appeared five years after the armistice.
James G. Randall's *Constitutional Problems under Lincoln* was, to a degree,
also a product of this new atmosphere. Writing in 1926, J. G. Randall could
hardly escape uttering some praise for the Supreme Court decision even though
it roundly criticized Lincoln. "One of the great doctrines of the Supreme Court,
as announced in the *Milligan* case, is that the Constitution is not suspended

during war," Randall said. In his influential text, *Civil War and Reconstruction*, published eleven years later, Randall wrote, echoing Charles Warren, that the *Milligan* decision "has become famous as one of the bulwarks of American civil liberty." Naturally, when a series of books on famous American trials was planned in the same decade, the first one featured was *The Milligan Case* by Samuel Klaus.[64]

Though the reputation of the decision was certainly on the rise with this new post–World War I outlook, the old cynical interpretation in the Burgess tradition of political science remained alive as well. Back in 1918, in an article that appeared in the *American Historical Review,* the Columbia political scientist William A. Dunning compared "Disloyalty in Two Wars," examining the records of Lincoln and Wilson. He did so without ever mentioning *Ex parte Milligan,* and he stressed the contrast between Wilson's reliance on congressional authorizations and Lincoln's exploitation of "the constitutional functions of the commander-in-chief of the army and navy." He concluded with an insight quite like Burgess's:

> The spirit and record of the Wilson administration must give much satisfaction to those who seek an abiding reign of law. It would, however, be a highly sanguine student of history who would assert that the normal course of justice would have been consistently maintained in our last war [the Great War] if the enemy had been as near Washington as he was in 1861, or if the conflict had lasted four years, or if great reverses had been experienced, or if our coasts had been threatened at close range by a high-seas fleet instead of by a lonely and furtive submarine.[65]

The growing reputation of the *Milligan* case in the 1920s was never great enough to lay to rest the tradition of cynical interpretation which survived, for example, in Samuel Klaus's book *The Milligan Case.* Klaus identified *Ex parte Milligan* as "the famous case in which the United States Supreme Court affirmed that the civil liberties guaranteed by the Constitution are to be safeguarded not less in the fever of civil war than in time of peace," but he began his long introduction to the case by saying: "Between Sumter and Appomattox, one is apt to infer, the opinion would simply have been irrelevant. As for the future, we have Burgess' prophecy." Klaus concluded that the case was irrelevant to the twentieth century:

> More than sixty years have passed since the decision....In all this period the Supreme Court of the United States has not found it necessary to reaffirm or to deny the doctrine of *Ex parte Milligan.* There simply has been no occasion for either the suspension of habeas corpus, whether by the president or by Congress, or the trial of civilians by military courts. In the few wartime cases in which *Ex parte Milligan* was urged to give an atmosphere of emotionality, the federal courts found no reason for either dwelling at length on the theory of the case or for limiting or broadening its application.
> ...a situation such as gave rise to *Ex parte Milligan* will not arise again.
> Civil conflict today is for the most part industrial.[66]

Even those who celebrated the *Milligan* decision in the 1920s had, in some cases, actually disregarded it rather cynically during the Great War. Charles

Warren, for example, testified in the United States Congress before the Committee on Military Affairs in 1917 that the *Milligan* decision was irrelevant to World War I. He assured the committeemen that certain classes of civilians could be subjected to military trials by calling the defendants "war spies" under Congress's constitutional power "to make rules for the government and regulation of the land and naval forces." Within President Wilson's Justice Department, where he worked as an assistant attorney general, Warren could be a good deal less circumspect in his language. "One man shot after court-martial is worth a hundred arrests by this Department," he wrote early in 1918. In April, he proposed to the Senate Military Affairs Committee a court-martial bill permitting military trials and capital punishment for persons interfering with the war effort. Woodrow Wilson opposed the legislation and the bill died quickly.[67]

By World War II, the *Milligan* decision, despite its enhanced historical reputation in the 1920s, had lapsed into disuse. President Franklin Delano Roosevelt, following a plan formulated by the Navy Department in 1936, placed about 120,000 Japanese-Americans in "relocation centers" (called "concentration camps" in the original plans) by means of a simple executive order to the secretary of war and through him to the military commanders in the western United States.[68]

One last attempt to resurrect the *Milligan* decision as a possible bulwark of American liberty followed in 1942. That year James Purcell petitioned for a writ of habeas corpus for an incarcerated Japanese-American named Mitsuye Endo. Purcell argued that *Ex parte Milligan* proved that "only by act of Congress could the writ of habeas corpus be suspended, and that no military action could be taken except under martial law conditions." The government's lawyer, Alfonso J. Zirpoli, brought to bear an influential article just published in the *Harvard Law Review*: Charles Fairman's "The Law of Martial Rule and the National Emergency." Fairman, a Stanford law professor, by this time actually a colonel on the judge advocate general's staff, minimized the relevance of *Ex parte Milligan*, saying, "When one considers certain characteristics of modern war—mobility on land, surprise from the air, sabotage, and the preparation of fifth columns—it must be apparent that the dictum that 'martial law cannot arise from a threatened invasion' is not an adequate definition of the extent of the war power of the United States." The judge in the case delayed his decision for almost a year and then dismissed the petition. When the chips were down—the only time when the *Milligan* decision could matter—the case usually proved irrelevant.[69]

Colonel Fairman was destined to write the volume of *The Oliver Wendell Holmes Devise History of the Supreme Court of the United States* dealing with the *Milligan* case, and he devoted over fifty pages to the subject. His analysis is more sophisticated, more deeply researched, more circumspect than that of the expansive Charles Warren. But Fairman still held that "Justice Davis' opinion is indeed a landmark of constitutional liberty," a judgment that is not exactly congruent with Fairman's wartime dismissal of the *Milligan* decision for failing to provide "an adequate definition of the extent of the war powers of the United States."[70]

Among the propagandists for twentieth-century war governments who at some time in their careers also wrote works on the constitutional history of the Lincoln administration, only sharp-tongued Edward S. Corwin, the Mc-Cormick Professor of Jurisprudence at Princeton University, consistently and openly maintained the crusty cynicism of John W. Burgess in interpreting the legacy of the *Milligan* decision. Already a man of forty during World War I, Corwin lent his considerable talents to the famous Committee on Public Information, serving as one of the three editors of the Red, White, and Blue Series's *War Encyclopedia* of 1918, which sought to provide definitions of terms like "freedom of speech" that would suit the mobilization efforts of the Wilson administration. Corwin took the view that the Great War came "at a peculiarly favorable moment for affecting lasting constitutional changes" and he constructed numerous arguments in defense of government power and the temporary circumscription of individual liberties. These he maintained into the era immediately after the war, coming into conflict with Zechariah Chafee and others concerned to carve out a more legally secure realm for civil liberties. Corwin's interpretation of *Milligan's* legacy remained a matter of withering cynicism:

> But what, it will be asked, of the decision in the *Milligan* case? It shows, to be sure, that two or three years after a grave emergency has been safely weathered and the country has reaped the full benefit of the extraordinary measures which it evoked, a judicial remedy may be forthcoming for some of the individual grievances which these produced, and a few scoundrels like Milligan himself escape a deserved hangman's noose—but it shows little more. And among judicial records it would be difficult to uncover a more evident piece of arrant hypocrisy than [one famous passage] from Justice Davis' opinion for the majority on that occasion.... To suppose that such fustian would be of greater influence in determining presidential procedure in a future great emergency than precedents backed by the monumental reputation of Lincoln would be merely childish.[71]

Ex parte Milligan lacked practical influence in protecting liberty, but it has had a powerful legacy in books on constitutional history. By and large, this decision made the sharp distinctions between trials by military commission and military arrests of civilians, and between martial law and the condition that follows a suspension of the writ of habeas corpus. Historians, not realizing how muddled the law was before 1866, have repeatedly misinterpreted the constitutional history of the Civil War mainly by making the choices seem clearer than they appeared to the protagonists at the time. Such distinctions were clearer to the justices in 1866 than to the hard-pressed Lincoln administration between 1861 and 1865. In fact, the real legacy of *Ex parte Milligan* is confined between the covers of the constitutional history books. The decision itself had little effect on history.

9

The Democratic Opposition

The *Milligan* decision provided a victory after the fact for opponents of the internal security measures taken by the Lincoln administration. It was all the more surprising and telling because it came from a Republican court and yet judged a Republican president harshly. It stands to reason that its sources must have included something other than Democratic denunciation of the Lincoln administration, and indeed a survey of the debate over the issue of civil liberties in the Civil War reveals that some doubts had been expressed on the Republican side all along. But naturally, the main opposition to the policy came from the Democratic party. They proved slow to criticize at first, lacking leadership and motivated by patriotic concern to preserve the Union. Eventually, their protests were heard, but a survey of Democratic criticism reveals a lack of depth and sincerity.

Early Criticism

Northern public debate on the question began in July 1861 only after Taney's pronouncement on the *Merryman* case was disseminated. Before that, little was heard. Even the rabidly anti-Republican *New York Freeman's Journal and Catholic Register* made no comment on the first order suspending the writ around Washington. On May 11, in fact, the paper stated: "The North stands as one man in saying that Washington, as the Capital of the country, shall be protected, and that *Whatever is necessary to this end must and shall be done.*"[1]

The situation was much the same in the West. When John C. Frémont declared martial law in St. Louis, Democratic newspapers in neighboring Illinois, which depended on the Missouri city's press for much of their material, uttered no dissent. Nor did Frémont's wider-reaching martial-law proclamation of August 30 prompt strenuous objections beyond the border states. On September 2, the *Illinois State Register,* the Democratic newspaper in Springfield

185

and the banner Democratic organ in the state, endorsed the limited use of martial law and even linked it to sturdy Democratic tradition:

> There is a great deal of reverence paid ... at present to the old maxim, that in the midst of war law should be silent. In a modified sense this is true. In the presence of actual hostilities, or in districts largely infected by treason, measures of a very stringent nature may be justified by a paramount public necessity. Thus Jackson was right at New Orleans, and it may be that martial law is essential to the peace and loyalty of St. Louis and Baltimore. But the proposition, so often made, that it should be extended all over the country, and particularly to the great cities, is not only foolish, but wicked. ... [2]

In the border states themselves, military arrests of civilians had immediate and genuinely practical impact, and resistance naturally flared up. Maryland was so politically traumatized by these early events of the Civil War that her state song even today is a protest against them. At the time, the state was replete with protests, many of them at the popular level of satirical songs and poems circulated in broadsides of palm-card size.[3] The state legislature became a major pamphleteer in August when it caused to be printed twenty-five thousand copies of its own *Resolutions of the Joint Committee... upon the Reports and Memorials of the Police Commissioners and the Mayor and City Council of Baltimore.* The pamphlet protested arrests in Baltimore "under arbitrary and frivolous pretexts" and the holding of "prisoners of State" at the "arbitrary pleasure of the President." This "gross and unconstitutional abuse of power" amounted to "a revolutionary subversion of the Federal compact."[4] Lincoln's actions attracted few defenders in Maryland, though Reverdy Johnson and Anna Ella Carroll wrote qualified defenses of the suspension of the writ of habeas corpus.[5]

The border states contained few Republicans in the early part of the war, but they held many Democrats of decidedly proslavery views. The latter, naturally, protested the suspension vigorously. One of the earliest formal protests came from the pen of Judge S. S. Nicholas of Louisville, Kentucky, who rushed a pamphlet into publication in 1861 by reprinting arguments he had used almost two decades before. In the 1840s, he had attacked John Quincy Adams, who had played a prominent role in the controversy over refunding Andrew Jackson's fine for ignoring the writ of habeas corpus in New Orleans and arresting the judge who issued it.[6] These prophetic arguments had shown that martial law, if it were ever imposed in the United States, would surely threaten slavery.

Nicholas's pamphlet remained a rare effort, however, and even the sensational arrests of the Maryland legislators in September 1861 did little to alarm Democrats outside the border states. In neighboring Pennsylvania, for example, Gettysburg's Democratic newspaper, the *Compiler,* charted the following course on the issue of civil liberties in 1861. The first sign of protest came in June, after Chief Justice Roger B. Taney had articulated an opposition position in the *Merryman* case. In nearby Harrisburg, the *Patriot and Union* cited precedents for locating the power to suspend the writ of habeas corpus in Congress rather than in the president's office. It defended Taney for doing his duty and warned against the "first steps toward the establishment of a military despotism." The Gettysburg paper clipped the story, the first of several on the issue printed early that summer. As state elections approached in the autumn, county conventions began to produce the customary ideological statements. The local Democrats did not mention civil liberties in their party resolutions. When the members of the Maryland legislature were arrested in mid-September, the Gettysburg *Compiler* ignored them. In fact, a reader informed of events in American political life only by that newspaper would not have known that the arrests occurred.

The western Democratic press seemed even more content with the events in Maryland. *The Crisis,* a sharply anti-administration Catholic newspaper in Columbus, Ohio, though it had earlier followed Taney's lead in criticizing the suspension of habeas corpus, did not mention the case of the Maryland legislature in the fall of 1861. When newspapers did take notice of the event, they sometimes offered bipartisan support. The Democratic Fort Wayne (Indiana) *Weekly Sentinel,* for example, stated in its issue of September 21, 1861, "The actions of the government in the matter of the arrests in Maryland..., meets our hearty approval.... It indicates that the day for trifling is past, and that rebels wherever found, and whatever their position, are to be treated as rebels." One article went on to say "that the present position of affairs in Maryland imperatively demands" that the writ of habeas corpus be set aside. The Chicago *Times,* also a Democratic newspaper, justified such arrests "upon the ground of overpowering necessity" because the "abolition congress... neglected their duty" and "did not last summer provide for the apprehension and detention of spies and traitors." Even after habeas corpus had become a partisan issue throughout the North, some Democrats accurately recalled the party's original acquiescence. Speaking in the New York State Assembly in 1863, J. S. Havens gave a brief history of Lincoln's tyranny, but he made an exception for the border states in 1861:

> I grant, sir, that there was a time when anarchy and confusion reigned in the Border Slave States, when a person guilty of crime, could not have had an impartial trial, and that it would not have been safe to have allowed him his liberty, but in my opinion there can be no such justification, no palliation, no excuse for this wanton violation of the rights of citizens of the loyal States of the North.[7]

By late 1861, opposition in the sensitive border states was beginning to spill over into the states of the Old Northwest. The protest of the Reverend Robert L. Breck of Maysville, Kentucky, was the first attack on the habeas-corpus suspension separately published in the Northwest. He drafted it for the December 1861 issue of the *Danville Quarterly Review,* but Kentucky Unionists like Robert Breckinridge would brook no anti-administration rhetoric in their journal. So Breck published the article himself as a pamphlet in Cincinnati early in 1862. It commended Taney's argument that only Congress could suspend the writ but made a more original contribution by denouncing "yet greater usurpations by the President and the military power":

> The suspension of the privilege of the *habeas corpus*, when it is legally accomplished, merely leaves under arrest and without a remedy for the time being, the citizen seized to prevent his aiding the rebellion or invasion; but leaves him the right when the exigency shall have passed, to a fair trial by a jury of his peers; and when abused, only subjects him to such tyranny as may be effected through the unjust confinement of his person. Martial law—which is nothing else than the enforcement of the arbitrary will of the commander-in-chief, or any of his subordinates...destroys the legal guarantees of all rights, and exposes them all to invasion.

Such distinctions would become important later, but early in the war, critics focused mostly on the question of presidential power to suspend the writ under the Constitution. Protestors from slave states, however, proved sensitive to the threatening possibilities that always lay in expansions of federal power.[8]

Republican Doubters and Defenders

Dissident Republicans, both to the left and right of the administration, at first prompted as much public debate on the issue as any Democrats outside the border states did. In the fall of 1861, a conservative Republican, writing a systematic critique of the *Merryman* decision, also put an extremely narrow construction on the president's acts. Joel Parker, Royal Professor of Law at Harvard University, had first aired his legalistic argument on June 1, 1861, before a law class, but the resulting article did not appear in the *North American Review* until October. Parker contended that "in time of actual war, whether foreign or domestic, there may be justifiable refusals to obey the command of the writ, without any act of Congress, or any order or authorization of the President, or any State legislation for that purpose; and the principle upon which such cases are based is, that the existence of martial law, so far as the operation of the law extends, is *ipso facto*, a suspension of the writ."

The question, then, narrowed to whether Fort McHenry, where Merryman had been held, lay under martial law when Taney's writ was served on the commander. Parker maintained that it did because the fort was an army installation, Merryman was a prisoner in wartime, and Fort McHenry's commander therefore need not heed a writ of habeas corpus for this prisoner. Martial law was the "military rule and authority which exists in time of war, and is conferred by the laws of war, in relation to persons and things under and within the scope of active military operations in carrying on the war, and which extinguishes or suspends civil rights, and the remedies founded upon them, for the time being." Professor Parker, an old Whig, was aware of John Quincy Adams's argument (so troubling to Southerners like S. S. Nicholas) that Jackson had definitely proved that American generals could declare martial law in times of war in theaters of war in America.[9]

The *Illinois State Register* took notice of the Harvard law professor's argument but chose to stress its assertion of "The Limited Range of Martial Law," as the headline expressed it. Such was not an unfair interpretation.[10]

The principal public debate over the policy was now occurring within the ranks of the Republican party. In the December session of Congress, Senator Lyman Trumbull of Illinois introduced a resolution asking the secretary of state "whether, in the loyal States of the Union, any person or persons have been arrested and imprisoned and are now held in confinement by orders from him or his Department; and, if so, under what law said arrests have been made." In the congressional debates that followed, Republicans mostly argued with each other. Democratic senators from the border states, and one Californian, spoke in favor of Trumbull's resolution. Northern Democrats were not yet eager to exploit the issue; they need not risk gaining an unpatriotic reputation when Republicans seemed willing enough to bring up this embarrassing issue on their own.[11]

Eventually, a Republican defense of Lincoln's policy provoked many Democratic responses and opened the serious partisan debate on civil liberties. In fact, it aroused more opposition of a systematic nature than any actual arrest or presidential order or proclamation on the subject before 1863. This was Horace Binney's pamphlet, *The Privilege of the Writ of Habeas Corpus under the Constitution,* published in Philadelphia shortly after New Year's Day 1862. Binney argued that the example of Great Britain, where the power to suspend lay with Parliament, was irrelevant to the United States because the habeas corpus in Great Britain mainly provided a buttress against the powerful monarchy. There was no corresponding need in republican America. A president empowered to suspend the writ was more readily accountable to the people than was Congress, where responsibility was necessarily divided.[12]

Argumentative responses immediately poured forth from lawyer pamphleteers, especially in Philadelphia. The net effect of Binney's defense, besides bracing the Republicans with learned arguments from an aged and respected jurist, was to galvanize Democratic opposition—and to focus it quite narrowly. The emphasis now was on Lincoln's suspension of the writ of habeas corpus, seen as executive usurpation of a power granted exclusively to Congress in the U.S. Constitution. John T. Montgomery, whom diarist and constitutional commentator Sidney George Fisher described as a "well-known Copperhead," was perhaps the first to assail Binney's arguments.[13] Another Philadelphia critic, perhaps the bitterest, Democrat Charles Ingersoll, entitled his pamphlet, published at the end of 1862, "An Undelivered Speech on *Executive* Arrests" (emphasis added). Louisville's S. S. Nicholas may have best expressed in style and substance the essence of the early Democratic argument:

> Mr. [John C.] Bullitt [one of the early Philadelphia pamphleteers who answered Binney] having produced . . . such a long roll of eminent judges, lawyers, and statesmen, including such names as Marshall and Story, who have expressly held the suspending power to be in Congress, justifies the belief that the roll, if completed, would amount to more than a hundred, whilst there are none, not one to the

contrary. These opinions, repeated without contradiction from the birth of the Constitution steadily down to the present day, amount to a fixed, settled construction, fully as authoritative as an express decision of the Supreme Court. The position of the President and Attorney General not having been indorsed by a single respectable lawyer, ... and Mr. Binney's construction, notwithstanding the great ability of its defense, having so signally failed of acceptance, it would seem to be the duty of all to acquiesce in the old, original construction of seventy years duration, as the only true one.[14]

Nicholas's own argument had an idiosyncratic proslavery twist not necessarily shared by the rest of the Democratic opposition. He thought that martial law simply could not be declared in the United States, by anyone, and for the sake of saving slavery, he rejected the Jackson precedent.[15]

Systematic thinkers among the Democrats, then, opposed military arrests for a number of reasons. The Kentuckian Nicholas, though a Unionist, feared the threat of martial law to slavery. Charles Ingersoll, the patrician son of a Democratic politician from Philadelphia, also had more than partisan reasons to oppose military arrests of civilians. Privately, he gave vent to his real hope that the South would win the war. He did not express such sentiments in public speeches.

On the whole, however, mainstream Democrats criticized Lincoln and contended that he usurped the constitutional power of Congress to suspend the writ of habeas corpus. They did not usually indulge in defenses of slavery, and they rarely expressed sympathy with the Confederacy. For example, George M. Wharton of Philadelphia, a Whig turned Democrat, kept his focus on Lincoln, though Wharton had publicly proclaimed the South justified during the secession crisis.[16]

In 1862, Democrats ceased publicly endorsing military arrests of civilians, however limited in scope. Yet their condemnation of Lincoln's policy remained sporadic and somewhat muted. Protesting executive usurpations was old Whig ground, a little unfamiliar to the party of Andrew Jackson. As yet, there were no martyrs or symbolic events nationally known that could be repeatedly and consistently referred to. Outside the troubled border states, even martyrs of local reputation proved scarce. Of the few arrests that occurred above the border areas, some were ignored or overlooked by the press, and others were reported only gingerly. Springfield's *Illinois State Register,* for example, published news of the arrest of some citizens in the county for aiding Confederate prisoners of war to escape.[17] Nothing else on the subject appeared in the newspaper after the initial story. Perhaps, the reports proved false, or maybe the Democrats did not wish to be associated with persons accused of a crime that clearly aided the rebellion.

The opposition party's leaders had not yet made it clear that acquiescence in military arrests of civilians was unacceptable. Editors in Illinois and elsewhere took their lead from Congress. Democrats in Washington had greater access to news of obnoxious arrests by the State or War departments, greater means for spreading such stories, and more opportunity to confer promptly with other influential Democrats. Only after Democrats in Congress attempted to embarrass the Republicans by asking for State Department documents in the near-arrest of Franklin Pierce did the *Register* publish an article about it.[18] Even the loosely drafted War Department orders of August 8, 1862, and the follow-up presidential proclamation of September 24 did surprisingly little to focus the Democratic party's response to the issue of military arrests. The Democrats in Illinois put up only a mild protest against the intensified internal security efforts of August, despite the irresponsible actions of U.S. Marshal Phillips in their own state. The *Register* complained about the appointments of provost marshals on the general grounds that there was "no necessity for this military surveillance in this state. Our state laws are ample to ferret out and punish disloyalty, and it is an imputation upon the integrity and loyalty of our executive and judicial officials to assume that, with peace reigning within our borders . . . our state should be infected with a corps of military police, to foster malicious espionage, and produce neighborhood bickering and distrust, and consequent general public injury."[19] Surely it is significant that the loosest and most poorly controlled arrests of the war, those following the August 8 orders, failed to provoke the most severe Democratic criticism—and this in a period when the very authority to suspend the writ remained in question, before Congress authorized the president's action.

Restiveness among Republican opinion leaders continued to cause nearly as much trouble for the administration as Democratic protest. Horace Greeley and his New York *Tribune* frequently criticized military arrests of civilians, and in the midst of the August 8 arrests, the *Tribune* complained:

> In our poor judgement, nine-tenths of the arbitrary arrests thus far had far better not been made; but we would not deprive the government of the power to make them in a crisis like the present. Let it be distinctly understood, however, that each arrest will be made the subject of rigorous and dispassionate inquiry after peace, and that, while no one should suffer for his innocent mistakes in honestly endeavoring to serve and save his country, it will go very hard with any one who is proved to have gratified his own malice or his love of exercising despotic power, without the warrant of public necessity.
>
> What we mainly apprehend, however, is not that individuals shall suffer—they always do and must suffer like this—but that the salutary popular instinct which regards with jealousy every infringement of constitutional safeguards for liberty may be debauched and broken down.[20]

Naturally, this *Tribune* article enjoyed the unusual fate of being reprinted in Democratic newspapers.

Despite embarrassing disclosures from uneasy Republicans, increased provocation from loosely controlled arrests, and ever bolder presidential proclamations, Democrats failed to concentrate on civil liberties in the off-year elections. The preliminary Emancipation Proclamation of September 22 and the continuing lack of a smashing Union military victory provided more important issues. The habeas-corpus and martial-law proclamation of September 24 went hardly noticed, overshadowed by the startling news of emancipation.

Besides, there was nothing new in the habeas-corpus proclamation, as the *Illinois State Register* commented:

> Nothing from the president surprises anyone, in these days, but the inquiry naturally suggests itself—where the necessity of such a proclamation when the secretary of war, for months, has been acting upon a policy which is now enunciated from the executive head of the nation? Martial law and suspension of the *habeas corpus* have been practically in force for months.

When the Democrats carried Lincoln's own congressional district, the *Register's* telltale epitaph for the Republicans was: "Such is the result in response to Mr. Lincoln's proclamation." But there had been two proclamations by Lincoln in September. Illinois Democrats had voted mainly against Lincoln's "proclamation"—singular—that is, the Emancipation Proclamation.[21]

The Role of Horatio Seymour

The opposition press paid more attention to the civil-liberties issue *after* the elections, in the winter of 1862–1863. Congress showed the way, as usual. There, the debates over a habeas-corpus bill and several resolutions inquiring about arrests in different Northern states kept the issue before the people. Democrats played a major role from the first day of the session that began in December 1862, when Ohio's S. S. Cox denounced military arrests.[22]

Yet things had not changed for the worse in the actual administration of the Republican policy. If anything, arrests diminished in number and significance after filling of quotas in the first draft in the fall and the consequent relaxation of the August 8 orders. Nor was interest in the subject spurred by the arrival of new Democratic members—this was a lame duck session of the old Congress.

Whatever force invigorated the congressional Democrats of the North, their decision to do battle on the subject of civil liberties did not stem from increased provocation. But their decision in Washington to protest did, in turn, provide the party throughout the nation with information and new-found indignation against military arrests of civilians. In December, articles on the subject began to appear every week in some Democratic papers.

By January 1863, the party seemed ripe for a major reformulation of Democratic policy on the issue, and one of their ablest politicians, Horatio Seymour of New York, provided it. Conveniently out of office at war's outbreak when the crisis caused consternation in Maryland, Seymour was unhampered by a personal public record on the arrests in Maryland. When his election as governor in the fall of 1862 made him a likely national leader of the still somewhat distracted Democracy, Seymour began to focus on the constitutional issues raised by Lincoln's proclamations of September 22 and 24. For his first annual message, of January 7, 1863, he consulted lawyer John V. L. Pruyn, especially on habeas corpus and martial law. Pruyn, a railroad executive and former chancery lawyer of distinguished reputation, was serving in the state legislature.

After his discussion with Pruyn, the governor devoted over two-thirds of his speech to national rather than state issues and much of it to the subject of military arrests.[23]

Seymour slighted the sterile and soon to be irrelevant issue of executive usurpation of congressional authority to suspend the writ of habeas corpus. He looked squarely at the more embarrassing issue for the Republicans, martial law. "'Martial Law' defines itself to be a law where war is," Seymour said. "It limits its own jurisdiction by its very term." He had no tolerance for what he called "this new doctrine...that the loyal North lost their constitutional rights when the South rebelled." Then he seized on what would one day become a vital distinction in the *Milligan* decision: "More than two centuries since, that bold defender of English liberty, that honest and independent judge, Lord Coke, declared: 'Where courts of law are open, martial law cannot be executed.'"[24]

New York assemblyman J. S. Havens, in a February speech, described the impact of Seymour's message:

> When in response to the voice of the people, Horatio Seymour...was called to occupy the highest office in the gift of [this state's] citizens, and denounced the arbitrary acts of the government, that the liberties of the people were endangered—that the government had falsified its trust—it found a glad echo in the heart of every citizen that loved his country for her institutions, and the blessings of a good government which they had so long enjoyed, but were now deprived of. Men whose lips had been sealed for months (except in their family circles) upon the great questions agitating the country, and dared not express their honest convictions in public for fear of arrest felt that the fetters had dropped from their limbs—that their tongues were unloosed.

The portion of Seymour's message devoted to national affairs was quickly and widely reprinted in Democratic newspapers across America. The Indiana legislature even thanked the New York governor in a joint resolution, passed 53–35. Other Democrats, like the newly inaugurated New Jersey governor, Joel Parker (not to be confused with the Royall Professor of Law at Harvard), followed Seymour's lead. "It is a clear principle," said Parker, "that what is called martial law cannot rightly be extended beyond the field of active operations of the commander."[25]

Articles and editorials on military arrests continued to appear every week. Occasionally, Democrats attempted to deal with the awkward Democratic precedent set early in the century by their party's founder, Andrew Jackson. The *Illinois State Register,* for example, addressed the subject of "Jackson and Lincoln—the Habeas Corpus" not long after printing Seymour's stirring message. Andrew Jackson, the article pointed out by way of contrast with Abraham Lincoln, "proclaimed and carried out martial law within the limits of a besieged city, whose inhabitants were all within his military lines."[26]

Despite an apparent increase in Democratic truculence and cohesion on the issue, congressional Republicans failed to close ranks. Members of Lincoln's own party—including the senior senator from his own state, Trumbull; the

radical John P. Hale of New Hampshire; and the moderate William Pitt Fes-
senden of Maine—insisted on creating some system of accountability for the
persons arrested. That kept debate open as surely as did the resolutions brought
in from Democrats' constituents demanding relief from arrests in their states.
Differences among Republicans on the issue stemmed from various causes,
including individual conscience and local political circumstance. Trumbull
appears from the start to have been uneasy with the prospect of constitutional
irregularities in fighting the rebellion. He labored for almost two years to create
legislation that might bring system to the arrests. Ohio's John Sherman, on the
other hand, felt the sting of political backlash against arrests that occurred in
his home state. "Of the few arrests made in Ohio," Sherman told the Senate
on December 9, 1862, "most...resulted only in evil to the party that made
the arrests." Edson B. Olds, he maintained, had been "a man comparatively
without influence" before his arrest. Though Olds had used "utterly disgraceful
and discreditable" language in the speech that caused his arrest, "it fell utterly
harmless and impotent." After his arrest, however, "he has been crowned a
martyr, and...has been elected a member of our State Legislature by a very
large party majority."[27]

These congressional debates were often rather uninformed on technical
jurisprudence. The congressmen may have been lawyers like Lincoln, but like
him, they too had little knowledge and no experience of military arrests or
martial law. Both those who supported the policy and those who criticized it,
and both Democrats and Republicans, shared the era's uncertainty about what
it all meant in practicality. For example, when John S. Carlisle, who represented
the nearly mythical constituency of Unionist Virginia, argued that the writ
should never be suspended in areas where the courts of justice were open,
he received a cool reply from Trumbull. The Illinois senator pointed out that
unless it were suspended where courts were open, it would never be suspended at all, because the writ would not be issued except from a court.[28]

Most of the debates in Congress in the winter of 1862–1863 remained
innocent of distinctions between martial law and the situation that obtains
when only the writ of habeas corpus is suspended. There were no special
objections raised to the use of trials by military commission, the only issue
that would later trouble the U.S. Supreme Court. Even Republicans, who were
surely in a better position than Democrats to know, possessed little knowledge
of who had been arrested and by whom. John Sherman, who was sharply
critical of the arrests, nevertheless suggested the following remedy:

> When the privileges of this writ are once suspended, every protection should be
> thrown around the accused. If it is not expedient for any reason to have an open
> trial upon a criminal accusation for an offense charged against the individual, then
> there ought to be a military examination by a board of officers, or by some one
> who could inquire into the reality of the charges and the ground for making them.[29]

This was virtually an invitation to use military commissions, but they were
already in use and would one day be regarded as the most abusive feature of

the whole dangerous system, certainly not as a remedy for abuses. Besides, War Department committees were already making occasional visits to the military prisons to review such cases.

Meanwhile, short and hastily compiled works of protest by former political prisoners began to appear as books and pamphlets. Francis Key Howard, the grandson of Francis Scott Key, had been among the Baltimoreans arrested in September of 1861. By December 1862, he had finished a manuscript about his prison experiences, and the little book made its appearance in print early in 1863. Like others in this genre of protest writing, Howard's work made a special point "to show...how men, who were guiltless..., were treated in this age, and in this country" and stressed the crowded conditions and spartan hardships of prison life. In a protest letter written to President Lincoln and reprinted in Howard's book, he and other prisoners of state likened the conditions in Fort Lafayette to those on "a slave-ship, on the middle passage."[30]

Sidney Cromwell answered Howard by making him look effete. Cromwell's pamphlet, called *Political Opinions in 1776 and 1863: A Letter to a Victim of Arbitrary Arrests and "American Bastiles",* stated:

> You complain of your sufferings, your lodging, and your fare; you whine...because you had "nothing but the ordinary rations of the soldier, which are of the coarsest kind," because "the dinners consisted of fat pork and beans, a cup of thin soup and bread, or of boiled beef and potatoes, and bread on alternate days...." But, sir, there are hundreds of thousands of men who, for two years, have lived on these rations of the common soldier; who, while enduring fatigue and braving the elements, and wounds and death at the hands of the supporters of that just cause of which you speak, have been glad to get those dinners of pork and beans, and beef and potatoes, and bread, which so offended your high stomach, and who cannot supply themselves with other fare at "one dollar per day," as you were allowed to do; and who have not been regaled with presents of "pheasants, chicken, tongues, pies, and other delicacies," as you admit that you were....Now do you think that the people of this country will weep much because you, who have seen fit to adhere to the public enemy, lived for fourteen months...upon the best fare which the nation can afford to those who are giving up their lives in defense of the country against that public enemy?

Cromwell also made a rare appeal to history, beyond Andrew Jackson, to the fathers of the republic. In the debate over habeas corpus, this was usually Democratic territory, the presidential suspension being unprecedented and it being generally believed that the Revolution and the War of 1812 were fought without suspending the writ—and, until the Battle of New Orleans, without declarations of martial law.

Cromwell, however, pointed to the Continental Congress's action in sending a military force to Queen's County on Long Island to disarm all and arrest some of the persons who had voted against sending delegates to the revolutionary convention. Those arrested were sent out of the province where the alleged crime was committed to be tried in Philadelphia, "the seat of the tyrannical government," as Cromwell expressed it with irony. The Continental

Congress, declaring that "those who refuse to defend their country should be excluded from its protection and be prevented from doing it injury," then essentially outlawed those Queens County men, putting them outside Congress's protection, ordering that trade with them cease, and making public enemies of any lawyers who should argue cases for them. Cromwell took particular glee in finding some of these historical facts in the work of George Ticknor Curtis, a Democratic pamphleteer for the Society for the Diffusion of Political Knowledge and brother of Benjamin R. Curtis, pamphleteer against Lincoln's use of executive power.[31]

Most of the arguments in the winter of 1862–1863 arose from the debates over the Habeas Corpus Act, which passed in March 1863. Despite its removal of what had once been the Democrats' principal formal complaint about the suspension, that is, its lack of congressional sanction, Democrats nevertheless remained discontented. As the Philadelphia *Inquirer* expressed it, "This act would seem to any candid mind to obviate most of the objections urged against the suspension of the writ in the earlier months of the present Rebellion. But so far from silencing the columns of the malcontents or satisfying their demands, they are now pouring out fiercer objurgations than ever."[32]

Yet opposition in the spring of 1863 was not particularly intense—national elections remained too far off. Even when news of the arrest of Clement Vallandigham hit the presses in May, the initial Democratic reaction in many areas of the country proved rather mild. The independent New York *Herald* observed that it was "the policy of the democracy to keep cool and be quiet. All such cases will tell in their favor in the next election." And the Democratic New York *World* said that if the obnoxious Vallandigham had been let alone, he would have done the Republican administration more good than anyone else in the country. Now the arrest had made a martyr of him and, as the *Illinois State Register* put it, "his illegal trial and outlandish punishment will only add thousands to the democratic strength in future elections."[33]

Democrats waited for the lead of Governor Seymour, now regarded by many as the party's major spokesman. His letter of May 16, 1863, set the stage for elaborate protest. It was read at an indignation rally in Albany and republished by Democratic newspapers in the North.[34]

The Albany indignation rally passed a series of resolutions which were sent by Erastus Corning to President Lincoln, whose response was the famous Corning letter. The Democrats' answer to that letter merits more fame than it now enjoys. Distributed as a pamphlet by the Society for the Diffusion of Political Knowledge, this *Reply to President Lincoln's Letter of 12th June, 1863* was signed by twenty-three committeemen and their chairman, John V. L. Pruyn, the distinguished lawyer who had advised Governor Seymour on questions having to do with military arrests for his inaugural address six months earlier.[35]

Nothing in the Albany *Reply* matched Lincoln's "simple-minded soldier boy" and "wiley agitator" for popular appeal, but the argument of Pruyn's committee was, if a bit too legalistic for maximum political effect, as precisely written and original as any Democratic argument against Lincoln's policy of military arrests of civilians:

> ...we can not acquiesce in your dogmas that arrests and imprisonment, without warrant or criminal accusation, in their nature lawless and arbitrary, opposed to the very letter of constitutional guarantees, can become in any sense rightful, by reason of a suspension of the writ of *habeas corpus.* We deny that the suspension of a single and peculiar remedy for such wrongs brings into existence new and unknown classes of offenses, or new causes for depriving men of their liberty. It is one of the most material purposes of that writ, to enlarge upon bail persons who, upon probable cause, are duly and legally charged with some known crime; and a suspension of the writ was never asked for in England or in this country, except to prevent such enlargement when the supposed offense was against the safety of the government. In the year 1807, at the time of Burr's alleged conspiracy, a bill was passed in the Senate of the United States, suspending the writ of *habeas corpus* for a limited time *in all cases where persons were charged on oath with treason or other high crime or misdemeanor,* endangering the peace or safety of the government. But your doctrine undisguisedly is, that suspension of this writ justifies arrests without warrant, without oath, and even without suspicion of treason or other crime.

Once Congress had sanctioned the suspension of the writ and absolved Lincoln of the old Democratic charge of "usurpation," the only recourse left to thoughtful Democrats was to insist that most traditional rights should remain unaffected. Democrats believed the Republicans used the suspension of this remedial legal device to invent new crimes unknown to law and new ways of suppressing them. The ironic effect of this was to minimize the importance of the "great writ" of myth.

At still another place in the text of the Democrats' *Reply,* habeas corpus was defined "as a remedial writ, issued by courts and magistrates, to inquire into the cause of any imprisonment or restraint of liberty; on the return of which and upon due examination, the person imprisoned is discharged, if the restraint is unlawful, or admitted to bail if he appears to have been lawfully arrested, and is held to answer a criminal accusation." This bears only a pale resemblance to "that great writ of right, that bulwark of personal liberty... which it cost the patriots and freemen of England six hundred years of labor and toil and blood to extort and to hold fast from venal judges and tyrant kings; written in the great charter at Runnymede"—of which Clement Vallandigham had spoken back in 1861.[36]

Pruyn and his committee were shaping a clever Democratic attack on the policy of military arrests of civilians, as that policy stood after the passage of the Habeas Corpus Act of March 3, 1863. They spoke not of a great writ but rather of one among many technical procedural remedies that should not be confused with the other abundant safeguards of personal liberty in America. They chided Lincoln:

> Inasmuch as this process may be suspended in time of war, you seem to think that every remedy for a false and unlawful imprisonment is abrogated; and from this

postulate you reach, at a single bound, the conclusion that there is no liberty under the Constitution which does not depend on the gracious indulgence of the Executive only. This great heresy once established, and by this mode of induction, there springs at once into existence a brood of crimes or offenses undefined by any rule, and hitherto unknown to the laws of this country; and this is followed by indiscriminate arrests, midnight seizures, military commissions, unheard-of-modes of trial and punishment, and all the machinery of terror and despotism.[37]

John V. L. Pruyn's grasp of the legal issues likely lay at the foundation of this interesting piece of political thought. The pamphlet enjoyed some success. There were at least two different printings by the Society for the Diffusion of Political Knowledge, one in English and one in German. Yet the modern world is more familiar with Lincoln's Corning letter than with the resolutions that provoked it or the response to it by the Albany protesters. The obvious reason for this—aside from the sheer power of Lincoln's posthumous reputation— was the legalistic language and dignified tone of the work of Pruyn's committee. A pamphlet that repeatedly used the word "enlarge" in its legal sense to mean "set at large" could hardly compete in the hearts of Americans with the "simple-minded soldier boy."[38]

The myth of the great writ was too powerful to be restrained by one partisan and legalistic pamphlet. To suggest that the suspension of the writ of habeas corpus amounted to little more than a measure to forbid bail to persons under indictment for treason was more than most people would believe. Some of the words of Pruyn's committee failed to square with the Democratic complaints of the past:

> We repeat, a suspension of the writ of *habeas corpus* merely dispenses with a single and peculiar remedy against an unlawful imprisonment; but if that remedy had never existed, the right to liberty would be the same, and every invasion of that right, would be condemned not only by the Constitution, but by principles of far greater antiquity than the writ itself. Our common law is not at all indebted to this writ for its action of false imprisonment, and the action would remain to the citizen if the writ were abolished forever. Again, every man when his life or liberty is threatened, without the warrant of law, may lawfully resist, and if necessary, in self-defense, may take the life of the aggressor. Moreover, the people of this country may demand the impeachment of the President himself for the exercise of arbitrary power. And when all these remedies shall prove inadequate for the protection of free institutions, there remains, in the last resort, the supreme right of revolution.[39]

When the president replied to a similar protest from the Ohio Democratic state convention about two weeks after the Corning letter, he showed himself to be a believer in the myth of the "great writ." One of the Ohio resolutions stated: "Expunge from the constitution this limitation upon the power of congress to suspend the writ of Habeas corpus, and yet the other guaranties of personal liberty would remain unchanged." Lincoln noted in passing that he did not think it a limitation on *Congress* and went on to say, "the benefit of the writ of Habeas corpus, is the great means through which the guaranties of personal liberty are conserved, and made available in the last resort; and corroborative of this view, is the fact that Mr. V[allandigham]. in the very case in question, under the advice of able lawyers, saw not where else to go but to the Habeas corpus."[40]

The Albany *Reply* retained some of the more routinely partisan aspects of political pamphleteering. Only one of these was marked by rhetorical genius, and that was the Pruyn committee's attack on what surely remains the most troubling paragraph of the Corning letter, Lincoln's broad attempt to impugn the loyalty of those who remained silent or who qualified their loyalty. The president had condemned those who stood silent while their country was in peril and those who maintained a loyalty qualified by "buts" and "ifs" and "ands." To this, the New York Democrats replied: "We think that men may be rightfully silent if they so choose, while clamorous and needy patriots proclaim the praises of those who wield power; and as to the 'buts,' the 'ifs,' and the 'ands,' these are Saxon words, and belong to the vocabulary of freemen."

While the New Yorkers honed such intellectual arguments, Democratic protest in the Old Northwest after the Vallandigham trial took more broadly popular forms. At a Democratic mass meeting in Lima, Ohio, in the fall of 1863, the central theme of the elaborate floats in the giant parade was "Eternal vigilance is the price of liberty." That particular slogan appeared on a wagon holding sixty-four ladies pulled by a sixteen-horse team. "Peace, and no Dictator!" proclaimed another float, while six four-horse teams pulled wagons with girls aged five to nine chanting "Vallandigham and Liberty." Over five hundred women rode horseback in the parade, and there were over three hundred wagons. Eight horses pulled a float called the "Lincoln Bastile," with eight old men representing prisoners in Ohio's different military prisons. Eight men in blackface, dressed in Union uniforms and armed with muskets, guarded the prisoners and occasionally pricked them with their bayonets.[41]

Lincoln and Andrew Jackson

Pruyn's committee did not bother to answer one part of Lincoln's celebrated letter to Erastus Corning and others, the part in which Lincoln reminded his critics of "a bit of pertinent history"—Andrew Jackson's imposition of martial law in New Orleans. The president supplied a dryly factual account of the events. Jackson had acted after the famous battle in 1815 was over and the peace treaty concluded, as was "well known in the city," though "official knowledge" had yet to arrive. The general arrested the author of a denunciatory newspaper article, the lawyer who then represented the writer, the judge who issued a writ of habeas corpus in the case, and still another critic of his actions. Presently, when "the ratification of the treaty of peace was regularly announced," Jackson released the prisoners. The judge then fined him $1000 for contempt of court, and Jackson paid the fine.

Some Democrats preferred not to be reminded of the incident. On March

13, 1863, for example, Harvey Palmer, a Republican assemblyman from New York, noted that in the "entire discussion" of Governor Seymour's message on military arrests and a resolution drafted in the Assembly to inquire into them, "the name of the distinguished statesman and hero of many battles, Andrew Jackson, was not even mentioned by a single democratic member." Indeed, in that debate, the first reference to him was made by a Republican.[42]

Lincoln was probably familiar with the history of the New Orleans incident from his days in the Whig party. In the early 1840s, Congress heatedly debated a proposal to refund Jackson's fine, and one of the leaders supporting the measure was Lincoln's old rival Stephen A. Douglas. Though surely not written purely from memory, Lincoln's account in the Corning letter sounded like an old piece of political ammunition. The president concluded his review of the facts in the Jackson episode by saying, "The late Senator Douglas, then in the House of Representatives, took a leading part in the debate, in which the constitutional question was much discussed. I am not prepared to say whom the Journals would show to have voted for the measure." Lincoln had not been forced to research deeply for this part of the Corning letter.

It may be remarked: First, that we had the same constitution then, as now. Secondly, that we then had a case of Invasion, and that now we have a case of Rebellion, and: Thirdly, that the permanent right of the people to public discussion, the liberty of speech and the press, the trial by jury, the law of evidence, and the Habeas Corpus, suffered no detriment whatever by that conduct of Gen. Jackson, or it's subsequent approval by the American congress.[43]

What Lincoln had thought of Jackson's conduct back in his youthful career in the Whig party, is unknown, though he did not let it mar his admiration for the old general as the hero of the Battle of New Orleans. Lincoln told his fellow legislators in the Illinois House on January 8, 1841, that he was "proud of the victory of New-Orleans, and the military fame of Gen. Jackson, though he could never find in his heart to support him as a politician." Lincoln approved a holiday observation of the event by the legislature and thought it should have "no view whatever to politics."[44]

Lincoln was not the only person to recognize the importance of Jackson's actions in New Orleans and their endorsement by Democrats in the 1840s. Daniel Gardner, for one, a New York lawyer and legal writer, published a so-called "Treatise" on the law of the war late in 1862. It was nothing more than a pamphlet, and a slim one at that, but it was chockablock with citations and references to earlier legal writers. Gardner managed to produce from them a sweeping argument for emancipation and the arming of black freedmen. Its mainstay was his belief that "the national and State courts have recognized the constitutional martial power of the President and his officers, in time of war, to supersede the civil code, including constitutional civil provisions to the extent of adjudged military necessities." And his principal citation as authority

was the "declaratory act of Congress, approving of General Jackson's declaration of martial law at New-Orleans, in 1815," This settled the legal question
for Gardner, whose argument was less a defense of Lincoln than a call for more
vigorous action against slaveholders. In fact, Lincoln hardly needed such allies
in his camp, for Gardner went so far as to say:

> The Roman maxim was *Leges silentur inter arma.* In war the martial supersedes
> the civil, as far as the Roman dictators judged necessary to the public safety. All
> nations have adopted it and acted on it. It is a doctrine essential to defend a nation
> from foreign and domestic foes.[45]

Lincoln never said that law was silent during war, and he surely did not want
his political allies to say so either.

Another person who recognized Jackson's importance was Charles Ingersoll
of Philadelphia, himself the victim of arrest in August 1862 after he gave a
speech denouncing the administration at a Democratic rally. He was quickly
released after Judge John Cadwalader issued a writ of habeas corpus. In December, Ingersoll published *An Undelivered Speech on Executive Arrests,* based
on notes assembled when he represented as counsel John H. Cook, arrested
by the Lincoln administration but released before Ingersoll had opportunity
to make his argument before a judge. The Democratic lawyer referred to "the
case of what is called General Jackson's fine," giving a brief description of the
general's actions and concluding:

> ... time brings with it oblivion of our faults. This fine was by the Congress of the
> United States repaid ... twenty-nine years after. It was repaid under circumstances
> to produce the extremest sanction of Jackson's conduct which it was possible for
> a grateful country to bestow: but without its occurring even to his most ardent
> admirers to go the length of pretending a legal justification of his act.

Ingersoll had good reason to be able to recall the affair, for his own father,
Charles J. Ingersoll, serving as a Democratic congressmen from Philadelphia,
had introduced the bill to repay Jackson's fine in December 1843. But what
the son said of the lack of arguments pleading the legal justification of Jackson's
act was patently false. Stephen A. Douglas, for one, as leader of the floor fight
for the bill in 1844, placed Jackson's defense on the broadest grounds of legal
justification rather than mere personal gratitude:

> The charge of exerting arbitrary power and lawless violence over courts, and
> legislatures, and civil institutions, in derogation of the Constitution and laws, and
> without the sanction of rightful authority, have been so often made (against Jack-
> son) and reiterated for political effect, that doubtless many candid men have been
> disposed to repose faith in their correctness, without taking the pains to examine
> carefully the grounds upon which they rest. A question involving the right of the
> country to use the means necessary to its defence from foreign invasion, in times
> of imminent and impending danger, is too vitally important to be yielded without
> an inquiry into the nature and source of that fatal restriction which is to deprive
> a nation of the power of self-preservation. ... For one, I maintain that, in the exercise
> of this power, General Jackson did not violate the Constitution, nor assume to

himself any authority which was not fully authorized and legalized by his position, his duty, and the unavoidable necessity of the case.[46]

The fact of the matter was that, in 1844 at least, the northern wing of the Democratic party was willing to argue that the military arrests of civilians by Andrew Jackson were "fully authorized and legalized by . . . [the] unavoidable necessity of the case." Democratic party traditions in no way specially prepared the Democracy of the Civil War era to defend civil liberties from martial assault.

Habeas Corpus after the Habeas Corpus Act

Republicans of a more theoretical bent than Lincoln breathed a sigh of relief when they saw the habeas-corpus proclamation of September 15, 1863, the first to appear after Congress's authorization of the suspension on March 3. Sidney George Fisher, for example, author of the *Trial of the Constitution*, noted in his diary on September 16:

> In the paper this morning appeared a proclamation by the President suspending the writ of habeas corpus throughout the U.S. *in accordance with the act of Congress* of March last, that is to say, *not* by executive but by legislative authority. This is a very important step as it settles the question whether the President or Congress has the power to suspend the privilege of the writ and settles it wisely in my judgement.[47]

Neither Lincoln nor the man who had drafted the proclamation, Seward, much cared, but Salmon P. Chase did. The proclamation of September 15, 1863, as Chapter 3 showed, bore a heavy impress from his thinking.

The ambiguity of the Habeas Corpus Act permitted an interpretation entirely different from Fisher's, namely, that Congress had endorsed the president's earlier unilateral suspensions. Thus, popular pamphleteers D. C. Cloud and John H. Power, both of Muscatine, Iowa, stated flatly in 1863 that the "power to make arrests by the political department of the Government in time of war, or to suspend the writ of *habeas corpus* under the law is vested in the President." And they pointed first to the law of March 3 as proof.[48]

Congressional action must have cut the ground from under some potential judicial obstruction. Before the legislation, when the pacifist Campbellite preacher Judson D. Benedict of Aurora, New York, was arrested for discouraging enlistments after the orders of August 8, 1862, a writ of habeas corpus had been issued by Judge Nathan K. Hall, of the United States District Court for the Northern District of New York. Hall, a former law partner of Millard Fillmore's who had served briefly as postmaster general in his cabinet, issued a writ to free Benedict on the grounds that "the President, without the authority of Congress, has no constitutional power to suspend the privilege of the writ of *habeas corpus* in the United States."[49] Now such legal reasoning would be impossible.

But political resistance on constitutional grounds to military arrests of ci-

vilians did not end. This was partly because of the imperatives of partisanship
for the Democratic party. And it was partly because General Ambrose Burnside
foolishly provided the opposition a new issue—trials by military commission—
by arresting and trying a celebrated politician, Clement L. Vallandigham. It was
due as well to Lincoln's decision to uphold Burnside's action and to justify it
in the widely circulated Corning letter without referring to the congressional
act of March 3, 1863.

There was another reason for the increase in criticism—the approach of
national elections in 1864. Attacking the Republican record on civil liberties
became a staple of the Democracy's campaign in 1864. A typical example is
Document No. 13 of a series issued for all Democratic newspaper offices in
1864: *Mr. Lincoln's Arbitrary Arrests.* This, of course, was less a systematic or
theoretical work than a hastily cobbled-together pamphlet meant to list as
many things as possible that might embarrass the Republican administration.
"Clergymen were seized while at prayer at the altar, on the morning of the
sabbath day," the Democrats charged—and "judges . . . for judicial decisions
rendered on the bench" and "ladies . . . subjected to nameless insults . . . and
most foully treated by the officers of Mr. Lincoln." "Mourners were seized at
funerals," and "young children . . . arrested and imprisoned for months, and
even years."[50]

As a critique of constitutional law, *Mr. Lincoln's Arbitrary Arrests* did not
amount to much. After repeating the Fourth, Fifth, Sixth, and Eighth amend-
ments to the Constitution—forbidding unreasonable searches and seizures;
guaranteeing indictment by grand jury for capital crime, due process of law,
and trial by jury; and outlawing cruel and unusual punishment—the Democratic pamphleteers gave a brief history of Lincoln's despotic acts. They praised
Taney's *Merryman* decision as the work of a "noble old judge," and they alluded
to the orders of August 8, 1862. They commented on the Habeas Corpus Act
of March 3, 1863, by saying that Lincoln had had "no color of authority" for
the arrest of Merryman in 1861 and "his own party admitted his usurpation
long afterward, and after hundreds of like cases had occurred, by passing an
act of Congress (*ex post facto*) to save him from the consequences of his
arbitrary use of power." Leaving aside the question whether the act legalized
such arrests after 1863 and failing to mention martial law, the Democrats
concluded by saying:

> During all these years the courts were open, and justice might have been admin-
> istered according to the Constitution. There was no possible excuse for substituting
> the will of the President for the arm of the law. However guilty a man was, he
> could and should have been dealt with according to the Constitution and laws of
> his land.[51]

By the time of the presidential campaign of 1864, Democrats had become
so well versed in denunciations of military arrests of civilians that even an
obscure Illinois politician in a rural district could ring interesting changes on
the theme. Thus, Andrew D. Duff of Benton, Illinois, addressed a Democratic

mass meeting in Williamson County on May 28, 1864, on *Footprints of Despotism, or the Analogy between Lincoln and Other Tyrants.* He knew what he was talking about, for Duff himself had been arrested in the summer of 1862 for discouraging enlistments, though he did not mention it in the speech. The speech was extreme but stopped well short of expressions of disloyalty, exalting instead a tradition of patriotic defenders of civil liberty "from Jefferson at the foundation of the republic to Vallandigham in banishment." Duff predicted that *"Abe Lincoln* will not allow us freedom of speech through the coming canvass, and will not permit us to vote at the next November election," but he did not counsel violence or unlawful resistance to government authority.

Duff's most original contribution lay in amassing a string of historical examples to prove the dangers of despotism that lay in Lincoln's acts. Using Kreighly's *History of England,* he recalled the times of the Stuarts, when "Sir Thomas Darnell, John Corbett, Walter Earl, John Heviningham and Edward Hampton . . . applied to the court of King's Bench for their writ of HABEAS CORPUS, the writ was granted but the warden of the fleet made return that the warrant of the privy council assigned no cause for their imprisonment, [more than ten thousand freemen have been imprisoned by Abe Lincoln and his minions in the same way.] [bracketed material is Duff's addition] the cause was argued on the 7th day of November before the court over which Sir Nicholas Hyde presided, Noy and Seldon eminent lawyers for the prisoners argued from MAGNA CHARTA that, 'NO FREEMAN SHALL BE TAKEN OR IMPRISONED UNLESS BY LAWFUL JUDGEMENT OF HIS PEERS OR THE LAW OF THE LAND.'" Duff recalled as well the specter of Oliver Cromwell.[52]

Though denunciations of "arbitrary arrests" escalated in volume and intensity over the course of the war and into the 1864 election year, they *never* became the *principal* focus of the Democratic party. The record of their presidential nominee would not permit it, for George B. McClellan had been a part of the scheme to arrest the members of the Maryland legislature back in 1861. This became an issue within the Democratic party before McClellan's nomination. Horatio Seymour, McClellan's closest rival for the nomination, had opportunity to make a sharp issue of civil liberties in the summer before the presidential nominating convention met. New York had experienced especially unsettling events. Memories of the bloody draft riots of the previous summer made both Democrats and Republicans nervous. The Lincoln administration's seizure of two newspapers in New York City, the *World* and *Journal of Commerce* in mid-May 1864, because of publishing a bogus draft proclamation, heightened tensions. Governor Seymour did not help soothe public anxieties when he called out 75,000 of the state militia, ostensibly to defend the state or to assist the national government in repelling future invasion. He warned ominously, "In addition to the dangers of invasion from without, and of popular dissentions at home, we have been warned by recent events of the STILL GREATER DANGER of arbitrary encroachments upon our liberties as citizens." The Democratic press hailed Seymour's determination "to show that there is one state in the Union prepared to resist tyranny."

Though no dramatic confrontation between state and federal forces in New York occurred, Seymour kept the issue alive by alluding to civil liberties in his speech renouncing the nomination at the Democratic convention. He threw New York's support to General McClellan and then attempted to excuse the nominee's record for the sake of Maryland's disgruntled delegates:

> I wish to say a few words with regard to the objections which have been urged against his nomination, and which have caused some excitement to this convention. I speak more particularly of the objections urged by the delegation from Maryland. ... With respect to the orders issued by General McClellan affecting the citizens of that State, I must say that I do not approve of them; but they must not be viewed in the light which events have since thrown upon the policy of the administration. At that time the wisest and best men of our country had confidence in its purposes. Then, the President denounced measures which he has since adopted; then the friends of the Union in the border States were listened to by him, with every appearance of respect and deference. The mask had not been thrown off, and obedience to his orders did not imply hostility to the rights of States. We must bear in mind how at that moment the public was convulsed by a condition of affairs without precedent, and by questions which were suddenly forced upon the public attention, and with regard to which, public men were compelled to act without time for reflection. What man can say after looking back over his own action during the past three years, that he had not fallen into many and grave errors with respect to his duty? God knows, I cannot.[53]

It must have pained Seymour to find excuses for McClellan, especially after the New York governor had blazed a clear trail for the Democrats on the issue of civil liberties in the North during the war.

Candidate and platform were notoriously unsynchronized in the Democratic canvas that followed this nominating convention in 1864. Despite McClellan's record, some Democrats took their cue from the Ohio gubernatorial campaign of the previous year. For example, a Democratic parade in Fort Wayne, Indiana, in October of 1864 included a "Lincoln Bastille" float, a cage with a man inside guarded by four men in blackface dressed as Union soldiers.[54]

No single arrest in 1864 dramatized or symbolized the civil-liberties issue, as Vallandigham's had in 1863. Valiant Val himself remained too controversial to provide the best symbol. After all, this was a year when the party's ticket was headed by a general running on an avowed promise to restore the Union by waging the war better than the president could. Mainstream Democratic editors, those opposed to the peace wing of the party, disavowed Vallandigham's principles even when they stuck up for his rights.[55] The more recent arrests of Lambdin P. Milligan and others in Indiana netted some Democrats deeply compromised by contacts with Confederate agents and may not have initially hurt the Republicans at all.

Governor Seymour attempted to blend the old civil-liberties issue with the 1864 campaign for McClellan. Delivering what was touted by some Democrats as "The Great Speech of the Campaign" in Philadelphia in October, the tired and ill New York governor adhered to what had become the main Democratic theme, namely, that the harsh way in which the Republican administration waged the war would make reunion of the two sections impossible. This was mostly a veiled reference to emancipation, and it recalled what Seymour had said at the nominating convention about Lincoln's early disavowal of "measures which he has since adopted." Emancipation would keep the South from the peace table and so would atrocities perpetrated on Southern civilians. He kept the focus on the border states for a long while in the speech, noting that incompetence in the handling of Missouri and Kentucky now made huge armies of occupation necessary. These unhappy states, which initially did not want to leave the Union, were now the scenes of atrocious applications of military power.

Maryland was another matter, at least now that the Democrats had chosen George B. McClellan as their standard bearer. Seymour did not say as boldly as he had in the summer that he disapproved of the early measures taken to hold Maryland in the Union. Instead, he admitted that, unlike the cases of the states of Kentucky and Missouri, "It is, perhaps, doubtful what course that state should have taken if left free to act for herself. But I ask you, if for the past three years the control of the government [in Washington]...has been of a character to bring that state back to its allegiance?"

Governor Seymour drew thunderous applause when he stressed the issue which had distinguished him as a politician in the Civil War:

> War does not extinguish liberty. We fought our war of the revolution to win, preserve and perpetuate liberty. War does not suspend the rights of men; and he who dares to say that Abraham Lincoln, at the head of his enormous armies, may rightfully do what George Washington would not do in the darkest hour of the revolution does not know what constitutional liberty is.[56]

Seymour had thus tailored his ideas to blend with McClellan's. For his part, the hero of Antietam did not enunciate many political ideas during his whole life, but those he did hold were fairly consistent and did not include any marked enthusiasm for libertarianism. McClellan's central idea on the Civil War was that it should be waged in a restrained way. This was prominent in his thought from the time of the famous Harrison's Landing letter of July 8, 1862, on. As he put it then, the Southern rebellion

> assumed the character of a war; as such it should be regarded, and it should be conducted upon the highest principles known to Christian civilization. It should not be a war looking to the subjugation of the people of any State in any event. It should not be a war at all upon population, but against armed forces and political organizations. Neither confiscation of property, political executions of persons, territorial organization of States, or forcible abolition of slavery should be contemplated for a moment....Military arrests should not be tolerated, except in places where actual hostilities exist; [military government should be] confined to the preservation of public order and the protection of political rights.

McClellan was thinking mainly of slavery, the abolition of which he opposed, at least on practical grounds. "A declaration of radical views," he added, "especially upon slavery, will rapidly disintegrate our present armies."[57]

These remained the general's political views to the end of the war. They were echoed in the Democratic press, mainly in their repeated assertions that emancipation would prevent a restoration of the Union by forcing the South to fight to its last man. Occasionally, the press complained of atrocities like the burning of Southern towns. As for the theme of military arrests, so well developed by McClellan's political rival Horatio Seymour, the general hardly pursued it further. As close as he came was in an ill-advised letter, written in the autumn of 1863, supporting Democrat George W. Woodward's candidacy for governor of Pennsylvania. Woodward, a Pennsylvania Supreme Court justice, well known for believing conscription unconstitutional, was hardly a darling of the more warlike Democrats. Nevertheless, McClellan's letter stated:

> I understand Judge Woodward to be in favor of the prosecution of the war with all the means at the command of the loyal states, until the military power of the rebellion is destroyed. I understand him to be of the opinion that, while the war is urged with all possible decision and energy, the policy directing it should be in consonance with the principles of humanity and civilization, working no injury to private rights and property not demanded by military necessity and recognized by military law among civilized nations. And, finally, I understand him to agree with me in the opinion that the sole great objects of this war are the restoration of the unity of the nation, the preservation of the constitution, and the supremacy of the laws of the country. Believing our opinions entirely agree upon these points, I would, were it in my power, give to Judge Woodward my voice and vote.[58]

McClellan's central message remained opposition to emancipation. Saying in late 1863 that the war's sole objects were the preservation of the Union and of the Constitution was pointedly to criticize the antislavery policies of the Republican administration. To be sure, the Constitution stood as a reminder of the allegedly unconstitutional methods of the Lincoln administration in policing the home front, but McClellan hardly went out of his way to stress that theme. Nowhere, after the Harrison's Landing letter, did he denounce any of the specific measures of the Lincoln administration affecting civil liberties.

Although the Democratic party of the Civil War era has been faulted for failing to move toward the ideological center, surely the party could hardly have made a more dramatic such move than in nominating McClellan. In fact, they followed the classic centrist pattern in American politics, nominating a military hero who stood for very little. By his own admission, McClellan had attempted to avoid political issues until 1863. When the Democrats repudiated the peace wing of the party in the 1864 nomination, they also turned their backs on the civil-liberties issue, for McClellan showed little interest in it and was personally tarnished by involvement with one of the most famous incidents involving military arrests of civilians in the whole war—the arrest of the Maryland legislators.

The Democratic party never ceased altogether to agitate the civil-liberties issue. The local press, here and there, stressed an arrest or a trial by military commission. But the protests became sporadic. The Democrats' opposition to military arrests of civilians during the Civil War was fairly short-lived and decidedly opportunistic. Even after Roger B. Taney led the way with the *Merryman* decision, Democrats did not follow *en masse* into the constitutional breach. Outside Maryland itself, there were few protests of the administration's rough handling of Maryland in 1861. Troubled Republicans did as much as Democrats to keep the civil-liberties issue alive until 1863.

Democratic opposition, led by the shrewd Horatio Seymour, proved intense and steady from early 1863 until the autumn of 1864, but McClellan's presidential nomination definitely blurred the party's focus on constitutional issues. Opposition to emancipation was the party's only consistent ideological theme that McClellan's campaign continued. Civil libertarianism was no more naturally Democratic ideological territory than it was Republican; the former had their Andrew Jackson and the latter, their Abraham Lincoln. When the wartime economy of high employment robbed the Democrats of any chance to use their traditional economic appeals, they simply floundered for a time and they switched—for a surprisingly brief period—to civil liberties. But they let that drop, more or less, late in 1864 to embrace the only issue left to them, and one they were familiar with using aggressively since the 1850s—race.

It is not entirely true to say, as does the leading historian of the Civil War Democratic party, that their "arguments grew out of an ideology rooted in their traditions and experiences and the perceptions developed in their past about the role and power of government, about the nature of the Constitution, and about the direction of racial and social policy in the nation." Nor is it altogether true to say, as does the leading constitutional historian of the Civil War, that Democrats "applied familiar static constitutional doctrines . . . to the novel assertions from Washington about war powers . . . they adopted an idiom of antipathy to martial ways and loyalty tests, of obeisance to the Bill of Rights, and . . . of respect for courts as defenders of individuals' liberties against internal-security policies." The Democrats had no more claim to the mantle of protector of civil liberties in wartime than did the Republicans. There was nothing "rooted in traditions" of the Democratic party or "familiar" to their party in their Civil War protests. That is why their protests started late and ended early. Moreover, Democrats thrashed about in search of fresh constitutional and legal ground on which to attack the administration's policies on civil liberties. Several Democrats began to express arguments that would have turned the "great writ" into an obscurely technical legal instrument.[59]

Only when they attempted to play down the importance of the writ after 1863, did the Democrats remain true to their antebellum views. Before the war, Northerners who were eager to protect blacks from slave-catchers had helped broaden the nature of the writ of habeas corpus. Gradually, Northern judges came to use it to review the evidence and merits of a case. They proved no longer satisfied with a return to the writ that would have been adequate in earlier times, one that showed that the person detained was held in pursuance of law. The judges would themselves determine the guilt or innocence of the person in the case. They would even decide whether the law under which the detainee was held was constitutional.[60] To the degree that Democrats protected the slave owners' interests in these matters, they too desired a narrowing of Northern judges' powers under habeas-corpus law. When Civil War came, Democratic judges acted differently. Taney, a judicial activist, thus interpreted the Constitution for the president when the writ in the *Merryman* case was ignored. The Pennsylvania judges who so disturbed Lincoln may have acted thus in the conscription cases. After 1863, Seymour's wing of the Democratic party, at least, began to retreat from such bold views of habeas corpus, to attempt to salvage freedom by belittling the liberties really protected by habeas corpus and to assert that other protections made most liberties secure. When Lincoln, on the other hand, said the great writ really was great, he was true to Republican traditions.

Otherwise, there was nothing especially sincere or characteristic of the party's traditional role in Democratic protests against the circumscription of civil liberties in the Civil War. Had the Democrats been in office and the Republicans in opposition during the war, their roles on this question would surely have been reversed, with the Democrats pointing proudly to the Jacksonian precedents for imposing martial law and the Republicans whining about military dictatorship.

This is not to say that Democratic opposition was wrong or impeded the war effort. Their opportunistic protests played a role crucial to a democratic country involved in war. They helped keep the army and the Republicans honest. They helped prevent the U.S. Army from an increasing reliance on military justice for the sake of convenience (for example, in the Indiana court decision on selling liquor to soldiers). Their protests about political abuse of the internal security system forced Abraham Lincoln, in his last habeas-corpus proclamation (the one for Kentucky issued in 1864) to disclaim any attempt to interfere with the electoral process. The Democratic party was a loyal opposition. If their rhetoric was calculated to goad Republicans, it also played a legitimate role in preserving civil liberties in wartime America.

10

Lincoln and the Constitution

The Democratic depiction of Lincoln as a tyrant was to have more influence on history than it merited, but like many political caricatures, it contained a certain element of truth. To be sure, there was nothing of the dictator in Lincoln, who stood for reelection in 1864 and, until General Sherman captured Atlanta, genuinely feared that he would lose the presidency. But he did not by habit think first of the constitutional aspect of most problems he faced. His impulse was to turn to the practical.

Whig Heritage

Lincoln had been a Whig for most of the life of that political party—twice as long as he was a Republican. And the Whigs generally took a broad view of what the Constitution allowed the federal government to do (create a national bank and fund the building of canals, roads, and railroads, for example). As a victim of rural isolation and lack of economic opportunity in his youth, Abraham Lincoln proved eager as a politician to provide the country with those things that seemed wanting in his hardscrabble past. His desire to get on with economic development made him impatient with Democratic arguments that internal improvements funded by the federal government were unconstitutional.

After years of political struggle to implement improvement schemes, Lincoln, as a congressman in the late 1840s, saw "the question of improvements . . . verging to a final crisis." The Democratic national platform in 1848 declared that "the constitution does not confer upon the general government the power to commence, and carry on a general system of internal improvements." Speaking on the subject in the House of Representatives, the 39-year-old Lincoln expressed plainly his mature judgment that "no man, who is clear on the questions of expediency, needs feel his conscience much pricked upon this."

Emphasis on the practical was characteristic of Lincoln, but his confidence in this instance stemmed in part from a belief that the constitutional arguments were also on his side. In the Civil War, Lincoln would again suggest practical reasons for action and then add assurances and proofs that the Constitution permitted it anyhow.

In his 1848 speech on the internal improvements crisis, Lincoln laid unusual emphasis on constitutional subject matter. Despite his assertion that practical demands for internal improvements should weigh heavily against constitutional doubt or controversy, Lincoln seemed preoccupied with constitutional questions in the speech, devoting eight of twenty-six paragraphs, almost a third of his time, to that issue. He began these arguments with a modest disclaimer:

> Mr. Chairman, on the . . . constitutional question, I have not much to say. Being the man I am, and speaking when I do, I feel, that in any attempt at an original constitutional argument, I should not be, and ought not to be, listened to patiently. The ablest, and the best of men, have gone over the whole ground long ago.

Lincoln followed this by quoting and summarizing at some length arguments from Kent's *Commentaries.*

The Democratic president, James K. Polk, had suggested that it would require a constitutional amendment to make such internal improvements possible. Lincoln did not much like this idea, in part no doubt because of its impracticability and time-consuming nature, but in the speech, he attacked it in the language of sweeping constitutional conservatism.

> I have already said that no one, who is satisfied of the expediency of making improvements, needs be much uneasy in his conscience about it's constitutionality. I wish now to submit a few remarks on the general proposition of amending the constitution. As a general rule, I think, we would [do] much better [to] let it alone. No slight occasion should tempt us to touch it. Better not take the first step, which may lead to a habit of altering it. Better, rather, habituate ourselves to think of it, as unalterable. It can scarcely be made better than it is. New provisions, would introduce new difficulties, and thus create, and increase appetite for still further change. No sir, let it stand as it is. New hands have never touched it. The men who made it, have done their work, and have passed away. Who shall improve, on what *they* did?[1]

Often quoted in later years, this passage had rather a different meaning in context from what constitutional conservatives since his day have imagined. What Congressman Lincoln was really saying was that no amendment was needed if a reasonably broad interpretation of the existing document were accepted.

Before this unusual speech, Lincoln had in his sixteen-year career in politics rarely gone on record on constitutional questions. In 1832, when he wrote at length on internal improvements for the Sangamon River in his first published

political platform, young Lincoln had dwelled exclusively on practical matters
of cost and navigability. In 1836, when he declared his candidacy for reelection
to the state legislature, the 27-year-old Whig again favored a plan to make
internal improvements possible and focused exclusively on financial questions:
"Whether elected or not, I go for distributing the proceeds of the sales of the
public lands to the several states, to enable our state, in common with others,
to dig canals and construct rail roads, without borrowing money and paying
interest on it." Except for a brief comment the next year on the legality of the
Illinois State Bank under the state constitution, Lincoln first spoke on a constitutional question in his speech on Martin Van Buren's subtreasury scheme
delivered December 26, 1839. He stressed the practical advantages of a national
bank—increased circulation of money as well as cheaper and safer operation—
over the Democrats' proposal. Then he addressed the question of constitutionality. Lincoln was satisfied that the U.S. Supreme Court had declared a
national bank constitutional, as had a majority of the country's famous founders.
Rather than go over that well-worn path again, however, he now wanted "to
take a view of the question . . . not . . . taken by anyone before. It is, that whatever
objection ever has or ever can be made to the constitutionality of a bank, will
apply with equal force in its whole length, breadth and proportions to the
Sub-Treasury." If there were no "express authority" in the Constitution to
establish a bank, he quipped, there was none to establish the subtreasury
either.[2]

Lincoln thought them both constitutional, of course. After all, the Constitution specified general authority "to make all laws necessary and proper"
for carrying into effect the powers expressly enumerated, and among those
enumerated was Congress's power to lay and collect taxes and to pay the
debts of the United States. To accomplish those things, the government had
to be able to collect, keep, transfer, and disburse revenues. The arguments
of his adversaries on this question made Lincoln so impatient that at one
point he dismissed them as "too absurd to need further comment." This
was not the tone he customarily used when arguing about matters of expediency. There he usually managed to find some sympathy for those who
held opposing beliefs, saying even of Southern slaveholders in 1854 that
they were "just what we would be in their situation." The practical legisla-
tor from Illinois was not comfortable on the high ground of inflexible constitutional principle.[3]

In the 1840s, Lincoln appeared to be marching steadily toward a position
of gruff and belittling impatience with constitutional arguments against the
beleaguered Whig program. A set of resolutions drafted by Lincoln and adopted
at a Whig meeting in Springfield in 1843 reiterated his position on the proven
constitutionality of a national bank and followed with this abrupt dismissal of
Democratic arguments against distribution of the proceeds from the sale of
the national lands: "Much incomprehensible jargon is often urged against the
constitutionality of this measure. We forbear, in this place, attempting to answer
it, simply because, in our opinion, those who urge it, are, through party zeal,

resolved not to see or acknowledge the truth." But Lincoln's movement away from constitutional modes of thinking was halted abruptly by the presidency of James K. Polk.[4]

When Lincoln spoke against the subtreasury in 1839, he devoted three of fifty-nine paragraphs to the constitutional issue; when he made his last-ditch defense of internal improvements (while Polk was president in 1848), he devoted eight of the twenty-six paragraphs to the constitutional question. What had changed was Lincoln's awareness of the importance of constitutional issues in general. And that heightened awareness was a result of the Mexican War.

Lincoln hated the war, which he considered "unconstitutional and unnecessary." He was not yet, if he ever became one, an internationally minded man. Lincoln did not worry much about Mexico for the sake of Mexicans. In fact, in a lecture on discoveries and inventions he gave years later, Lincoln celebrated the Yankee *"habit* of observation and reflection" which he thought responsible for the quick discovery of gold in California, "which had been trodden upon, and over-looked by indians and Mexican greasers, for centuries." The slavery issue was not the key to his opposition to the Mexican War either. While campaigning for Zachary Taylor in the summer of 1848, Lincoln stated that (as the press reported it) he "did not believe with many of his fellow citizens that this war was originated for the purpose of extending slave territory."[5]

Lincoln maintained instead that "it was a war of conquest brought into existence to catch votes." The war was unnecessary "inasmuch as Mexico was in no way molesting, or menacing the U.S." and unconstitutional "because the power of levying war is vested in Congress, and not in the President." Polk's motive for starting the war with Mexico "was to divert public attention from the surrender of 'Fifty-four forty or fight' to Great Brittain [*sic*], on the Oregon boundary question."[6]

When Lincoln's law partner, William H. Herndon, also an active Whig, disputed this interpretation of the origins of the Mexican War, Lincoln engaged in a rare exercise for him: a long letter, lecturing in tone, on a constitutional question. Herndon's letter, which has not survived, may have caused his partner to focus on the constitutional question, for Lincoln's letter begins, "Your letter of the 29th. Jany. was received last night. Being exclusively a constitutional argument, I wish to submit some reflections upon it." Whatever the cause, once focused, Lincoln's scrutiny proved close and intense:

The provision of the Constitution giving the war-making power to Congress, was dictated, as I understand it, by the following reasons. Kings had always been involving and impoverishing their people in wars, pretending generally, if not always, that the good of the people was the object. This, our Convention understood to be the most oppressive of all Kingly oppressions; and they resolved to so frame the Constitution that *no one man* should hold the power of bringing this oppression upon us. But your view destroys the whole matter, and places our President where kings have always stood.[7]

Slavery and the Constitution

At the very time that Abraham Lincoln's awareness of constitutional questions was on the rise, the issue of slavery in the territories was injected into American politics. The Wilmot Proviso, which would have forbidden slavery in any territory acquired as a result of the Mexican War, came up several times while Lincoln was in Congress, and he consistently voted for it. But the slavery controversy did not make a constitutional thinker of Lincoln, any more than the old economic issues of the 1830s and 1840s had.

This sets Lincoln apart from his era. A leading historian of the political ideas of the Whig party, Daniel Walker Howe, has criticized its era for a tendency to make every political question into a constitutional question: "The tendency to debate the constitutionality of issues rather than their expediency did little to temper the discussion; if anything, it exacerbated differences." This volatile constitutionalism became even more a factor in the era that followed. The 1850s, as Don E. Fehrenbacher has pointed out, witnessed the increasing "fashion of constitutionalizing debate on slavery." When pressed, Lincoln voiced an antislavery interpretation of the Constitution, but he was not one to constitutionalize the debate over slavery or anything else.[8]

Lincoln did think more about the Constitution after 1848 than in previous decades, but his ideas were quite unoriginal. He viewed the document as most antislavery moderates did, shunning the anti-Constitution "covenant-with-death" views of the abolitionists and their unconstitutional political positions as well. He embraced the interpretation of the Constitution as a reluctant guarantor of the slave interest existing at the time of the country's founding. The Constitution betrayed the basically antislavery sentiments of its authors by hiding slavery "away, . . . just as an afflicted man hides away a wen or a cancer, which he dares not cut out at once, lest he bleed to death; with the promise, nevertheless, that the cutting may begin at the end of a given time." Like many of his fellow Republicans, Lincoln attributed great importance to the absence of any explicit mention of slavery or the Negro race in the document. It seemed a sure sign that the founders looked forward to the day when, with slavery eradicated by time, there would be "nothing in the constitution to remind them of it."[9]

Lincoln was a lawyer, but antislavery sentiment and Whig tradition go farther than professional outlook to explain Lincoln's views of the Constitution. The influence of Lincoln's profession on his political ideas has been exaggerated in recent years: "the last Blackstone Lawyer to lead the nation," one writer calls him. Such views have been expressed especially by biographers and historians interested in what is widely regarded as "Lincoln's first speech of distinction," his address to the Young Men's Lyceum of Springfield, delivered January 27, 1838. This speech contained not so much constitutional views as cheerleading for the laws of the land and was widely quoted in later years:

Let reverence for the laws, be breathed by every American mother, to the lisping babe, that prattles on her lap—let it be taught in schools, in seminaries, and in colleges;—let it be written in Primmers, spelling books, and in Almanacs;—let it be preached from the pulpit, proclaimed in legislative halls, and enforced in courts of justice. And, in short, let it become the *political religion* of the nation; and let the old and the young, the rich and the poor, the grave and the gay, of all sexes and tongues, and colors and conditions, sacrifice unceasingly upon its altars.

Lincoln mentioned the Constitution itself at the end of the speech, when he invoked "Reason, cold calculating, unimpassioned reason" to "furnish all the materials for our future support and defense. Let those [materials] be moulded into *general intelligence,* [sound] *morality* and, in particular, *a reverence for the constitution and laws.*"[10]

Thus, Lincoln gave "eloquent expression to the developing ideology of his profession," according to historian George M. Fredrickson, who sees "Lincoln's early speeches as an aspiring young lawyer and Whig politician" as part of a "'conservative' response to the unruly and aggressive democracy spawned by the age of Jackson." Indeed Fredrickson finds this conservative law-and-order strain in Lincoln's political thought substantially unshaken until the *Dred Scott* decision of 1857 undermined "Lincoln's faith in the bench and bar as the ultimate arbiters of constitutional issues." The problem with such an interpretation stems mainly from its approach, that of "intellectual history," for Abraham Lincoln was neither an intellectual nor a systematic political thinker. He was a politician, and historians ignore the instrumental side of his political thought only at great peril. He rarely thought abstractly about the Constitution and the laws. He usually thought about them when a particularly pressing political problem arose. At the time of the Lyceum speech in 1838, Lincoln's recent admission to the Illinois bar was surely a less important circumstance than the political situation. The purpose of the speech was to urge the protection of unpopular minorities. Lincoln mentioned recent headlines describing mob violence and vigilante justice that victimized Mississippi gamblers and unfortunate black people. Most other interpreters of Lincoln's speech in modern times have assumed that the real shadow hanging over it was that of the martyred Elijah Lovejoy, killed just prior to the address by an anti-abolition mob in Alton, Illinois. To say that Lincoln here was at odds with democracy in the age of Jackson is either unfair to Lincoln or paints a dark caricature of Jacksonian democracy.[11]

Lincoln was not searching so much for order and community as for usable arguments and instruments. That is not to say that his constitutional thinking was nakedly opportunistic or embarrassingly shallow, but only that he changed his mind from time to time and that he did not characteristically reach first for a copy of the U.S. Constitution when confronted with a political or social problem. Even to survey Lincoln's ideas on the Constitution is to run the risk of overemphasizing his constitutional concerns, because thinking in constitutional ways did not come naturally to him. It was more often forced on him—by a rigid Democratic president like Polk or by his own law partner's arguments against the articles of their party faith.

Whatever the Lyceum address may seem to mean, it is, in fact, difficult to find any tough threads of legalistic, procedural, or constitutional conservatism woven into Lincoln's political thought of the 1850s, even before the *Dred Scott* decision. Lincoln quickly embraced a moralistic antislavery ideology that pointed to the Declaration of Independence and the political libertarianism of Thomas Jefferson as its fundamental source while relegating the Constitution and the laws to a rather pale secondary role. Shortly after the passage of the Kansas-Nebraska Act in 1854, Lincoln told an audience in Springfield on October 4 that the "theory of our government is Universal Freedom. 'All men are created free and equal,' says the Declaration of Independence. The word 'Slavery' is not found in the Constitution." His political message varied little from this until 1861.[12]

Jefferson and the Declaration of Independence assumed a conspicuous place in Lincoln's political imagery in this period. Nevertheless, careful readers of the previous paragraph will have noted, perhaps a little impatiently, that the Declaration did *not* say that all men were created "free" and equal. Lincoln did not take his political ideas straight and by rote from any single printed source, but of their many sources, the slogans of Thomas Jefferson proved to be of greater importance than the words of the Constitution and of increasing importance to Lincoln after 1854. On October 16, 1854, he spoke of "Mr. Jefferson, the author of the Declaration of Independence," as "the most distinguished politician of our history." He justified his anti-Nebraska stand by saying that "the policy of prohibiting slavery in new territory originated" with Jefferson and the Northwest Ordinance. The policy began "away back of the constitution, in the pure fresh, free breath of the revolution." The hint in this statement that the Constitution may have smothered the free spirit of the revolutionary era never was carried further in Lincoln's thought. He was not a systematic thinker. He was a politician, and few mainstream political candidates wanted to be placed in opposition to the government's founding document.

Instead, antislavery Republicans like Lincoln embraced an antislavery interpretation of (or created an antislavery myth about) the Constitution. "This same generation of men," he said, "mostly the same individuals . . . who declared this principle [of self-government]—who declared independence—who fought the war of the revolution through—who afterwards made the constitution under which we still live—these same men passed the ordinance of '87, declaring that slavery should never go to the north-west territory." With such language as this, Lincoln made of the founders a single cohort of heroes who drafted the Declaration of Independence, won the Revolution, and wrote the Constitution.[13]

In truth, the Constitution stood as an embarrassment to the antislavery cause. It protected slavery in the states as surely as it did anything, and all politicians, Republican and Democrat alike, knew it. The best the antislavery politicians could do was to find antislavery tendencies in the document. In building a mythical past for his political platform, Lincoln preferred to state the antislavery interpretation of the Constitution and get on quickly past that document to the Declaration of Independence. In a speech in Chicago on July 10, 1858, for example, he said, "We had slavery among us, we could not get our constitution unless we permitted them to remain in slavery, we could not secure the good we did secure if we grasped for more, and having by necessity submitted to that much, it does not destroy the principle that is the charter of our liberties. Let that charter stand as our standard." The spirit of the Constitution, properly and carefully looked at, was antagonistic to the Kansas-Nebraska bill, Lincoln could say after elaborate argument, but it was easier to say that the "spirit of seventy-six" and "the spirit of Nebraska" were "utter antagonisms."[14]

After the *Dred Scott* decision, Lincoln's constitutional views changed little, and his overall political thought, less. The Taney court's decision may have accelerated his rush to the Declaration of Independence. In a Springfield speech after the decision, Lincoln asked: "I should like to know if taking this old Declaration of Independence, which declares that all men are equal upon principle and making exceptions to it where will it stop.... If that declaration is not the truth, let us get the Statute book, in which we find it and tear it out!" Of course, the Declaration of Independence was not law, as the Constitution was, and could not properly be located in a "statute book." Lincoln knew this and, when not on the stump, could write about it in more lawyerly fashion. In an 1858 letter to an Illinois politician named James N. Brown, Lincoln said more soberly: "I believe the declara[tion] that 'all men are created equal' is the great fundamental principle upon which our free institutions rest; that negro slavery is violative of the principle; but that, by our frame of government, that principle has not been made one of legal obligation." The *Dred Scott* decision merely forced Lincoln to articulate his view of what makes a lasting Supreme Court decision, which he did with characteristic avoidance of Latinate distinctions. Speaking rhetorically to the Southern people early in 1860, Lincoln pointed out what he thought were good reasons for doubting the force of this Supreme Court decision:

> Perhaps you will say the Supreme Court has decided the disputed Constitutional question in your favor. Not quite so. But waiving the lawyer's distinction between dictum and decision, the Court have decided the question for you in a sort of way. The Court have substantially said, it is your Constitutional right to take slaves into the federal territories, and to hold them there as property. When I say the decision was made in a sort of way, I mean it was made in a divided Court, by a bare majority of the Judges, and they not quite agreeing with one another in the reasons for making it; that it is so made as that its avowed supporters disagree with one another about its meaning, and that it was mainly based upon a mistaken statement of fact—the statement in the opinion that "the right of property in a slave is distinctly and expressly affirmed in the Constitution."[15]

Here, in challenging Taney's careless opinion, the Republicans' emphasis on the absence of the words "slave," "slavery," or "property" in connection with the idea of slavery, was all the constitutional doctrine Lincoln needed.

War and the Constitution

Once Lincoln became president and faced civil war, his clear record on the Constitution became paradoxical and unclear. To be sure, his constitutional outlook all along had left room for the argument of "Necessity." Lincoln demonstrated this years before the Civil War, and not only in recognition of the founders' necessary protection of slavery in the U.S. Constitution. In 1854, while reviewing the history of the slavery-expansion controversy in a speech in Bloomington, Illinois, Lincoln stated matter-of-factly: "Jefferson saw the necessity of our government possessing the whole valley of the Mississippi; and though he acknowledged that our Constitution made no provision for the purchasing of territory, yet he thought the exigency of the case would justify the measure, and the purchase was made." As a Whig and a critic of the Mexican War, Lincoln's record was not as pro-expansion as that of most western politicians, but he admired Jefferson and agreed that the Louisiana Purchase was too good an opportunity to miss, no matter what the Constitution said. The prompt development of an attitude of indifference to the niceties of constitutional interpretation involved in suspending the writ of habeas corpus might have been predicted from the unpricked constitutional conscience of Lincoln's pre-presidential career. He naturally responded vigorously to the exigency of civil war. But on the question of emancipation, Lincoln appeared to some antislavery advocates at the time and to many historians since to have been strangely stricken with a paralyzing constitutional scrupulousness. When it came to putting the spirit of seventy-six into action, Lincoln as president grew suspiciously reluctant.[16]

The most telling event was his revocation of Frémont's emancipation proclamation in Missouri in 1861. In a moment of pique reminiscent of his argument with fellow Whig Herndon over the Mexican War thirteen years earlier, President Lincoln now lectured fellow Republican Orville Hickman Browning on the constitutional issues involved. As Chapter 2 noted, Lincoln had not bothered to mention the Constitution or to dwell on law in first reprimanding Frémont, but once Browning brought it up, the president waded in:

> Genl. Frémont's proclamation, as to confiscation of property, and the liberation of slaves, is *purely political,* and not within the range of *military* law, or necessity. If a commanding General finds a necessity to seize the farm of a private owner, for a pasture, an encampment, or a fortification, he has the right to do so, and to so hold it, as long as the necessity lasts; and this is within military law, because within military necessity. But to say the farm shall no longer belong to the owner, or his heirs forever; and this as well when the farm is not needed for military purposes as when it is, is purely political, without the savor of military law about it. And the same is true of slaves. If the General needs them, he can seize them, and use them; but when the need is past, it is not for him to fix their permanent future condition. That must be settled according to laws made by law-makers, and not by military proclamations. The proclamation in the point in question, is simply "dictatorship." It assumes that the general may do *anything* he pleases—confiscate

the lands and free the slaves of *loyal* people, as well as of disloyal ones. I cannot assume this reckless position; nor allow others to assume it on my responsibility. You speak of it as being the only means of *saving* the government. On the contrary it is itself the surrender of the government. Can it be pretended that it is any longer the government of the U.S.—any government of Constitution and laws,— wherein a General, or a President, may make permanent rules of property by proclamation?

I do not say Congress might not with propriety pass a law, on the point, just such as General Frémont proclaimed. I do not say I might not, as a member of Congress, vote for it. What I object to, is, that I as President, shall expressly or impliedly seize and exercise the permanent legislative functions of the government.

When he finished that part of the letter, Lincoln wrote, "So much for principle. Now as to policy." And then he proceeded to talk about Kentucky. It seems striking that when delaying freedom for the slave, Lincoln thought first of constitutional principle, then of policy. But policy considerations came first with him in dealing with the crisis following the firing on Fort Sumter. Was he willing to go farther to save the Union than to free the slaves? Did he value the Union more than liberty after all?[17]

To answer those questions will require a quick review of American thinking on the subject of emancipation and war before Abraham Lincoln faced both as live subjects rather than abstract possibilities. The review can be brief because there had been little thought on the subject and because what little thought there was had been clearly and succinctly put by an intelligent politician, John Quincy Adams. After his return to Washington as a member of the House of Representatives, "Old Man Eloquent" attempted to avenge his loss of reelection to the presidency in 1828 by attacking Southerners and the "Slave Power." In a debate in Congress as early as 1836, Adams expressed the belief that from "the instant that our slaveholding states become the theater of war, civil, servile or foreign... the war powers of Congress extend to interference with the institution of slavery in every way by which it can be interfered with." During the early rounds of the Texas controversy in 1842, when war was much spoken of, Adams again warned Southerners that Congress would have "full and plenary power" over slavery in a state at war.[18] Finally, galvanized by the vigorous example of his old nemesis Andrew Jackson at New Orleans, Adams, in the congressional debate over refunding the general's fine, suddenly declared that the president and even his subordinate commander of the army have the power to abolish slavery:

When your country is actually in war, whether it be war of invasion or a war of insurrection, Congress has power to carry on the war, and it must carry it on according to the laws of war; and, by the laws of war, an invaded country has all its laws and municipal regulations swept by the board, and martial law takes the place of them.

This power in Congress has, perhaps, never been called into exercise under the present constitution. But when the laws of war are in force, what, I ask, is one of those laws? Is it this: that when a country is invaded, and two hostile armies are met in martial array, the commanders of both armies have power to emancipate all the slaves in the invaded territory.

And here, I recur again to the example of General Jackson.... You are ... passing a law to refund to General Jackson the amount of a certain fine imposed upon him by a judge under the laws of Louisiana. You are going to refund him the money with interest, and this you are going to do because the imposition of the fine was unjust. And why was it unjust? Because General Jackson was acting under the laws of war, and because the moment you place a military commander in a district that is the theatre of war, the laws of war apply to that district.

... I lay this down as the law of nations. I say the military authority takes for the time the place of all municipal institutions and of slavery among the rest, and that under that state of things, so far from its being true, that the States where slavery exists have the exclusive management of the subject, not only the President of the United States, but the commander of the army, has power to order the universal emancipation of slaves.[19]

Lincoln's constitutional journey was similar to Adams's.[20] When President Lincoln revoked the Missouri proclamation in September 1861, he had already abandoned any belief that slavery, because of the Constitution, could not be touched in the states where it then existed. Generals could not fix the "permanent future condition" of the slaves in Missouri but the slaves' status could "be settled according to laws made by law-makers." Such an idea, daring though it was, did not fully anticipate the constitutional grounds of Lincoln's actions a year later when he announced his own Emancipation Proclamation, because Lincoln was no "law-maker." He was head of the executive branch; Congress made the laws. Lincoln still thought in 1861 that if he did it himself, it would constitute "dictatorship," but Lincoln would "not say Congress might not with propriety pass a law ... just such as General Frémont proclaimed." He even hinted that if he were in Congress as Browning was, he might vote for it. He simply did not think he should, "as President ... expressly or impliedly seize and exercise the permanent legislative functions of the government."

What has never been noticed in the furor over the Frémont proclamation is how far Lincoln already had traveled down John Quincy Adams's constitutional road in 1861. Although Lincoln would never reach the point where he believed a general could proclaim emancipation, within less than a year of the Frémont episode, he had reached Adams's view that the president could do so. What Lincoln learned, between September 1861 and the drafting of the Emancipation Proclamation in July 1862, was not Adams's view that war threatened slavery; Lincoln knew that already. He learned, perhaps from William Whiting's *War Powers of the President,* published in 1862, that war gave the president and not law-makers only the power to abolish slavery in enemy territory.[21]

Lincoln moved faster in adjusting his prewar ideas about the Constitution and slavery than most historians have previously believed. What he said to Browning, admittedly, was said in private. In public, Lincoln's utterances sounded more skeptical, but his constitutional doubts had clearly been dispersed well before the public announcement of the Emancipation Proclamation on September 22, 1862, and he wanted the public to know it. His famous letter of August 22, 1862, to Horace Greeley, counseling patience on the slavery question, said nothing of constitutional obstacles to action and expressed a willingness to free slaves, if such action would save the Union. On September 13, he explicitly told a delegation of Chicago Christians urging emancipation: "I raise no objections against it on legal or constitutional grounds; for, as commander-in-chief of the army and navy, in time of war, I suppose I have a right to take any measure which may best subdue the enemy." Emancipation was for him "a practical war measure" and as soon as military circumstances at the front seemed to require it and political circumstances in the border states seemed to permit it, Lincoln acted to end slavery in the Confederacy.[22]

Lincoln did worry more about the consequences of emancipation than the consequences of suspending the writ of habeas corpus—and for good reason. Lincoln regarded the suspension of the writ as an exception for a temporary emergency, and he felt sure that the American people would never want to continue the condition when the emergency was over. He put it more vividly, of course, comparing such an unimaginable course to that of a man fed emetics in illness and then insisting on "feeding upon them the remainder of his heathful life." About this, Lincoln proved essentially correct.

Emancipation was different. Though it might be adopted as a practical measure to end the war, it could not be reversed when the crisis was over. As Lincoln put it in his letter to Browning about the slaves in Missouri, "If the General needs them, he can seize them, and use them; but when the need is past, it is not for him to fix their permanent future condition. That must be settled according to laws." Emancipation, though perhaps a matter of situational ethics in the midst of war, would necessarily affect American society for all time to come. Lincoln was a practical man, all right, but he did occasionally think about the country's "permanent future condition." He saw no danger in the temporary suspension of habeas corpus during rebellion or invasion, but the case of black people was clearly different. Only rigid safeguards would protect freedmen from popular race prejudice and possible reenslavement. Black freedom might prove as temporary and situational as the whites' brief loss of customary liberties during the Civil War. So Lincoln's thoughts necessarily turned to a constitutional amendment to end slavery in the United States.

This was a major change in his constitutional thinking. The Constitution was last amended five years before Abraham Lincoln was born. He was on record in a speech in Congress recommending that the document be left alone and that the American people not get into the habit of changing it. In the desperate throes of the secession crisis, he did agree to a proposed amendment that would have explicitly guaranteed slavery where it already existed. But this was redundant in Lincoln's view, merely reassuring the South of what it already had. In 1864, he wanted an amendment to guarantee that there would be nothing temporary about emancipation.

This ability to balance short-term practicality and long-term ideals is perhaps the essence of statesmanship. In Lincoln's case, the one helped preserve the Constitution as the law of the land, and the other brought such changes as made it worth preserving "thoughout the indefinite peaceful future."

Epilogue

Abraham Lincoln never discussed most of the arrests described in this book. His statements on this policy dealt with sensational instances of abridgement of freedom of speech or freedom of the press. This anomaly brings to mind historian Phillip Paludan's observation that President Lincoln made a more extreme defense of military arrests of civilians than necessary[1]

Instead of saying that the administration seldom arrested individuals merely for criticizing the war, Lincoln maintained that it was quite all right to do so. He told his critics in the Corning letter of June 18, 1863, that he possessed the necessary constitutional power, and he explained why dissent could not safely be tolerated. He did not point to the dissent he did tolerate, nor did he argue that his power was most often used against persons who had done something other than criticize the war in words. Instead, he suggested that even silence could be a crime:

> . . . arrests by process of courts, and arrests in cases of rebellion, do not proceed altogether upon the same basis. The former is directed at the small per centage of ordinary and continuous perpetration of crime; while the latter is directed at sudden and extensive uprisings against the government, which, at most, will succeed or fail, in no great length of time. In the latter case, arrests are made, not so much for what has been done, as for what probably would be done. The latter is more for the preventive, and less for the vindictive, than the former. In such cases the purposes of men are much more easily understood, than in cases of ordinary crime. The man who stands by and says nothing, when the peril of his government is discussed, can not be misunderstood. If not hindered, he is sure to help the enemy. Much more, if he talks ambiguously—talks for his country with "buts" and "ifs" and "ands."[2]

Why did Lincoln, as Paludan put it, establish so "broad" a "definition of dangerous speech"? The explanation may lie in the mythical aspects of habeas corpus. Whenever Lincoln spoke of it, in public pronouncements at any rate, he spoke more of a symbol of American freedom than of a specific legal instrument. In the crucial Corning letter, he twice invoked habeas corpus in a litany of traditional American freedoms, saying the first time that he did not really fear that the American people would, "by means of military arrests during the rebellion, lose the right of public discussion, the liberty of speech and the press, the law of evidence, trial by jury, and Habeas corpus, throughout the indefinite peaceful future." He even put himself in the strange position of chiding his Democratic critics for belittling the importance of the writ.[3]

Although the privilege of the writ of habeas corpus is surely helpful in maintaining free speech, a free press, and trial by jury, it is not identical with them, and its relation to maintaining the laws of evidence seems downright obscure. Lincoln was not speaking of the actual writ, something that could be issued by Democratic judges or Republican judges, something that could aid the freedom of fugitive slaves or help slave-catchers. He was speaking of a mythical writ that is ever liberating and that is always and everywhere a cornerstone of freedom. His loose equation of it with such freedoms as freedom of speech and with such procedural safeguards as the laws of evidence led him to defend military arrests of civilians as though they necessarily assaulted such freedoms and safeguards in every instance. The mythical rather than legalistic image of the writ in Lincoln's mind is further revealed in his reiteration of the familiar litany of freedoms later in the Corning letter. After pointing out that in 1815 Andrew Jackson had established martial law in New Orleans and defied a writ of habeas corpus, Lincoln said, "that the permanent right of the people to public discussion, the liberty of speech and the press, the trial by jury, the law of evidence, and the Habeas Corpus, suffered no detriment whatever by the conduct of Gen. Jackson, or it's subsequent approval by the American congress."[4] For the most part, such was the plane on which the debate over habeas corpus was conducted—at a broad, politically symbolic level rather than at a jurisprudentially profound and precise one. That is true of both sides in the debate, and it is true of much of the debate over the issue conducted in history books written since the war.

Historiography

For a subject of such obvious importance, the historical literature on civil liberties in the Civil War is strangely unsatisfying. For one thing, it is astonishingly meager. James G. Randall wrote the only book-length scholarly treatment, *Constitutional Problems under Lincoln,* more than sixty years ago. At the time, he could not find another book or comprehensive article on the subject worth citing in his bibliography. "So little has been done by historians in searching the voluminous legal material of the Civil War period," Randall stated without exaggeration, "that I have felt myself to be breaking new ground."[5]

Randall referred, of course, to scholarly literature; some books and articles had appeared outside the realm of scholarship. From the Civil War to World War I, Democrats noisily denounced Lincoln's record on the Constitution. The most important example of that partisan impulse, John A. Marshall's *American Bastile*, first published in 1869, may have gone through as many as thirty-four printings in at least two editions.[6]

Despite its partisan animus, *American Bastile* enjoyed a long life and had considerable influence on historical writing. It was essentially a book of martyrs, offering its readers dozens of sketches of Northern Democrats arrested by military authority during the Civil War. Some of these provided important anecdotal evidence. But most important, the author of *American Bastile* committed the fatal errors of method and interpretation that crippled the literature on the subject afterward. First, Marshall provided the total number of arrests "as variously estimated." He accepted the estimates of others and did not pursue his own independent investigation of the question. Second, he assumed that these tens of thousands of civilian prisoners were "FREE citizens of FREE STATES," that is, of the "Free Northern States." The prisoners of state did number in the tens of thousands, probably, and they were indeed civilians, but most of them were not from free Northern states. The simple assumption that citizens in military prisons were Northerners—or to use the old-fashioned term, victims of "arbitrary arrest"—dogged the work of every subsequent historian.[7]

Denunciations of Republican tyranny under Lincoln, like those in *American Bastile*, proved lastingly popular with Democrats, but they were not without their drawbacks. Republican orators after the war liked nothing better than waving the bloody shirt to remind the electorate of alleged Democratic disloyalty during the Civil War. For the Democrats to bring up the tyranny issue was to open themselves to questions about their past loyalty. It raised the very subject the Republicans wanted to discuss. Gradually, partisan writers shifted the focus. The 1885 edition of *American Bastile* carried a longer subtitle and additional contents: *A History of the Arbitrary Arrests and Imprisonment of American Citizens in the Northern and Border States, on Account of their Political Opinions, during the late Civil War, together with a Full Report of the Illegal Trial and Execution of Mrs. Mary E. Surratt, by a Military Commission, and a Review of the Testimony, showing her Entire Innocence.* Shifting the focus to the martyrdom of Mrs. Surratt, who had been charged with conspiring to murder Abraham Lincoln and tried by a military commission rather than a civil court, proved forward-looking. Democratic lawyer David M. Dewitt followed the lead in 1895 with *The Judicial Murder of Mrs. Surratt*, which narrowed the focus to this one case of special interest (because Mary Surratt had been a Roman Catholic) to one bloc of Democratic voters.[8]

Other factors militated against serious study of the issue beyond the level of partisan charges. In the two decades immediately following the Civil War, the regnant view of the Constitution—as a static decree, handed down in 1787 to be preserved, rather than as something that grows, changes, or adapts—proved inimical to assessing its recent history under the Lincoln administration. The centennial of the Constitution in 1887 saw no books or articles published on the subject of Lincoln and the Constitution. The next year, Sydney George

Fisher published "The Suspension of Habeas Corpus during the War of the Rebellion" in the *Political Science Quarterly,* a fledgling journal in only its third year of publication. Fisher's work hardly celebrated the great founding document; instead, he defended Lincoln's conduct during the Civil War by saying that the Constitution was defective, and it remained so in 1888 because it did not grant the government enough power to defend itself.[9] Although Fisher was an advanced constitutional thinker, hardly given to uncritical Constitution worship, neither he nor other scholars or jurists had yet adopted the broad answer to the question "What Is a Constitutional History of the United States?" to be given by Francis N. Thorpe in 1902: "a history of the origin and growth of a civil system embodied in a constitution, whether written or unwritten."[10]

At the end of the nineteenth century, scholars, especially political scientists, grew more willing to tout Lincoln's record on the Constitution. Reformers of the Progressive Era generally admired executive leadership and vigorous nationalism, and for perhaps the first time since the 1860s, Lincoln's handling of dissent during the Civil War seemed appealing for its very toughness. This viewpoint was epitomized by Nathaniel W. Stephenson, the first academic historian to write a biography of Lincoln. Stephenson praised Lincoln's defense of the "right of the President to assume in emergency vast authority" and ridiculed his enemies as "rhetorical visionaries," "fanatics," and "parasites" who were not "fully conscious of the Nation as a whole."[11]

The modern constitutional historian Herman Belz credits political scientist William Archibald Dunning with establishing "the dictatorship question as a theme in historical writing on the Civil War." In his 1897 *Essays on the Civil War and Reconstruction,* Dunning pointed to the appearance of "a new principle . . . in the constitutional system of the United States" in the period between April 12 and July 4, 1861, "namely, that of a temporary dictatorship." Many of the scholars of the Progressive Era not only discerned but also somewhat admired this feature of the Civil War, and Lincoln's model served them in good stead when the United States entered World War I. When he compared the "Disloyalty in Two Wars" in the October 1918 issue of the *American Historical Review,* Dunning himself found in the Civil War and World War I "a far-reaching parallelism in incidents and ideas." Excesses were committed in the name of patriotism in both wars, Dunning admitted. It was often difficult to tell exactly where the iconoclastic Dunning stood, but he nevertheless appeared to give Wilson high marks from the legal point of view and to have some reservations about Lincoln's record.[12]

The Progressive Era was the cradle of the most important work ever written on the subject of civil liberties in the Civil War, Randall's *Constitutional Problems under Lincoln.* And here one encounters, even more than in Dunning, the second remarkable aspect of the literature, its ambivalence. Although not published until 1926, the book had its inception in 1911, and by 1918 Randall had completed a draft manuscript called "Constitutional Problems of the Civil War." Before the Great War, Randall had shown some predisposition to condemn the methods of the North. His doctoral dissertation was entitled "The Confiscation of Property during the Civil War," and an early article based on it, which appeared in the *American Historical Review* in October 1913, focused on "the use of extreme methods in crippling an enemy." In this pre–World War period, Randall adopted a denunciatory tone in discussing Civil War issues:

> On both sides the methods of conducting the war were of questionable reputableness, and this was true not only in unauthorized orders and in breaches of discipline, but in many measures which received the full sanction of government. The humanizing effect of modern international law has been nowhere more strikingly revealed than in the guarantees which have been introduced for the security of the lives and property of non-combatants, and the principle of the inviolability of private property on land has been thoroughly established. Yet the thoughtless repetition of "Marching through Georgia" is but a glorification over the harshness of Sherman's most famous campaign, and the failure of this sort of warfare to produce a sentiment of condemnation is but an evidence of callousness due to the frequency of such outrages.

Randall's sentiments doubtless derived in part from his Southern-born dissertation adviser, William E. Dodd, but since Randall condemned both Union and Confederate methods, they may have been influenced by his personal political beliefs as well. Randall was a deeply religious young man with somewhat of a reformer's heart, a prohibitionist, and a critic of "Railroad Frauds, bought elections, unspeakable graft, . . . vice.[. . .] Labor War." He was critical of war, too, as "An Outworn World Idea," not justifiable on social Darwinian grounds or on any other as a really useful instrument of policy.[13]

Despite such ideas, during the Great War, like many other American academics and intellectuals, Randall enlisted his talents wholeheartedly in the war effort once the United States intervened in Europe. There was no coercion of any sort. While teaching at tiny Roanoke College in Salem, Virginia, Randall volunteered to write propaganda for the Wilson administration's Committee on Public Information (CPI) and National Board for Historical Service. He accused the enemy of having the worst possible record on civil liberties. "Germany," Randall wrote in 1918, "seems to hold that any otherwise illegal thing may be done on the plea of military necessity, that inter arma silent leges, etc., whereas the American and Anglo-Saxon point of view is much more restrictive [,] that war does not countenance extreme illegality, and that even in the most desperate times the rule of law should prevail."[14] In a manuscript of World War I vintage or later, Randall reflected on America's experience in the Civil War:

> It is true that dangerous possibilities lurked in the executive "suspension of the writ," that civilians were made prisoners by the thous[ands] and without judicial process, that some of the Union military officers out of touch with Lincoln's spirit had the erroneous notion that war breaks down the rule of law and substitutes the rule of force, that as a consequence of imperfect central control over subordinate officers many frivolous arrests were made and unwarranted orders executed.

The anti-administration alarm raised by such an agitator as Vallandigham and his "peace party" may even have had, here and there, some foundation. Yet in the main, and viewing the whole period not merely the first year of the war, it is evident that the limitations of governmental power were carefully heeded—so carefully that at times it did seem that war was actually being conducted *in vinculis*, which may, after all, be the best way for it to be conducted.

Randall added, with feelings typical for persons with his Midwestern Republican roots, "Lincoln's intention, it must be remembered, was often milder than the temper of the officers who carried out the Union policy."[15]

Randall's position was a far cry from the extreme "presidential-dictatorship" thesis of William A. Dunning, and the propaganda needs of World War I were pulling him still further away. American scholars suddenly had reason to see the American Civil War in a different light, as a war conducted rather benignly when compared with Germany's conduct in World War I. This could lead an author to justify Lincoln's record on civil liberties and to deny that he was a dictator, so that America would seem consistent in upholding law and the Constitution amidst war efforts throughout her history. Before the war, Randall had focused on "extreme methods" of fighting the Civil War and found them of "questionable reputableness...not only in unauthorized orders, and in breaches of discipline, but in many measures which received the full sanction of government." After the United States entry into World War I, he saw Civil War abuses of power mainly as "a consequence of imperfect control over subordinate officers." The change was subtle, a matter of emphasis only, and yet in a scholar as careful as James G. Randall, it was surely important.

Modern readers of Randall's *Constitutional Problems under Lincoln* are sure to be struck by its curious ambivalence. It seems on the one hand to condemn many Union war policies and on the other hand to praise Lincoln's record overall. The problem was that the book retained some of the antiwar views of the prewar dissertation, whereas other parts were informed by an apologetic outlook induced by Randall's experiences during the Great War. His contacts with the Wilson administration had led to a job as historian for the United States Shipping Board until 1919 and to hopes that he might become a historian for the general staff. Soon, however, he returned to teaching and to work on *Constitutional Problems under Lincoln.*[16]

Randall's work for the government left its mark on his writing. Not long before publication of the book, an old friend of Randall's, political scientist Edwin P. Tanner, read the manuscript and commented on it in a letter:

I do differ a little from your general interpretation. Personally I think the whole idea of the War Power contains enormous dangers which it seems to me you together with most writers on the subject minimize. You show that you realize them of course. But somehow you give the impression that Lincoln was nearly always right. Now, I yield to nobody in my admiration for the sense and moderation of Lincoln.... Nevertheless, I think his theory of the War Power was dangerous in the extreme.

Randall's reply was quite revealing:

> I may have gone too far in justifying the extreme war powers. My real convictions are similar to yours,—that many dangers lurk in the war power theory. Possibly my admiration for Lincoln has carried me too far.[17]

The publishing history of *Constitutional Problems under Lincoln* illustrated another factor that has kept to a surprising minimum the amount of scholarly literature on this subject: the extremely limited audience for serious constitutional history in the United States. Randall at first thought his manuscript had commercial possibilities, but Harcourt Brace refused even to read it. Macmillan read it and rejected it, as did Little, Brown and Houghton Mifflin—all on the grounds that the book, though admirable for its scholarship, simply had no commercial appeal. When Yale University Press dragged its feet an unconscionably long time in evaluating the manuscript, Randall decided to pay for publication himself. Appleton's finally brought the book out in 1926, but the author paid them the costs of production for the 1,500 copies published.[18]

After the appearance of Randall's book, developments in world politics, political theory, and historical scholarship served to erode any aggressive confidence in praising the more vigorous aspects of Lincoln's wartime leadership. In the realm of Civil War scholarship, the most important change came in the way historians depicted Northern opposition to Lincoln. Nathaniel Stephenson's eager Progressive Era defense of Lincoln had been premised on the assumption that the Civil War president had encountered an enormous fifth-column threat. Stephenson pointed in particular to "those extensive secret societies which all through the war seem always on the verge of rebellion in the Middle West." In 1926, Randall showed himself to be of the same mind. Disloyalty was "widespread," he said, and in "view of such extensive disloyalty, the number of political arrests is comprehensible."[19] By the mid-1930s, careful students had their doubts about the extent of disloyalty in the North. Constitutional historian Andrew C. McLaughlin, for example, soberly pointed out in an important 1936 textbook that "arbitrary arrests were frequent. They took place not alone in areas where many persons were known to be secretly disloyal and engaged in more or less active conspiracies, but also in regions where there was no evidence of widespread disaffection or of dangerous combinations.[20] Revisionist historians, writing in the 1930s, proved that the Democratic party had played the role of a strictly loyal opposition during the Civil War, and eventually the works of Frank L. Klement in the 1950s, 1960s, and 1970s would demolish the myth of a large, secret, well-organized disloyal Northern opposition to the Lincoln administration.[21] Once the Copperhead menace melted away in the works of the revisionist historians, Lincoln's record of "frequent" arrests began to look quite embarrassing.

Important developments in political theory in the United States after World War I affected the interpretation of civil liberties in the Civil War as well. A new awareness of the fragility of freedom of speech and the growth of "First Amendment law" practice came from the work of Harvard law professor Zechariah Chafee, Jr. Disturbed by World War I's Espionage Act, he wrote *Freedom of Speech,* published in 1920, and warned in an article six years later—the very year Randall's book appeared—that in the next war all civilian attempts to question the wisdom of the war "will . . . become perilous."[22]

The specter of dictatorship loomed in world politics after the war, too. Its unsettling effect on Lincoln scholarship is apparent in the work of constitutional historian Andrew C. McLaughlin. When he delivered the banquet address at the annual meeting of the Abraham Lincoln Association in Springfield on February 12, 1936, his paper seemed curiously tentative and sober for such an essentially celebratory occasion. The reason was best expressed by McLaughlin himself in his Pulitzer prize-winning *Constitutional History of the United States,* also published that year: "That a president armed with the 'war power,' may some day wreck the whole constitutional system is theoretically possible, and the dictator, if he ever appears, may discover precedents in the conduct of Lincoln."[23]

Although such factors tended to make historians after the Progressive Era uneasy about Lincoln's record, the old static view of the Constitution no longer impeded understanding. The era of the Great Depression and the New Deal saw the triumph of a more flexible view of constitutional history, expressed in McLaughlin's speech to the Abraham Lincoln Association:

> A constitution need not be looked upon as only a piece of parchment stored away in a safe, free from the prying eyes of the multitude, consigned to the clever exposition of politicians and subjected to the astute argument of jurists. In a very real sense the actual structure of a nation . . . is something more than a document and all the incrustations of statutes and judicial interpretations; it includes the common and conventional attitudes of the citizens, the principles which animate them, their substantial concepts of justice, liberty, and safety, their readiness to be ruled by others or their determination to compel their rulers to serve them.[24]

But this more flexible view failed to aid Lincoln's reputation as a constitutional leader. The legacy of fear left over from the security measures taken in World War I and the ominous rise of dictatorships in Europe left scholars nervous about Lincoln's record.

So nervous were they, in fact, that the ground Randall broke in 1926 was not plowed again in a major way until the 1970s. By then, however, changes in the academy caused the central focus of inquiry to shift radically. For one thing, Reconstruction, because of its obvious lessons for the modern civil-rights movement, grew increasingly important to modern academic historical scholarship. Books would now treat the Constitution in Civil War *and* Reconstruction, thus halving the amount of space devoted to Lincoln's administration and placing less emphasis on the civil liberties of white men than on the civil rights of blacks. Constitutional history changed, too. When *A More Perfect Union: The Impact of the Civil War and Reconstruction on the Constitution* appeared in 1973—almost fifty years after Randall's book—its author, Harold Hyman, deliberately treated topics which "had little or nothing to do with the

traditional concerns of Civil War and Reconstruction constitutional history."
He wrote, instead, about subjects ranging from public-health law to legal profes-
sionalism. The result was an original and genuinely ground-breaking study. A
necessary concomitant, however, of the new focus was a lack of lengthy reconsideration of the older themes Randall had covered. In a more recent book,
written with William M. Wiecek, *Equal Justice under Law: Constitutional
Development, 1835–1875,* Professor Hyman devotes only forty-six pages to
the Civil War.[25]

Because of the notion (born of Seward's notorious reputation) that the first
year was the worst for civil liberty during the Civil War and because sources
in published form are more readily available for that period, study of military
arrests of civilians before February 15, 1862, has been greater than for the
remainder of the war. Dean Sprague's *Freedom under Lincoln,* which appeared
in 1965, relied on the *Official Records* as well as newspaper accounts, for a
rather anecdotal narrative critical of Lincoln. Kenneth A. Bernard's "Lincoln
and Civil Liberties," an article published in 1951, had relied on similar published sources for a defense of Lincoln's record. Neither advanced the argument
beyond Randall's.[26]

Sprague's book, though footnoted, was essentially a popular history, and
Bernard's article was brief. Academic historians wrote surprisingly little on the
subject. But others have rushed in where historians feared to tread. As was
the case with the history of the Lincoln assassination, which academic historians
deliberately ignored for many years, abdication of scholarship led to the
triumph of irresponsibly cynical and iconoclastic popular theories.

The dominant popular view today has been forged outside the historical
profession, where literary critics, political scientists, and historical novelists
unembarrassedly assume that Abraham Lincoln was a dictator. Such an assumption underlies the influential works of the eminent literary critic Edmund
Wilson and of his less-able intellectual offspring, Dwight C. Anderson, Gore
Vidal, and William Safire.

Wilson set the stage in 1962 in his widely read book *Patriotic Gore,* which
compared Lincoln to Bismarck and Lenin. Like them, Wilson argued, Lincoln
united a great national power, became an uncompromising dictator in the
process, and was "succeeded by agencies which continued to exercise this
power and to manipulate the peoples he had been unifying in a stupid, despotic
and unscrupulous fashion." Wilson was fascinated by Freudian psychology, and
in a remarkably influential section of the book, he argued that Lincoln foresaw
the tyrant he would become in the speech (discussed in Chapter 10) that the
youthful Illinois lawyer delivered in 1838 before the Young Men's Lyceum of
Springfield. Psychohistorians found Wilson's work stimulating and later focused
obsessively on the warning against a tyrant in the Lyceum speech. The root
of Wilson's interest in the speech was his personal fear of the growth of the
power of the federal government. The ultimate source of his view that the
Civil War president wielded Bismarckian or Leninist powers was Wilson's own
extremist theories of individual freedom. These brought him perilously near

prosecution for income-tax evasion by the Internal Revenue Service. So closely related were the origins of *Patriotic Gore* and the writer's refusal to pay his income taxes that when he wrote *The Cold War and the Income Tax* in 1963, he considered it a supplement to the introduction to *Patriotic Gore.*[27]

Dwight Anderson's psychohistorical interpretation in his symptomatic book *Abraham Lincoln: The Quest for Immortality* (1982) started where Wilson left off, and he went on to argue that Lincoln was "demonic," a "Robespierre," and "a tyrant who would preside over the destruction of the Constitution in order to gratify his own ambition."[28]

Gore Vidal's novel *Lincoln,* published in 1984, owed a heavy debt to Wilson as well. In a key scene Vidal wrote:

> For the first time, [Secretary of State] Seward understood the nature of Lincoln's political genius. [Lincoln] had been able to make himself absolute dictator without ever letting anyone suspect that he was anything more than a joking, timid back-woods lawyer, given to fits of humility in the presence of all the strutting military and political peacocks that flocked about him.[29]

Finally, political columnist William Safire's long novel *Freedom* likewise takes the view, expressed in its ironic title, that Lincoln willfully crushed civil liberties during the war and played fast and loose with the Constitution of the United States. Its debt to Edmund Wilson is apparent not only in its message but also in its telltale language. The original title of Wilson's essay on Lincoln was "The Union as Religious Mysticism," a phrase the literary critic took from Confederate vice president Alexander H. Stephens, who said of Lincoln that "the Union with him in sentiment, rose to the sublimity of a religious mysti cism." At least twice in Safire's book, characters comment on the curiously "mystical" nature of Lincoln's attachment to the Union.[30]

The reader should not be misled by the amount of time and space devoted here to earlier historical theories. It is important for perspective to keep in mind how often these theories appeared as only small parts of more general histories or as mere articles or banquet speeches. All in all, the most remarkable feature of the scholarly literature on this subject is how little of it exists. Its scantiness forces even this brief historiographical essay to discuss novels.

The skimpiness of the serious literature suggests that historians have been more or less embarrassed by Lincoln's record on the Constitution. It is as though historians reacted as William Herndon did when a pious Eastern biographer asked him about Lincoln's religious ideas. "The less said, the better," was Herndon's reply.[31] Historians have, to put it simply, shied away from the subject.

The Historical Record

Ignoring Lincoln's record on civil liberties will not cause it to go away, and a continuing focus on the habeas-corpus cases of mythic proportions—*Vallan-*

digham and *Milligan*—will tell a misleading story and cause historians to continue the fruitless debate over "dictatorship." Of more interest is to examine the hard social realities of the individual identities of persons arrested by Union military authority. The old ways of understanding military arrests of civilians under the Lincoln administration must be altered. Were there really 13,535 such arrests from February 1862 on? Was the first year of the war (Seward's regime in internal security) the worst for civil liberties? Were the persons who were arrested "political prisoners" in the modern sense of the term, like politically disaffected victims of a totalitarian regime?

The answers to all these question could not be affirmative because such large numbers of arrests, if they were "political" in any sense, would surely have caused the Lincoln administration more trouble than they did. Conscription had the power to shake American society, but civil liberties, as the Lincoln administration handled them, did not. Conscription provoked numerous incidents of violent, even murderous resistance, most notably in New York City in July 1863 but also day-to-day in bloody conflicts with provost marshals and enrolling officers. Tens of thousands of men broke the law, refused to report, and became, technically, deserters. Military arrests of civilians, on the other hand, caused some mass indignation rallies and much heated rhetoric but little violent resistance or law-breaking and no violent resistance in the first year of the war, previously alleged to be the worst. If the much-written-about *Vallandigham* and *Milligan* cases were at all representative, it would seem hardly possible that the Lincoln administration could have faced so little resistance.

It was not difficult to see why arrests under Seward caused so few political problems for the administration. Few of them had anything to do with conventional politics. The discovery that many of the prisoners were citizens of the Confederacy, blockade-runners, foreign nationals, returning Southern sea captains, and the like explained the lack of violent political reaction. Such people were not voters in the North.

Research in the National Archives to discover the sources of F. C. Ainsworth's influential 13,535 figure for arrests in the period after February 1862 (when Stanton was in control of internal security) began to uncover another problem with older interpretations. A letter about a civilian arrested by the military in New Mexico territory in 1861 provided an early clue. After all, the president had not suspended the writ of habeas corpus in New Mexico. This arrest did not have anything to do with the suspension of the writ.

A majority of the arrests would have occurred whether the writ was suspended or not. They were caused by the mere incidents or friction of war, which produced refugees, informers, guides, Confederate defectors, carriers of contraband goods, and other such persons as came between or in the wake of large armies. They may have been civilians, but their political views were irrelevant, even perverse—some were black men and others came into Union lines for the purpose of declaring their loyalty to the U.S. rather than to the Confederacy.

Such prisoners were hardly "political"; nor did their arrests seem "arbitrary" in the dictionary sense of the word. But did the term as employed in the Civil War era have some other technical meaning? What was an "arbitrary arrest," anyhow? Upon examination of the records, it proved to be neither very precise, technical, nor legally well-defined. In this term lay the major conceptual problem with previous interpretations: the total number of civilians in military prisons in the North, whatever that was, would not equal the number of "arbitrary arrests" (that is, what historians had always meant by the term "arbitrary arrests").

Continuing research in records from the National Archives revealed that the 13,535 figure was not only unverifiable but erroneous. There were many more arrests than Ainsworth found, but they were of less significance in the history of civil liberties than anyone ever imagined. As the count made for this book neared 14,000, no new patterns in the arrests were emerging. Many more records were available, but the time had come to abandon the count and focus on the meaning of the statistics.

Even counting every last relevant scrap of paper remaining in the National Archives would not tell a historian how many civilians were arrested. It is not always easy to tell whether a person was a civilian or not. It is not always easy to tell whether a person had actually been arrested or not. The incidents of war simply did not leave the kinds of records that made an exhaustive study possible.

The only large group of arrests that appeared in any way to resemble what most Democrats complained about at the time, the sort that really disturbed historians and civil libertarians, were those that followed the August 8, 1862, orders. And those were sincerely meant to enforce conscription rather than stifle dissent—which explains their relaxation in one month's time after most draft quotas had been filled.

The later history of the war revealed incidents and eventually whole patterns of arrests never before glimpsed by Civil War historians. At its worst, the system tended to spread, imposing military justice and discipline to new areas often merely for convenience' sake—in the ill-governed western territories, in policing liquor-sellers near the Union camps, or in punishing fraudulent government contractors. At its very worst, the system even indulged in routine torture of civilian prisoners well behind the lines. In enemy territory, when conscription made every male citizen a potential soldier and when guerrilla warfare made it difficult to distinguish soldier from civilian, there was a tendency in the Union army to destroy the age-old distinction between combatants and noncombatants. But this occurred only late in the war and, more important, mostly only in theory.

Competing theories, decidedly more humanitarian, clearly held sway as well. There was a surprising adherence to international law, even when the law contradicted U.S. interests. In fact, interest in such laws grew as the Civil War went on. Instead of offering a spectacle of degeneration, with modern warfare obliterating nice points of law, the Civil War saw the U.S. draft a legal code to govern its armies in the very middle of the conflict—the first such code in history to be promulgated in the midst of war. With the Dix commission in 1864, even the army attempted to declare itself ineligible to try civilians in military courts.

Through it all, there appeared an astonishing variety of complicated cases. Even more astonishing was the confused inability of the protagonists to explain what legal conditions obtained in the war. It was far from being the case that Abraham Lincoln moved with decisiveness in 1861 from that familiar situation in which the writ of habeas corpus was secure to an opposite situation, equally well understood though far more fearful or deplorable, in which the writ was suspended. What happened after suspension was clear to no one. Lincoln's steadily growing confidence or decisiveness was as much a function of his indifference to constitutional scruple as anything else—except his sure sense of the purpose of the government to win the war and keep the country whole so that democracy could not be said to have failed. The government and its critics often used the terms "martial law" and "suspension of the writ of habeas corpus" interchangeably. Even at war's end, the finer practical points remained without definition. The question of proper notice, for example, was never resolved to the satisfaction of the judiciary, on the one hand, or of the executive, on the other, for Lincoln's government issued redundant orders and proclamations throughout the war.

There may be many lessons for the present in this book, but none as clear as that one. If a situation were to arise again in the United States when the writ of habeas corpus were suspended, government would probably be as ill-prepared to define the legal situation as it was in 1861. The clearest lesson is that there is no clear lesson in the Civil War—no neat precedents, no ground rules, no map. War and its effect on civil liberties remain a frightening unknown.

Notes

Introduction

1. Alexander H. Stephens, *A Constitutional View of the Late War Between the States,* II (Philadelphia: National Publishing Co., 1868–70), 448.

2. The technical term is "suspension of the privilege of the writ of habeas corpus," but I refer to it with briefer phrases throughout the book, as is common practice among constitutional historians.

3. For a summation see Frank L. Klement, *Dark Lanterns: Secret Political Societies, Conspiracies, and Treason Trials in the Civil War* (Baton Rouge: Louisiana State Univ. Press, 1984); Joel H. Silbey, *A Respectable Minority: The Democratic Party in the Civil War Era, 1860–1868* (New York: W.W. Norton, 1977).

4. William Starr Myers, *A Study in Personality: General George Brinton McClellan* (New York: D. Appleton-Century, 1934), 464.

5. Milton Cantor, "The Writ of Habeas Corpus: Early American Origins and Development," in Harold M. Hyman and Leonard W. Levy, eds., *Freedom and Reform: Essays in Honor of Henry Steele Commager* (New York: Harper & Row, 1967), 74.

6. Frederick J. Blue, *Salmon P. Chase: A Life in Politics* (Kent: Kent State Univ. Press, 1987), 31–40. The famous Matson slave case, argued in 1847, is discussed in John A. Duff, *A. Lincoln, Prairie Lawyer* (New York: Rinehart, 1960), 130–45.

7. John Codman Hurd, *The Law of Freedom and Bondage in the United States,* II (Boston: Little, Brown, 1858–62), 195n.

8. Thomas D. Morris, *Free Men All: The Personal Liberty Laws of the North, 1780–1861* (Baltimore: Johns Hopkins Univ. Press, 1974), 56; Dallin H. Oaks, "Habeas Corpus in the States, 1776–1865," *University of Chicago Law Review,* XXXIII (1965), 268–69, 278–79.

9. Morris, *Free Men All,* 112.

10. Hurd, *Law of Freedom and Bondage,* II, 745. President Millard Fillmore worried that the law might violate the constitutional guarantee of the writ of habeas corpus, but Attorney General John J. Crittenden ruled that the provision properly carried out Congress's constitutional duty to "deliver up" fugitives. See Albert Kirwan, *John J. Crittenden: The Struggle for the Union* ([Lexington]: Univ. of Kentucky Press, 1962), 267.

11. Morris, *Free Men All,* 178–79, 187; Robert M. Cover, *Justice Accused: Antislavery and the Judicial Process* (New Haven: Yale Univ. Press, 1975), 183–87.

12. Morris, *Free Men All,* xi; Roy P. Basler et al., eds., *The Collected Works of Abraham Lincoln,* III (New Brunswick: Rutgers Univ. Press, 1953–55), 384, 386, 391; *Coll. Works of Lincoln,* IV, 152, 157.

1. Actions Without Precedent

1. *Coll. Works of Lincoln,* IV, 195, 269.

2. Ibid., 261n, 271.

3. See, for example, James Henderson to Abraham Lincoln, Apr. 16, 1861, Abraham Lincoln Papers, Library of Congress (microfilm).

4. The language, presumably Lincoln's, comes from the memorandum answering the president's request.

5. On Coffey, see Harry J. Carmen and Reinhard H. Luthin, *Lincoln and the Patronage* (New York: Columbia Univ. Press, 1943), 56, and *The Diary of Edward Bates, 1859–1866,* ed. Howard K. Beale, *Annual Report of the American Historical Association for the Year 1930,* IV, 107n, 182.

6. Edward Bates, Memorandum, Apr. 20, 1861, Abraham Lincoln Papers. Coffey found Roger B. Taney's seemingly relevant opinion in *Luther v. Borden* that the "government of *a State,* by its legislature has the power to protect itself from destruction by armed rebellion, by declaring Martial law, and the Legislature is the sole judge of the existence of the necessary exigency." As much as the federal government appeared to be in need of self-defense, the president could hardly embrace that opinion, given Taney's clear indication that such powers lay in the legislative branch. By underlining the word "State," Coffey perhaps emphasized the irrelevance of the opinion to the government in Washington.

7. James Kent, *Commentaries on American Law,* 6th ed., II (New York: William Kent, 1848), 26–27; *Coll. Works of Lincoln,* VI, 268. Lincoln knew Kent's work and quoted it in 1848 (*Coll. Works of Lincoln,* I, 485–86). Besides Blackstone, that is the only book that mentioned the subject that one can be certain Lincoln read. See David C. Mearns, "Mr. Lincoln and the Books He Read," in Arthur Bestor, David Mearns, and Jonathan Daniels, *Three Presidents and Their Books* (orig. pub. 1955; Urbana: Univ. of Illinois Press, 1963), 58–62.

8. Joseph Story, *Commentaries on the Constitution of the United States* (orig. pub. 1833; Durham: Carolina Academic Press, 1987), 483. One Republican polemicist later laid emphasis on Story's awkward language, recalling: "Justice Story in his Commentaries *does* incidentally say: 'That the power to suspend the writ would *seem* to be a legislative power.'" This was clearly a misrepresentation of Story's meaning. See P.E. Havens, *Arbitrary Arrests. Speech of..., March 5, 1863* (n.p.: New York State Union Central Committee, [1863]), 2. Rollin Carlos Hurd's *Treatise on the Right of Personal Liberty, and on the Writ of Habeas Corpus....* (Albany: H.C. Little, 1858) also assumed that Congress wielded the suspending power: "Rebellion and invasion are eminently matters of national concern; and charged as Congress is, with the duty of preserving the United States from both these evils, it is fit that it should possess the power to make effectual such measures as it may deem expedient to adopt for their suppression" (p. 133).

9. Lincoln first used the anecdote in 1863 (*Coll. Works of Lincoln,* VI, 268–69). It was not mentioned in the Bates-Coffey digest, and Lincoln doubtless knew of it from the dispute in the U.S. Congress over remitting Jackson's fine in 1844 if not from reading

history or biography. *Herndon's Life of Lincoln,* ed. by Paul M. Angle (Cleveland: World, 1949), 386; *Coll. Works of Lincoln,* IV, 249n, 341–42.

10. John G. Nicolay and John Hay, *Abraham Lincoln: A History,* IV (New York: Century, 1890), 121–22; George William Brown, *Baltimore and the Nineteenth of April, 1861: A Study of the War* (orig. pub. 1887; Baltimore: Maclay & Associates, 1932), 58–59.

11. Lyman Trumbull to Abraham Lincoln, Apr. 21, 1861; Orville Hickman Browning to Abraham Lincoln, Apr. 22, 1861, Abraham Lincoln Papers.

12. Andrew H. Reeder to Simon Cameron, Apr. 24, 1861; James Watson Webb to Abraham Lincoln, Apr. 24, 1861, Abraham Lincoln Papers.

13. Salmon P. Chase to Abraham Lincoln, Apr. 25, 1861; Edward Bates, Memorandum, Apr. 23, 1861, Abraham Lincoln Papers; Nicolay and Hay, *Abraham Lincoln: A History,* IV, 166.

14. Charles L. Wagandt, *The Mighty Revolution: Negro Emancipation in Maryland, 1862–1864* (Baltimore: Johns Hopkins Univ. Press, 1964), 12–13; Benjamin F. Butler, *Butler's Book* (Boston: A.M. Thayer, 1892), 210.

15. *Coll. Works of Lincoln,* IV, 344.

16. Jean H. Baker, *The Politics of Continuity: Maryland Political Parties from 1858 to 1870* (Baltimore: Johns Hopkins Univ. Press, 1973), 55–58. Scott's orders to Butler nevertheless said that "arrested individuals, notorious for their hostility to the United States," should not be surrendered except on Scott's order. Winfield Scott to Benjamin F. Butler, Apr. 26, 1861, in Jessie Ames Marshall, ed., *Private and Official Correspondence of Gen. Benjamin F. Butler During the Period of the Civil War,* I (Norwood: privately published, 1917), 43.

17. Abraham Lincoln to Winfield Scott, Apr. 25, 1861, Abraham Lincoln Papers. The most recent work on the history of habeas corpus says that it is "interesting that Lincoln viewed suspension as a step to be taken only if absolutely and indispensably necessary; a step even more extreme than bombardment." See William F. Duker, *A Constitutional History of Habeas Corpus* (Westport: Greenwood, 1980), 110, 167n.

18. Nicolay and Hay, *Abraham Lincoln: A History,* IV, 151–52. Nicolay and Hay forgot how the troops were to arrive by the new route—on trains rather than ships.

19. Benjamin Franklin Cooling, *Symbol, Sword, and Shield: Defending Washington during the Civil War* (Hamden: Archon Books, 1975), 33–38.

20. Winfield Scott, General Order No. 4 and Daily Report No. 20, Apr. 26, 1861; Mr. Cooper, Report, Apr. 26, 1861, Abraham Lincoln Papers; Winfield Scott to Benjamin F. Butler, Apr. 29, 1861, *Private and Official Correspondence of Gen. Benjamin F. Butler,* I, 54.

21. *Coll. Works of Lincoln,* IV, 347.

22. *The War of the Rebellion: A Compilation of the Official Records of the Union and Confederate Armies,* ser. II, vol. 2 (Washington: Government Printing Office, 1880–1902), 18–19 (hereafter cited as *O.R.,* with series no. in Roman numerals, volume in Arabic).

23. *O.R.,* II, 1, pp. 567–68. The order suspending the writ was not a general order of the War Department or the adjutant general's office. See *General Orders of the War Department, . . . 1861, 1862 & 1863* (New York: Derby & Miller, 1864); *General Orders Affecting the Volunteer Force. Adjutant General's Office, 1861, 1862, 1863* (Washington: Government Printing Office, 1864).

24. Carl B. Swisher, *The Oliver Wendell Holmes Devise History of the Supreme Court of the United States,* Volume V: *The Taney Period, 1836–64* (New York: Mac-

millan, 1974), 842–43; Taney is quoted from *Ex parte Merryman,* reprinted in *O.R.,* II, 1, p. 578.

25. *Coll. Works of Lincoln,* IV, 364–65. Frederick W. Seward states that "when it was found that the courts in Florida were being used to reduce the little garrison of United States troops by bringing the soldiers, one by one, before the local judge on a writ of *habeas corpus,* and then taking care that they did not get back to the fort again," this proclamation was "sent down by the expedition to reinforce Fort Pickens." However, he misdates the proclamation as having been issued in Apr. *Seward at Washington, as Senator and Secretary of State* (New York: Derby and Miller, 1891), 574.

26. William H. French to Lorenzo Thomas, May 5, 1861, in *O.R.,* I, 1, p. 406.

27. William H. French to George L. Hartsuff, May 8, 1861, in *O.R.,* I, 1, pp. 411–12. William Watson Davis, *The Civil War and Reconstruction in Florida* (New York: Columbia Univ. Press, 1913), 248–49, says French's actions were permitted by Lincoln's proclamation, but this cannot be true since the actions preceded the proclamation.

28. See, for example, *The Opinion of the Hon. Roger Brooke Taney, in the Habeas Corpus Case of John Merryman* (Baltimore: Lucas Brothers, 1861); *Decision of Chief Justice Taney in the Merryman Case...* (Philadelphia: John Campbell, 1862); *The Merryman Habeas Corpus Case, Baltimore.... The United States a Military Despotism* (Jackson: J.L. Power, 1861); *Opinion of Chief Justice Taney, in the Case of ex parte, John Merryman....* (New Orleans: George Ellis, 1861); "The Merryman Case," *New York Freeman's Journal and Catholic Register,* June 15, 1861; "The Merryman Case," *The Crisis* (Columbus, Ohio), June 13, 1861.

29. *Coll. Works of Lincoln,* IV, 390.

30. Ibid., IV, 414 and VI, 263. The Chase order is misdated in *O.R.,* II, 2, p. 193, as the *Coll. Works of Lincoln* points out (IV, 414n). William F. Duker says that the Chase order "amounted to a bill of attainder." *Constitutional History of Habeas Corpus,* 110, 168n.

31. *Coll. Works of Lincoln,* IV, 419.

32. Ibid., 429–31; James G. Randall, *Constitutional Problems under Lincoln* (New York: D. Appleton, 1926), 121, 123.

33. *Coll. Works of Lincoln,* V, 242. As late as May 27, 1862, Lincoln was admitting in a public document that "some" of the measures taken in the "crisis" after the fall of Fort Sumter "were without any authority of law."

34. *O.R.,* II, 2. pp. 18–19; the Senate, on the other hand, knew about the order of Apr. 27 and the proclamation of May 10; see George Clark Sellery, *Lincoln's Suspension of Habeas Corpus as Viewed by Congress,* Bulletin of the Univ. of Wisconsin, no. 149 (Madison: n.p., 1907), 12–14.

35. *Coll. Works of Lincoln,* V, 554.

36. William H. Seward to Winfield Scott, Oct. 14, 1861 (draft), Papers of William H. Seward, Univ. of Rochester (microfilm reel 66). Lincoln allowed Seward to draft for him such routine documents as proclamations of days of prayer and thanksgiving.

37. Frederick W. Seward, *Reminiscences of a War-Time Statesman and Diplomat, 1830–1915* (New York: G.P. Putnam's Sons, 1916), 175–78; Earl Schenck Miers, ed., *Lincoln Day by Day: A Chronology, 1809–1865,* III (Washington: Lincoln Sesquicentennial Commission, 1960), 62.

38. Dean Sprague, *Freedom under Lincoln* (Boston: Houghton Mifflin, 1965), esp. 87.

39. George B. McClellan, *McClellan's Own Story: The War for the Union...* (New

York: Charles L. Webster & Co., 1887), 146–47. Government detective Lafayette C. Baker maintained in later years that he penetrated the conspiracy to vote Maryland out of the Union at Frederick, but his reminiscences are not always trustworthy. See his *History of the United States Secret Service* (Philadelphia: L.C. Baker, 1867), 85–86. Actually, Allan Pinkerton did much of the dirty work; see *O.R.,* II, 1, pp. 678–88.

40. Jean H. Baker says Cameron and McClellan overestimated rumors of a secession conspiracy in *The Politics of Continuity: Maryland Political Parties from 1858 to 1870,* 58. Allan Nevins pronounced Maryland "entirely safe" for the Union by June 1861 in *The War For the Union,* Volume I: *The Improvised War, 1861–1862* (New York: Charles Scribner's Sons, 1959), 139. For Hicks's letter see *O.R.,* II, 1, p. 685.

41. Nathaniel P. Banks to Abraham Lincoln, Aug. 22, 1861, Abraham Lincoln Papers.

42. Ward Hill Lamon to Abraham Lincoln, Aug. 23, 1861; Nathaniel P. Banks to Lincoln, Aug. 28, 1861, Abraham Lincoln Papers.

43. William H. Seward to Abraham Lincoln, Aug. 22, 1861, Abraham Lincoln Papers.

44. *Correspondence between S. Teackle Wallis... of Baltimore, and... John Sherman... Concerning the Arrest of Members of the Maryland Legislature....* (n.p., 1863), 7. Among those arrested were some avowed secessionists, for example, the reading clerk of the Maryland Senate, John M. Brewer. See John M. Brewer, *Prison Life* (n.p., n.d.), 25. On the elections see Baker, *Politics of Continuity,* 69–74.

45. *Coll. Works of Lincoln,* IV, 523.

46. Seward's biographers have pulled few punches on this point. See Glyndon Van Deusen, *William Henry Seward* (New York: Oxford Univ. Press, 1967), 290; Frederic Bancroft, *Life of William H. Seward,* II (New York: Harper & Brothers, 1900), 280; Frederick W. Seward, *Seward at Washington, as Senator and Secretary of State,* p. 608.

47. Dean Sprague, *Freedom under Lincoln,* pp. 302–3.

48. Ibid., 158, 159.

49. Ibid., 159. Frederick W. Seward mentions the bell in *Seward at Washington,* p. 608, as does Frederic Bancroft in *Life of William H. Seward,* II, 280. Glyndon Van Deusen ignores the story in *William Henry Seward.*

50. W.A. Dunning stated that the "energetic" Seward "above all" was responsible "for filling the prisons." Dunning, "Disloyalty in Two Wars," *American Historical Review,* XXIV (Oct. 1918), 626.

51. *O.R.,* II, 2, pp. 300, 301; see also the case of David R. Mister, p. 305.

52. Bancroft, *Life of William H. Seward,* II, 276–77.

53. *O.R.,* II, 2, pp. 305, 311–12.

54. Ibid., II, 2, pp. 305–19.

55. See the cases of Michael Reilly and George Shannon as well as that of George W. Barnard, *O.R.,* II, 2, pp. 318, 300.

56. Paul C. Nagel, *One Nation Indivisible: The Union in American Thought, 1776–1861* (New York: Oxford Univ. Press, 1964), 88, 106. Nagel concludes that Seward's political ideas constituted a prime example of belief in the Union as "absolute." Major L. Wilson agrees in *Space, Time, and Freedom: The Quest for Nationality and the Irrepressible Conflict, 1815–1861* (Westport: Greenwood Press, 1974), 212–34.

57. Donn Piatt, *Memories of the Men Who Saved the Union* (New York: Belford, Clarke, 1887), 159. In the context of the Civil War, his willingness to embrace a "higher law" than the Constitution would be criticized as the sort of attitude certain to bring despotism. One of the most articulate victims of arbitrary arrest would identify this as the root of the problem: "With its first breach, Abolitionism repudiated the Constitution,

and appealed from its obligations to the higher law." D.A. Mahony, *The Prisoner of State* (New York: Carleton, 1863), 13.

58. William H. Seward to Abraham Lincoln, Mar. 15, 1861, Abraham Lincoln Papers.

59. Thomas D. Morris, *Free Men All,* p. 133; Frederick Seward, *Seward at Washington,* pp. 608–9. After the war, reporter Donn Piatt recalled that Seward told him before the war began that "a written Constitution is a superstition that presupposes certain impossibilities. The first is that it can express all the wisdom of the past, and anticipate all the wants of the future. It supposes that its creators were both saints and sages." Piatt, *Memories of the Men Who Saved the Union,* p. 136.

60. This figure and the statistics which follow vary slightly from those published in my earlier article on this subject, "The Lincoln Administration and Arbitrary Arrests: A Reconsideration," *Papers of the Abraham Lincoln Association,* V (Springfield: Abraham Lincoln Assn., 1983), 7–24. Statistics for the article were compiled in the winter of 1981–82, and information found in other sources since that time has dictated minor adjustments. The most notable difference in number of cases appears in the category of residence. Principal sources are *O.R.,* II, 2, pp. 290–1557; *Democratic Almanac for 1867* (New York: Van Evrie, Horton, & Co., 1866), 24–33; and the *American Annual Cyclopaedia . . . of 1861* (New York: D. Appleton, 1862), 361. Index cards for each case and tally sheets for the statistics for this chapter and all others are deposited in the Louis A. Warren Lincoln Library and Museum, Fort Wayne, Indiana.

61. [Lawrence Sangston], *The Bastiles of the North* (Baltimore: Kelly, Hedian & Piet, 1863), 57–58.

62. *Coll. Works of Lincoln,* VI, 265; Sangston, *Bastiles of the North,* p. 76n.

63. E.M. Archibald to Lord Lyons, Sept. 20, 1861, in *O.R.,* II, 2, p. 546.

64. Sangston, *Bastiles of the North,* p. 54.

65. Robert Murray to William H. Seward, Oct. 10, 1861, *O.R.,* II, 2, p. 96.

66. Percentages are calculated from those cases where records note residence and nationality.

67. The District of Columbia qualifies as a border area, as it bordered the Confederacy and slavery was legal there. Lincoln included it in his famous estimate of the cost of compensated emancipation in the border states. See *Coll. Works of Lincoln,* V, 160.

68. The estimate was arrived at by extrapolating for the whole population of persons arrested under Seward from the percentage calculated for the cases where residence is noted in the records. Some 85.6 percent of the arrests containing information about residence involved people who did not live in the North or who owed allegiance to a foreign government; 14.4 percent, then, came from the states above Maryland, Delaware, Missouri, and Kentucky. Taking 14.4 percent of the total number of arrests under Seward yields the figure of 125.

69. James G. Randall, *Lincoln the President: Midstream* (New York: Dodd, Mead, 1953), 195–97.

70. *O.R.,* II, 2, pp. 1021–23. I am indebted to Howard Westwood for the constitutional point here.

71. *O.R.,* II, 2, p. 331 (Hawley, patriotic envelopes), p. 750 (Corlies), p. 348 (Stewart), p. 801 (Cory), p. 302 (Stannard), p. 703 (Davis), p. 348 (Jones), p. 298 (Deckart), pp. 805–29 (Durrett); p. 938 (Flanders), p. 802 (McMaster), p. 787 (Mills), p. 788 (Piggott), p. 665 (Reeves), p. 771 (Wall). See Chapter 2 for the early Missouri cases.

72. *Coll. Works of Lincoln,* VI, 326.

73. *O.R.,* II, 2, pp. 1076 (Mason & Slidell), 561, 858; Thomas A. Jones, *J. Wilkes Booth: An Account of His Sojourn in Southern Maryland after the Assassination of Abraham Lincoln...* (Chicago: Laird & Lee, 1893).

74. *O.R.,* II, 2, p. 296; Sangston, *Bastiles of the North,* pp. 101–2. Details of the anecdote vary, but the story is corroborated in John M. Brewer, *Prison Life* (n.p., n.d.), 22–23.

75. Fred Harvey Harrington, *Fighting Politician: Major General N.P. Banks* (Philadelphia: Univ. of Pennsylvania Press, 1948), 60.

76. J. Heermans, *War Power of the President,* Loyal Publication Society, no. 32 (New York: C.S. Westcott & Co., Printers, 1863), 5. On the Democrats' quiet agreement on Maryland, see Chapter 9.

77. Sangston, *Bastiles of the North,* p. 38n.

78. Howard K. Beale, ed., *Diary of Gideon Welles, Secretary of the Navy under Lincoln and Johnson,* I (New York: W.W. Norton, 1960), 549. Seward's views were confirmed by the artist himself in Francis B. Carpenter, *Six Months at the White House with Abraham Lincoln* (New York: Hurd and Houghton, 1866), 73.

2. Missouri and Martial Law

1. *O.R.,* II, 1, pp. 106–15. Dean Sprague hints at the relative insignificance of the West in the national awareness of legal and constitutional matters in *Freedom under Lincoln,* p. 77. The McDonald case was mentioned in the article on "Habeas Corpus" in *The American Annual Cyclopaedia... of the Year 1861* (New York: D. Appleton & Co., 1862), 356. The *Illinois State Register* defended Judge Treat from "malicious slander" in its issue of June 11, 1861, but no mention was made of the writ of habeas corpus or of the merits of the McDonald case.

2. Mark E. Neely, Jr., "The Perils of Running the Blockade: The Influence of International Law in an Era of Total War," *Civil War History,* XXXII (June 1986), 116–17.

3. John Y. Simon, ed., *The Papers of Ulysses S. Grant,* II (Carbondale: Southern Illinois Univ. Press, 1967–85), 136–37. Eleven days earlier, General John C. Frémont, commanding the Western Department, had declared martial law in St. Louis County, but Grant was operating many miles and several counties away from St. Louis.

4. *Papers of Ulysses S. Grant,* II, 139.

5. William F. Swindler, "The Southern Press in Missouri, 1861–1864," *Missouri Historical Review,* XXXV (Apr. 1941), 398. By the end of 1861, Union authorities had silenced newspapers in Cape Girardeau, Hannibal, Lexington, Warrensburg, Platte City, Troy, Osceola, Oregon, and Washington, Missouri, without asking authorities in Washington, D.C., for policy guidelines on the state's pro-Southern press.

6. *Papers of Ulysses S. Grant,* II, 102 and III, 4.

7. William E. Parrish, *A History of Missouri, Volume III: 1860 to 1875* (Columbia: Univ. of Missouri Press, 1973), 64.

8. James G. Randall, *Lincoln the President: Springfield to Gettysburg,* II (New York: Dodd, Mead, 1945), 7, 16. See also James A. Rawley, *Turning Points of the Civil War* (Lincoln: Univ. of Nebraska Press, 1966), 37–38.

9. *Coll. Works of Lincoln,* IV, 506. Frémont's proclamation appears in *O.R.,* II, 1, pp. 221–22.

10. *O.R.,* II, 1, pp. 221–22; *Coll. Works of Lincoln,* IV, 532. See *O.R.,* II, 1, p. 283,

for the case of Joseph Aubuchon, tried by military commission in Ironton, Missouri, on Sept. 5, 1861, and other early cases.

11. *Ex parte Milligan,* 4 Wallace 297. See especially Randall, *Constitutional Problems under Lincoln,* pp. 140–47, for a discussion that implies that the distinction was blurred during the Civil War but should not have been.

12. *O.R.,* II, 1, p. 181.

13. Stephen E. Ambrose, *Halleck: Lincoln's Chief of Staff* (Baton Rouge: Louisiana State Univ. Press, 1962), 13; *O.R.,* II, 1, p. 230; Richard S. Brownlee, *Gray Ghosts of the Confederacy: Guerrilla Warfare in the West, 1861–1865* (Baton Rouge: Louisiana State Univ. Press, 1958), 146.

14. *O.R.,* I, 8, p. 817.

15. *O.R.,* II, 1, p. 231.

16. *Coll. Works of Lincoln,* V, 27.

17. *O.R.,* II, 1, pp. 232–33.

18. Seward's undated draft order to Henry W. Halleck appears (misfiled) at Dec. 6, 1861, on reel 67 of the Papers of William H. Seward. For the text printed in the *Coll. Works of Lincoln,* V, 35, Basler used a document signed by Lincoln but not otherwise written in his hand.

19. *O.R.,* II, 1, pp. 247–48.

20. Halleck declared martial law on Dec. 26, 1861. See *O.R.,* II, 1, p. 155.

21. *O.R.,* II, 1, p. 270.

22. *Papers of Ulysses S. Grant,* IV, 32, 40.

23. *O.R.,* II, 1, p. 174.

24. *O.R.,* II, 1, pp. 125–26.

25. *O.R.,* II, 1, pp. 139–40, 175–76.

26. Henry W. Halleck to T. Ewing, January 1, 1862, in *O.R.,* II, 1, p. 247.

27. Winfield Scott, *Memoirs of Lieut. General Scott, LL. D.* (New York: Sheldon & Co., 1864), 393, 541.

28. Charles Winslow Elliott, *Winfield Scott: The Soldier and the Man* (New York: Macmillan, 1937), 448; Scott, *Memoirs,* pp. 392–93, 394, 395.

29. Older studies give Americans high marks for their behavior during the occupation of Mexico, especially Justin H. Smith, "American Rule in Mexico," *American Historical Review,* XXIII (Jan. 1918), 287–302. By contrast, most existing studies of the Union occupation of Missouri give Lincoln's forces low marks. As for the applicability of the precedent, Scott himself expressed no doubts. While still general-in-chief during the Civil War, he eagerly instituted military arrests of civilians in the East. He was still holding overall command when trials by military commission began in Missouri.

30. Charles Fairman, "The Law of Martial Rule and the National Emergency," *Harvard Law Review,* LV (June 1942), 1259.

31. For Halleck's definition of the role of military commissions see *O.R.,* II, 1, pp. 247–48.

32. See Chapter 8.

33. *O.R.,* II, 1, p. 283.

34. Randall, *Constitutional Problems under Lincoln,* pp. 77–78. Congress had passed a conspiracies act on July 31, 1861, providing punishments for conspiring to overthrow or oppose the government, but the crimes triable by military commissions were defined by the army rather than Congress.

35. *O.R.,* II, 1, pp. 283, 284. Higginbotham had apparently been entrapped, "inveigled into an expression of... sympathies against the United States."

36. See for example, *O.R.*, II, 1, p. 284.

37. *O.R.*, II, 1, pp. 386–89.

38. *O.R.*, II, 1, p. 405.

39. *O.R.*, II, 1, pp. 443–45.

40. *O.R.*, II, 1, pp. 467–68.

41. Based on trial records in *O.R.*, II, 1, pp. 282–504.

42. *O.R.*, II, 1, p. 453.

43. *O.R.*, II, 2, p. 327. Though residence and place of arrest were rarely noted in these records, the few instances where such information is available show that a few prisoners were present from elsewhere in the West, Kansas for example. And Alton, unlike the prisons in St. Louis, often held prisoners who were not from Missouri and who had been arrested elsewhere.

44. Comparisons of arrest records with separate records of trials by military commission reveal few duplicate names. As those tried would also constitute political prisoners, the absence of their names from the lists in Washington indicates the incompleteness of the Missouri records.

45. Union Provost Marshals' File of Papers Relating to Two or More Civilians, RG 109, National Archives, (microcopy 416), reel 90 (hereafter cited PMG 416 with reel no. after slash, as PMG 416/90); Parrish, *A History of Missouri*, III, 28, 33–34, 47.

46. War Department Collection of Confederate Records, RG 109 (94 reels).

47. Richard D. Cutts to Henry W. Halleck, Jan. 31, 1862, PMG 416/93.

48. PMG 416/90.

49. Cases of Thomas Beard, PMG 416/90; Clinton Burbidge, PMG 416/90.

50. Gratiot Street records appear on PMG 416/90, 416/91, 416/92, and 416/93.

51. PMG 416/90; W.B. Hesseltine, "Military Prisons of St. Louis, 1861–1865," *Missouri Historical Review*, XXIII (Apr. 1929), 381–84, 386–87.

52. Charles R. Mink, "General Orders, No. 11: The Forced Evacuation of Civilians during the Civil War," *Military Affairs*, XXXIV (Dec. 1980), 132; *O.R.*, I, 22, pt. 2, p. 472. On Missouri in general, see Michael Fellman's *Inside War: The Guerrilla Conflict in Missouri During the American Civil War* (New York: Oxford Univ. Press, 1989).

53. Ann Davis Niepman, "General Orders No. 11 and Border Warfare During the Civil War," *Missouri Historical Review*, LXVI (Jan. 1972), 198; Ewing to C.W. Marsh, in *O.R.*, I, 22, pt. 2, p. 428; William H. Wherry to Ewing, in *O.R.*, I, 22, pt. 2, pp. 450–51.

54. *O.R.*, I, 22, pt. 2, p. 461.

55. *Coll. Works of Lincoln*, VI, 492 and VIII, 308. For a succinct description of Lincoln's hopeless policy in Missouri see Michael Fellman, *Inside War:*, pp. 84–86.

56. *Coll. Works of Lincoln*, VI, 492.

3. Low Tide for Liberty

1. On Aug. 8, 1861, Colonel E.R.S. Canby announced that the writ of habeas corpus had been suspended in the Department of New Mexico, far from any area where the president had thus far authorized its suspension. *O.R.*, II, 2, p. 40.

2. Howard K. Beale, ed., *Diary of Gideon Welles*, I, 150. Lincoln used the proper phrase consistently in his message to Congress of July 4, 1861. *Coll. Works of Lincoln*, IV, 429–31.

3. Frederick W. Seward, *Seward at Washington, as Senator and Secretary of State*, p. 574.

4. David Donald, ed., *Inside Lincoln's Cabinet: The Civil War Diaries of Salmon P. Chase* (New York: Longmans, Green, 1954), 156.

5. Eugene C. Murdock, *One Million Men: The Civil War Draft in the North* (orig. pub. 1971; Westport: Greenwood Press, 1980), 6.

6. *O.R.,* II, 4, pp. 358–59; undated draft affidavit filed with documents of Aug. 1862 and at the end of 1862's documents in the Stanton Papers, Library of Congress (microfilm). Stanton's affidavit was evidence in a suit for wrongful arrest made under the Aug. 8 orders.

7. *O.R.,* III, 2, pp. 321–22. On their being overlooked by history, see Robert J. Chandler, "Crushing Dissent: The Pacific Coast Tests Lincoln's Policy of Suppression, 1862," *Civil War History,* XXX (Sept. 1984), 1, 2n.

8. David Donald, ed., *Inside Lincoln's Cabinet,* p. 111.

9. D.W. Alvord to L.C. Turner, Aug. 11, 1862, and Turner to Alvord, Aug. 15, 1862 (draft), Turner-Baker Papers (microfilm) Case #213. William Osgood of Hornellsville, New York, wrote on Aug. 8 to ask a similar question about constables and justices of the peace in towns and villages. William Osgood to L.C. Turner, Aug. 8, 1862, Turner-Baker Papers (microfilm) Case #2602.

10. Logan to Lincoln, Aug. 30, 1862, and Lincoln endorsement, Oct. 9, 1862, Turner-Baker Papers (microfilm) Case #142. In the discussion of Blanchard's case in John A. Marshall, *American Bastile* (Philadelphia: Thomas W. Hartley, 1869) there is no mention of the roles of either Logan or Lincoln in gaining his release from prison.

11. Allan Nevins, "The Case of the Copperhead Conspirator," in John A. Garraty, ed., *Quarrels That Have Shaped the Constitution* (New York: Harper & Row, 1962), 106. Writing at about the same time, David Donald put the number arrested at "at least 15,000." See "Died of Democracy," in David Donald, ed., *Why the North Won the Civil War* (orig. pub. 1960; New York: Collier, 1962), 87.

12. Joseph Holt, endorsement, Sept. 9, 1862, on note from Edwin M. Stanton; Daniel Clark to Levi C. Turner, Sept. 12, 1862; Jacob H. Eler to Holt, Sept. 22, 1862, Turner-Baker Papers (microfilm) Case #200.

13. Turner-Baker Papers (microfilm) Case #600.

14. Vol. I, no. 3 is the Aug. 1862 issue; vol. I, no. 4 is the Apr. 1863 issue.

15. Joseph George, Jr., "'Abraham Africanus I': President Abraham Lincoln through the Eyes of a Copperhead Editor," *Civil War History,* XIV (Sept. 1968), 226–28.

16. A. Ricketts to Levi C. Turner, Aug. 28, 1862; Andrew G. Curtin to Turner, Sept. 2, 1862; Ricketts to Turner, Oct. 11, 1862, Turner-Baker Papers (microfilm) Case #385.

17. Frank L. Klement, *The Copperheads in the Middle West* (Chicago: Univ. of Chicago Press, 1960), 19; John Hughes to William H. Seward, Aug. 27, 1862, Turner-Baker Papers (microfilm) Case #413.

18. See Chapter 9.

19. William Millward to Edwin M. Stanton, Aug. 28, 1862, Turner-Baker Papers (microfilm) Case #14; William Dusinberre, *Civil War Issues in Philadelphia, 1861–1865* (Philadelphia: Univ. of Pennsylvania Press, 1965), 142–43.

20. Turner-Baker Papers (microfilm) Case #191 (Anderson), Case #94 (Anthony), Case #565 (Bond), Case #604 (Goheen), Case #388 (Strantzenheimer), Case #539 (Bobson).

21. Turner-Baker Papers (microfilm) Case #201 (Bush), Case #601 (Grier), Case #196 (Slagel).

22. Turner-Baker Papers (microfilm) Case #542 (Barney), Case #300 (Fraleys),

Case #12 (Gray), Case #689 (Lyon), Case #197 (Palmer), Case #157 (Peters), Case #151 (Hamilton, Ohio) Case #409 (Wright).

23. Turner-Baker Papers (microfilm) Case #751 (Kershaw), Case #4 (King), Case #534 (Meredith), Case #312, Case #381.

24. Mahony, *Prisoner of State,* pp. 117–19.

25. Turner-Baker Papers (microfilm) Case #3344.

26. Turner to John A. Kennedy, Aug. 29, 1862, Turner-Baker Papers, Case #159.

27. *Coll. Works of Lincoln,* V, 370; Stanton to Boyle, Aug. 13, 1862, in *O.R.,* II, 4, pp. 378, 380.

28. Chandler, "Crushing Dissent" pp. 239–40.

29. James Brooks, *The Two Proclamations. Speech of the Hon. . . . before the Democratic Union Association. Sept. 29th, 1862* (New York: Van Evrie, Horton, & Co., [1862]), 6. Though Democrats opposed both proclamations, not all of them linked them in the way Chandler and Brooks suggest. Rather than saying that the proclamation of the 24th was necessary to suppress opposition to the proclamation of the 22nd, Congressman Aaron Harding of Kentucky linked the two only in their derivation from the same despotic impulse: "when an absolute and despotic power is assumed over all the slave property of whole States and communities, without any regard to the guilt or innocence, loyalty or disloyalty of the owner—sweeping away, in thousands of instances, from loyal citizens, from aged men and women, and from helpless infants, all the property they have, and reducing them to beggary and want; and when all this is done without *any process of law* at all, the citizen would naturally conclude, however loyal and upright he may be, that his own personal liberty was no longer safe. And . . . he could have remained in doubt or suspense on that question *only two days;* for the most abundant and conclusive evidence of the correctness of such a conclusion was furnished by the President, in his second proclamation." *Speech of A. Harding, of Ky., on the President's Two Proclamations and the Two Rebellions* (Washington, D.C.: McGill & Withcrow, Printers, 1863), 7–8.

30. Turner-Baker Papers (microfilm), Case #470.

31. Turner-Baker Papers (microfilm), Case #37.

32. Turner-Baker Papers (microfilm) Case #677 and Case #676.

33. *Coll. Works of Lincoln,* V, 436–37.

34. Frank L. Klement, *The Limits of Dissent: Clement L. Vallandigham & The Civil War* ([Lexington]: Univ. Press of Kentucky, 1970), 173–77.

35. [Edwin M. Stanton], "Maj. Genl Burnside," draft order, May 13, 1863; "General Orders No., 1" May 13, 1863, Abraham Lincoln Papers.

36. Klement, *The Limits of Dissent,* p. 166; *Coll. Works of Lincoln,* VI, 215. Humphrey H. Leavitt heard the case. See Randall, *Lincoln the President: Midstream,* pp. 221–24.

37. Nicolay and Hay, *Abraham Lincoln: A History,* VII, 343.

38. Lincoln was an old hand at using conspiracy charges, as were most American politicians of the era; see David M. Potter, *The Impending Crisis, 1848–1861* (New York: Harper & Row, 1976), 289.

39. For further discussion of this part of Lincoln's argument see Chapter 11.

40. *Coll. Works of Lincoln,* VI, 260–69; Nicolay and Hay, *Abraham Lincoln: A History,* VII, 349.

41. Randall, *Constitutional Problems under Lincoln,* pp. 130–31.

42. *Coll. Works of Lincoln,* VI, 303.

43. James G. Randall and David Donald, *The Civil War and Reconstruction*, 2nd ed. (Boston: D. C. Heath, 1961), 316.

44. *House Executive Documents*, 39 Cong., 1 sess., vol. IV, pt. 1, pp. 32, 34.

45. Howard K. Beale, ed., *The Diary of Edward Bates, 1859–1866* (Washington: Government Printing Office, 1933), 306; Howard K. Beale, ed., *Diary of Gideon Welles*, I, 432.

46. Beale, ed., *Diary of Gideon Welles*, I, 432–33; *Diary of Edward Bates*, p. 307. Welles was probably right about Stanton's instigation of the subject, for Chase found Lincoln huddled with his secretary of war in the president's office just after breakfast on the 14th. See David Donald, ed., *Inside Lincoln's Cabinet*, pp. 192–93.

47. Donald, ed., *Inside Lincoln's Cabinet*, p. 194.

48. Dallin H. Oaks, "Habeas Corpus in the States, 1776–1865," *University of Chicago Law Review*, XXXII (1965), 274–75; Morris, *Free Men All*, p. 35.

49. *Coll. Works of Lincoln*, VI, 460.

50. Donald, ed., *Inside Lincoln's Cabinet*, pp. 195–96.

51. *Coll. Works of Lincoln*, VI, 451–52.

52. Beale, ed., *Diary of Gideon Welles*, I, 434, 435.

53. Donald, ed., *Inside Lincoln's Cabinet*, p. 199.

54. Ibid., p. 196.

55. A further mark of Lincoln's extreme agitation over this issue was his manuscript opinion on the draft, unissued in Lincoln's day but used by Nicolay and Hay and found in Lincoln's papers at the Library of Congress when they were opened in 1947. The undated manuscript began in a moderate tone, hopeful of avoiding "misunderstanding between the public and the public servant" and admitting that some who were opposed to conscription were nevertheless "sincerely devoted to the republican institutions, and territorial integrity of our country." Soon, though, Lincoln was saying that opposition to conscription on constitutional grounds marked "the first instance . . . in which the power of congress to do a thing has ever been questioned, in a case when the power is given by the constitution in express terms." Lincoln grew increasingly disgusted:

> The principle of the draft, which simply is involuntary, or enforced service, is not new. It has been practiced in all ages of the world. It was well known to the framers of our constitution as one of the modes of raising armies, at the time they placed in that instrument the provision that "the congress shall have power to raise and support armies." It had been used, just before, in establishing our independence; and it was also used under the constitution in 1812. Wherein is the peculiar hardship now? Shall we shrink from the necessary means to maintain our free government, which our grand-fathers employed to establish it, and our own fathers have already employed once to maintain it? Are we degenerate? Has the manhood of our race run out?

In truth, this manuscript dealt with the constitutionality of conscription and not with judicial obstructions to its operation. It never mentioned habeas corpus and is almost certainly misdated in the *Coll. Works of Lincoln* as stemming from the controversy with the Pennsylvania judges over habeas corpus in Sept. 1863. Nicolay and Hay date it in Aug., from the time when President Lincoln, in the aftermath of the draft riots, was embroiled in controversies over continuing the draft and over the inequities of the enrollment. In fact, it seems even more likely to have been provoked later, after the Pennsylvania Supreme Court declared the federal conscription law unconstitutional in Nov. *Coll. Works of Lincoln*, VI, 444–48.

56. Andrew G. Curtin to Abraham Lincoln, Sept. 4, 1863, Abraham Lincoln Papers,

Library of Congress (microfilm); A.G. Curtin, *Message of…, Governor of Pennsylvania, Relative to Military Arrests, February 12, 1863* (n.p., 1863), 4.

57. Andrew G. Curtin to Abraham Lincoln, Sept. 18, 1863, Abraham Lincoln Papers. See also James A. Rawley, *The Politics of Union: Northern Politics during the Civil War* (Hinsdale, Ill.: Dryden Press, 1974), 130.

4. Arrests Move South

1. *Coll. Works of Lincoln,* VI, 29.

2. Again, the figure is as good as the records will allow. As in all periods, some other arrests were doubtless made.

3. For a spectacular instance, see Chapter 5.

4. Turner-Baker Papers (microfilm) Case #3303.

5. See Paddy Griffith, *Battle Tactics of the Civil War* (New Haven: Yale Univ. Press, 1989).

6. T. Harry Williams, *Lincoln and His Generals* (New York: Alfred A. Knopf, 1952), 291–97, 304–10; U.S. Grant to George G. Meade, Apr. 9, 1864 and to Benjamin F. Butler, Apr. 2, 1864, in *Papers of Ulysses S. Grant,* X, 246, 274.

7. U.S. Grant to William Hoffman, Aug. 11, 1863, in *Papers of Ulysses S. Grant,* IX, 170.

8. U.S. Grant to Phillip Sheridan, Aug. 16, 1864, in *Papers of Ulysses S. Grant,* XII, 15, 13.

9. *Personal Memoirs of P. H. Sheridan,* II (New York: Charles L. Webster & Co., 1888), 99–100; *Loudoun County and the Civil War: A History and Guide* (Leesburg: Civil War Centennial Commission, 1961), 47–50; James V. Forsythe to Wesley Merritt, Nov. 27, 1864, in *O.R.,* I, 43, pt. 2, p. 679; Merritt to William Russell, Dec. 6, 1864, in *O.R.,* I, 32, pt. 1, p. 671. Jefferson Davis denounced the scorched-earth policy in the Valley in his message to Congress of Nov. 7, 1864, in Dunbar Rowland, ed., *Jefferson Davis, Constitutionalist: His Letters, Papers and Speeches,* VI (Jackson: Miss. Department of Archives and History, 1923), 385.

10. Charles Wells Russell, ed., *The Memoirs of Colonel John S. Mosby* (Boston: Little, Brown, 1917), 152–55. Hatton was imprisoned first in Old Capitol and then in Fort Warren. He was exchanged in Feb. 1864.

11. In addition to the listing in Record of Prisoners of State, Turner-Baker Papers (unmicrofilmed), National Archives (hereafter cited as RPS, NA), see Turner-Baker Papers (microfilm) Case #2641.

12. Case of Elijah Beach, RPS, NA, and Turner-Baker Papers (microfilm) Cases #2406, #3250, #3845.

13. Ould to Seddon, Nov. 1, 1864, *O.R.,* II, 7, p. 1080.

14. For such a caricature of Grant see the chapter called "A Strategy of Annihilation: U.S. Grant and the Union," in Russell F. Weigley, *The American Way of War: A History of United States Military Strategy and Policy* (orig. pub. 1972; Bloomington: Indiana Univ. Press, 1977), 128–52.

15. U.S. Grant to Phillip Sheridan, Aug. 21, 1864, Nov. 11, 1864, in *Papers of Ulysses S. Grant,* XII, 63, 127, 397.

16. *Papers of Ulysses S. Grant,* IX, 133–34.

17. Ulysses S. Grant to Henry W. Halleck, Aug. 11, 1863, in ibid., IX, 173–74; U.S. Grant to William T. Sherman, Aug. 6, 1863, in ibid., IX, 155; Sherman to Grant, Aug. 6, 1863, in ibid., IX, 156n.

18. Ulysses S. Grant to Henry W. Halleck, Sept. 4, 1864, in *Papers of Ulysses S. Grant,* XII, 124, 125n; E. Merton Coulter, *The Civil War and Readjustment in Kentucky* (Chapel Hill: Univ. of North Carolina Press, 1926), 221–22.

19. Ulysses S. Grant to Henry W. Halleck, July 1, 1864, in *Papers of Ulysses S. Grant,* XI, 155.

20. William Hoffman to Henry Dent, Feb. 10, 1863, *O.R.,* II, 5, p. 263.

21. Horatio G. Wright to Julius White, Feb. 26, 1863, *O.R.,* II, 5, p. 299.

22. William Hoffman to S.E. Jones, Dec. 3, 1862, *O.R.,* II, 5, p. 19; Wright to White, ibid., p. 299; B.F. Kelly endorsement, May 26, 1864, Turner-Baker Papers (microfilm) Case #3044; C.W. Nelson, "How we Live...," Turner-Baker Papers (microfilm) Case #1419.

23. Wright to White, *O.R.,* II, 5, pp. 299–300.

24. Turner-Baker Papers (microfilm) Cases #589, #47; Turner-Baker Papers, RPS, NA, Cases of Margaret Augiers, Columbus C. Fearson, William M. Harrison, James McGee, and Hugh McMahon.

25. Emma Lou Thornbrough, *Indiana in the Civil War Era* (Indianapolis: Indiana Historical Bureau & Indiana Historical Society, 1965), 68, 194–95, 208–9.

26. *INDIANA Daily State Sentinel,* Feb. 1, 1864.

27. Harold M. Hyman, *A More Perfect Union: The Impact of the Civil War and Reconstruction on the Constitution* (New York: Alfred A. Knopf, 1973), 249.

28. *INDIANA Daily State Sentinel,* Feb. 1, 1864; Emma Lou Thornbrough, "Judge Perkins, the Indiana Supreme Court, and the Civil War," *Indiana Magazine of History,* LX (Mar. 1964), 88–90, 95–96.

29. *INDIANA Daily State Sentinel,* Mar. 8, 1864; Thornbrough, *Indiana in the Civil War Era,* pp. 68–69.

30. Don E. Fehrenbacher, "The Paradoxes of Freedom," in *Lincoln in Text and Context: Collected Essays* (Stanford: Stanford Univ. Press, 1987), 135. Randall offered no explanation for the change in 1862 or for the particular limitations of the various proclamations of 1861 in *Constitutional Problems under Lincoln,* pp. 149, 149n., 151–52. Randall cited the first three proclamations but confused the chronological order in which they appeared in *Lincoln the President: Springfield to Gettysburg,* I, 373–74, and mentioned one more in II, 174. In *Lincoln the President: Midstream,* pp. 157–61, Randall doubled back to deal more fully with the first order and later mentioned the last (p. 168). Charles Sellery mentioned the first two in *Lincoln's Suspension of Habeas Corpus as Viewed by Congress,* p. 220. Then he glibly dismissed the rest, saying: "It is not necessary to refer specifically to any of the subsequent orders." He mentioned one other proclamation (p. 56n.). Dean Sprague, though he tells anecdotes of many arrests of civilians that could not have come under either order, mentions only two suspensions in *Freedom Under Lincoln,* pp. 25, 116. Harold M. Hyman mentions only two in *A More Perfect Union: The Impact of the Civil War and Reconstruction on the Constitution,* pp. 82, 217. Clarence A. Berdahl enumerated seven in *War Powers of the Executive in the United States,* Univ. of Illinois Studies in the Social Sciences, IX, Mar.–June 1920 (Urbana: Univ. of Illinois, 1921), 190–91. Kenneth A. Bernard mentions three, the first applying to "limited areas" and the last two, which "broadened the scope of suspension," in Lincoln and Civil Liberties," *Abraham Lincoln Quarterly,* VI (Sept. 1951), 377. The only complete enumeration previous to this book came in William F. Duker, *A Constitutional History of Habeas Corpus,* pp. 143–70, but Duker failed to mention the unissued suspension in the Vallandigham case.

31. *Coll. Works of Lincoln,* VII, 425–26; William H. Seward, Draft Proclamation,

misdated c. 1861, Papers of William H. Seward (no. 6441, reel 87); Seward to Edwin M. Stanton, July 4, 1864, Stanton Papers. The proclamation, as quoted in the text above, is considerably shortened from the originals cited in this note.

32. Edwin M. Stanton to C.C. Augur, July 5, 1864, Stanton Papers; Earl Schenck Miers, ed., *Lincoln Day by Day: A Chronology, 1809–1865,* III, 268–70.

33. *Diary of Gideon Welles,* I, 433–35.

34. Frank L. Klement says that arbitrary arrests "backfired." He assumes they were meant to control elections, but the election results showed they did not. See for example *The Copperheads in the Middle West,* pp. 18–23.

5. The Dark Side of the Civil War

1. Levi C. Turner to William Whiting, Feb. 6, 1863, Turner-Baker Papers (microfilm) Case #1729, National Archives.

2. *Coll. Works of Lincoln,* IV, 495, 521.

3. Turner-Baker Papers (microfilm) Cases #2742 and #3642.

4. See especially *O.R.,* III, 5, p. 792.

5. Eugene C. Murdock, *Patriotism Limited, 1862–1865: The Civil War Draft and the Bounty System* ([Kent, Ohio]: Kent State Univ. Press, 1967), 107–9.

6. Turner-Baker Papers (microfilm) Case #2147.

7. See Howard K. Beale, ed., *Diary of Gideon Welles, Secretary of the Navy under Lincoln and Johnson,* II, 240.

8. Lafayette C. Baker, *History of the United States Secret Service,* p. 399.

9. Turner-Baker Papers (microfilm) Case #2147.

10. Charles A. Dana, *Recollections of the Civil War: With the Leaders at Washington and in the Field in the Sixties* (New York: D. Appleton, 1899), 161–162. The money figures are for the fiscal year June 1863 to June 1864. Harold M. Hyman, *Stanton: The Life and Times of Lincoln's Secretary of War* (New York: Alfred A. Knopf, 1962), 125–26, 137, 152–53.

11. Dana, *Recollections,* pp. 162–64. Dana himself concluded years later "that the moderation of the President was wiser than the unrelenting justice of the Assistant Secretary would have been."

12. J. Thomas Scharf, *History of Maryland from the Earliest Period to the Present Day* (orig. pub. 1879; Hatboro: Tradition Press, 1967).

13. *Coll. Works of Lincoln,* VIII, 224n.

14. "To the President," petition signed by 25 Baltimore merchants, ca. Feb. 8, 1865, Louis A. Warren Lincoln Library and Museum, Fort Wayne, Indiana.

15. *Coll. Works of Lincoln,* VIII, 303.

16. Ibid., VIII, 303n.

17. Dana, *Recollections,* pp. 235–38.

18. *O.R.,* III, 5, pp. 791, 793.

19. Ibid., pp. 801–2.

20. *Coll. Works of Lincoln,* V, 241–42. The cabinet meeting at which were made the controversial decisions described by Lincoln in this message to Congress, was the same one seared into the memories of Seward and Welles.

21. Robert S. Harper, *Lincoln and the Press* (New York: McGraw-Hill, 1951), 289–303; *Diary of Gideon Welles,* II, 38.

22. Harper, *Lincoln and the Press,* pp. 289–303.

23. *Diary of Gideon Welles,* II, 35.

24. Gabor S. Boritt, *Lincoln and the Economics of the American Dream* (Memphis: Memphis State Univ. Press, 1978), 225.

25. *Cong. Globe,* 38 Cong., 1 sess., pt. 1, p. 974.

26. *Corruptions and Frauds of Lincoln's Administration,* Papers from the Society for the diffusion of Political Knowledge, no. 14 (New York: no. pub., 1864), 3–4; Leonard P. Curry, *Blueprint for Modern America: Nonmilitary Legislation of the First Civil War Congress* (Nashville: Vanderbilt Univ. Press, 1968), 234–40, esp. 239n.; Shannon, *The Organization and Administration of the Union Army, 1861–1865,* I (Cleveland: Arthur H. Clark, 1928), 58, 69, 74.

27. Shannon, *Organization and Administration of the Union Army,* I, 71, 66–67, 72; Fort Wayne *Weekly Sentinel,* May 18, 1861, Aug. 1, 1863, Jan. 30, 1864; *Oxford English Dictionary,* p. 2793.

28. *Cong. Globe,* 37 Cong., 3 sess., pt. 2, pp. 348, 952–58. A bill in 1865, also spearheaded by Wilson and Howard would have subjected substitute brokers to military justice. Again Cowan criticized it, and this time he prevailed. Allan G. Bogue, *The Earnest Men: Republicans of the Civil War Senate* (Ithaca: Cornell Univ. Press, 1981), 267–68.

29. Historians have failed to explore the issue as thoroughly as it deserves, since Shannon's path-breaking work. Curry showed interest in the issue of corruption only as what he imagined to be a tool of the Radical Republicans to usurp executive power. When, thirteen years later, Bogue wrote on the Senate, he ignored fraud and corruption as political issues in the Civil War.

30. Shannon, *Organization and Administration of the Union Army,* I, 73.

31. Turner-Baker Papers (microfilm) Case #333. See also Henry Snyder, Cases #1171 and #1623; Louis Solomon, Case #893; David Stauerclonz, Case #884; Victor Wolff, Case #1056; Barrett Woolf, Case #1292; Meyer G. Wyenberg, Case #836; Benjamin Jackson, Case #893.

32. Turner-Baker Papers (microfilm) Case #31.

33. Turner-Baker Papers (microfilm) Case #3047.

34. John Y. Simon, ed., *Papers of Ulysses S. Grant,* VII, 50, 51n–56n.

35. Ibid., p. 54n; Bates to Lincoln, Jan. 12, 1863, Abraham Lincoln Papers, Library of Congress (microfilm). Chase was not keeping a diary in this period; only Welles and Bates among cabinet members were keeping diaries. Bertram W. Korn's singling out Lafayette Baker for particular denunciation for his anti-Semitism may be a little unfair. It was equaled by many other officers and, of course, exceeded by men like Benjamin Butler, who richly deserves Korn's uncompromising denunciation. See Bertram W. Korn, *American Jewry and the Civil War,* 2nd ed., (Cleveland: World, 1961), 164–66, 169–70.

36. Byron Farwell, *Mr. Kipling's Army: All the Queen's Men* (New York: W.W. Norton, 1981), 110, 218.

37. Bertram W. Korn, *American Jewry and the Civil War,* p. 77. The Democratic opposition performed well on this human-rights issue. Peace Democrat Clement L. Vallandigham agitated the issue of Jewish chaplains well before Republicans moved to do anything about it in Congress. And after Grant's General Order No. 11, two conservative Democrats, George H. Pendleton of Ohio and Lazarus W. Powell of Kentucky, introduced resolutions of condemnation. Simon, ed., *Papers of Ulysses S. Grant,* VII, 55n.

38. Turner-Baker Papers (microfilm) Case #2998.

39. Lyons to William H. Seward, July 25, 1864; G.B. Case to [Levi C. Turner], July 31, 1864, Turner-Baker Papers (microfilm) Case #3023.

40. Lyons to William H. Seward, Aug. 22, 1864; Levi C. Turner to Seward, Oct. 7, 1864, Turner-Baker Papers (microfilm) Case #3023.

41. Turner-Baker Papers (microfilm) Case #3027; R.G. Rutherford to Levi C. Turner, Aug. 20, 1864, Turner-Baker Papers (microfilm) Case #3027.

42. Turner-Baker Papers (microfilm) Case #2919 (Buckley); Case #3053 (Williams); Ira Berry to Levi C. Turner, Jan. 7, 1865, Turner-Baker Papers (microfilm) Case #3053. Williams was arrested at the Washington railroad station, as were many such prisoners. He was suspect because he wore a shirt and drawers with regimental markings.

43. Turner-Baker Papers (microfilm) Cases #2919, #2928 (2), #3023, #3027, and #3053.

6. Numbers and Definitions

1. *The American Annual Cyclopaedia and Register of Important Events of the Year 1865,* vol. 5 (New York: D. Appleton, 1866), 414.

2. *Executive Document No. 1,* House of Representatives, 39 Cong., 1 sess., vol. IV, pts. 1 & 2; I also checked the control copy in RG110, National Archives. The Illinois report appears in pt. 2, p. 50. A report for Maryland, similarly included, does not mention arrests except of deserters. See also the Historical Reports of the State Acting Assistant Provost Marshals General and District Provost Marshals, 1865, Records of the Provost Marshal General's Bureau (Civil War), RG110, National Archives.

3. John A. Marshall, *American Bastile,* p. 713; 1885 ed., p. 843.

4. James Ford Rhodes, *History of the United States from the Compromise of 1850,* Volume IV: *1862–1864* (orig. pub. 1899; New York: Macmillan, 1904), 234; James Bryce, *The American Commonwealth,* rev. ed., I (New York: Macmillan, 1899), 61; Alexander Johnston, "Habeas Corpus (In U.S. History)," in John J. Lalor, ed., *Cyclopaedia of Political Science* II (orig. pub. 1881; New York: Charles E. Merrill & Co., 1893), 433–34. For a bound compilation called *Circulars, Provost Marshal General's Office,* see RG110, National Archives. James A. Woodburn stated in 1905 that "Lalor's *Cyclopaedia* has, since its first publication a quarter of a century back, been recognized by teachers and students as the most valuable compendium of information on the various subject matters considered." See Alexander Johnston, *American Political History, 1763–1876,* ed. by James Albert Woodburn, I (New York: G. P. Putnam & Sons, 1905), iii.

5. Thomas J. Pressly, *Americans Interpret Their Civil War* (orig. pub. 1954; New York: Free Press, 1965), 169–70.

6. F.C. Ainsworth to James Ford Rhodes, May 29, 1897, and June 1, 1897, (carbons) Adjutant General's Office, RG94, War Department Record and Pensions request no. 481397, National Archives. Ainsworth sent two letters containing the same information.

7. The figure covers the period after Feb. 15, 1862, when the War Department assumed jurisdiction over prisoners of state. Modern authorities generally add an estimate of the number arrested in the State Department period. The most recent estimate, 18,000, appears in Harold M. Hyman and William M. Wiecek, *Equal Justice under Law: Constitutional Development, 1835–1875* (New York: Harper & Row, 1982), p. 233; the authors do not explain how they arrived at the figure. Another recent estimate is

James M. McPherson's of "at least 15,000 civilians" in *Ordeal by Fire: The Civil War and Reconstruction* (New York: Alfred A. Knopf, 1982), 295. McPherson thus duplicates an older estimate by David Donald in "Died of Democracy," in Donald, ed., *Why the North Won the Civil War,* p. 87.

8. James G. Randall, "Has the Lincoln Theme Been Exhausted?" *American Historical Review,* LXI (1936), 270.

9. Randall, *Constitutional Problems under Lincoln,* p. 152n. Major General Davis spent, at most, four days on the problem for Randall. See Robert C. Davis to James G. Randall, June 26, 1925 (carbon), Adjutant General's Office, RG94, General Information Index, Prisons and Prisoners of War, National Archives.

10. Rhodes, *History of the United States from the Compromise of 1850,* Vol. IV, p. 234; Randall, *Constitutional Problems under Lincoln,* pp. 519–20. These authors knew of Lincoln's legendary penchant for pardoning soldiers and doubtless assumed he applied it to civilians. But their circumstances were not necessarily analogous, especially in Lincoln's mind.

11. Kenneth A. Bernard, "Lincoln and Civil Liberties," *Abraham Lincoln Quarterly,* VI (Sept. 1951), 393–95, 399; Dean Sprague, *Freedom under Lincoln,* pp. 301, 302.

12. Based on fragmentary records of Old Capitol prison found in the unmicrofilmed portion of the Turner-Baker Papers, National Archives.

13. See, for example, W. Winthrop, ed., *Digest of Opinions of the Judge Advocate General of the Army...,* 3rd ed. (Washington: Government Printing Office, 1868).

14. Frank L. Klement, *The Copperheads in the Middle West,* p. 331.

15. Turner-Baker Papers (microfilm) Case #2963.

16. Record and Pensions Index, Request No. 481397, June 1, 1897.

17. S.S. Nicholas, *Martial Law* (n.p., [1861]), 1.

18. *Speech of Hon. S.A. Douglas, of Illinois, on Refunding the Fine Assessed upon Gen. Jackson by Judge Hall in 1815* (Washington: Towers, [1844]).

19. *Military Despotism! Arbitrary Arrest of a Judge!!,* broadside published by P.W. Derham in New York. Derham first appears in New York City directories in 1864. The original broadside is located in the Abraham Lincoln Library and Museum, Lincoln Memorial Univ., Harrogate, Tennessee.

20. *Coll. Works of Lincoln,* IV, 372.

21. *O.R.,* II, 2, p. 223; William Whiting, *The War Powers of the President, Military Arrests, and Reconstruction of the Union,* 8th ed. (Boston: John L. Shorey, 1864), 184. Lincoln's ablest defender outside the administration itself, Philadelphia lawyer Horace Binney, referred to "Discretionary imprisonment," admitting in the same sentence that it constituted "an arbitrary *ouster* from all the benefits of Government." See Binney, *The Privilege of the Writ of Habeas Corpus under the Constitution* (Philadelphia: C. Sherman and Sons, 1862), 56.

22. J. Heermans, *War Power of the President,* p. 10; *Coll. Works of Lincoln,* VI, 269. Weeks later, Lincoln noted "phraseology" in Democratic protests of Vallandigham's arrest "calculated to represent me as struggling for an arbitrary personal prerogative" (*Coll. Works of Lincoln* VI, 303). He was alert to the political usage of the term.

23. "The Writ of Habeas Corpus," Gettysburg *Compiler,* June 10, 1861; "Freedom of Opinion and the Privilege of Habeas Corpus Inalienable Rights," Gettysburg *Compiler,* July 22, 1861.

24. "Democratic Platform," Gettysburg *Compiler,* Sep. 29, 1862 (printing the platform adopted on July 4, 1862); "Wholesale Military Arrests," Gettysburg *Compiler,* Aug. 11, 1862; George M. Wharton, *Remarks on Mr. Binney's Treatise on the Writ of*

Habeas Corpus, 2nd ed. (Philadelphia: John Campbell, 1862), 4; "Illegal arrests," Gettysburg *Compiler,* Dec. 29, 1862; "Dr. Olds on Arbitrary Arrest," Gettysburg *Compiler,* Jan. 19, 1863; "Arbitrary Arrest," Gettysburg *Compiler,* Feb. 2, 1863; and many articles thereafter.

25. *Coll. Works of Lincoln,* V, 318.

26. *O.R.,* I, 5, pp. 414, 417, 418.

27. Ibid., pp. 419–20.

28. *O.R.,* II, 3, p. 213; *O.R.,* II, 5, p. 26. On the nature of arrest, see Kent, *Commentaries on American Law,* II, 26.

29. *O.R.,* II, 4, pp. 270, 271.

30. Though available for almost a century in printed form, they were never previously counted or analyzed statistically. James G. Randall devoted less than 2 pages of his 580-page book to them, though they were the best records of military arrests of civilians available to him. See Randall, *Constitutional Problems under Lincoln,* pp. 149–50. The State Department records received more attention from Dean Sprague, who supplemented his research with newspaper sources as well. Yet he failed to count them and arrived at sweeping conclusions about the severity of William H. Seward's tenure as head of internal security that will not stand up. Finally, Kenneth Bernard, who made the most exhaustive study of the State Department records, nevertheless ignored their most salient characteristic as documents (their slipshod quality), did not bother to total them, and in the end could not formulate a conclusion that went beyond those already available in the literature.

31. Lucille H. Pendell and Elizabeth Bethel, "Preliminary Inventory of the Adjutant General's Office" (Washington: National Archives, 1949) 42–43.

32. Turner to Stanton, Oct. 30, 1865, in Turner-Baker Papers (microfilm) Case #3969.

33. Ibid.

34. *O.R.,* III, 5, p. 600.

35. *O.R.,* III, 5, p. 109.

36. Turner to Stanton, Oct. 30, 1865, in Turner-Baker Papers (microfilm) Case #3969.

37. The acting assistant provost marshal general's report for Illinois included 443 persons arrested who were not deserters. Among these, 49 were identified as "suspected deserters," 107 resisted enrollment or conscription, and 71 harbored deserters. Thus 227, just under half the total, were conscription-related arrests. *House Executive Documents,* 39 Cong., 1 sess., vol. IV, pt. 1, p. 50.

38. Turner-Baker Papers (microfilm) Case #3419.

39. Turner-Baker Papers (microfilm) Case #3418 (Andrews) and #1233 (Boileau).

40. William Dusinberre, *Civil War Issues in Philadelphia, 1856–1865* (Philadelphia: Univ. of Pennsylvania Press, 1965), 154–56; Harper, *Lincoln and the Press,* pp. 233–35.

41. Adrian Cook, *The Armies of the Streets: The New York City Draft Riots of 1863* ([Lexington]: Univ. Press of Kentucky, 1974), 60–61, 184–87.

42. *Report of Select Committee on Military Arrests. Ohio...* (n.p., 1863); *Report and Evidence of the Committee on Arbitrary Arrests, in... Indiana...* (Indianapolis: Joseph J. Bingham, 1863).

43. *O.R.,* II, 8, pp. 998–99.

44. *O.R.,* II, 7, p. 1106; *O.R.,* II, 8, p. 256.

45. *O.R.,* II, 7, pp. 898–99.

46. *O.R.,* II, 7, p. 145. (The POW's were later forwarded to Elmira, New York—see p. 502.)

47. *O.R.,* II, 8, p. 832; *O.R.,* II, 6, p. 24.

48. Hesseltine, *Civil War Prisons: A Study in War Psychology* (orig. pub. 1930; New York: Frederick Ungar, 1978), 42; *O.R.,* II, 7, p. 563; LeRoy P. Graf and Ralph W. Haskins, eds., *Papers of Andrew Johnson,* VI (Knoxville: Univ. of Tennessee Press, 1983), 499.

49. Randall, *Constitutional Problems under Lincoln,* p. 155.

7. *The Revival of International Law*

1. Davis quoted in Quincy Wright, "The American Civil War, 1861–1865," in Richard A. Falk, ed., *The International Law of Civil War* (Baltimore: Johns Hopkins Univ. Press, 1971), 54.

2. Frank Lawrence Owsley, *King Cotton Diplomacy: Foreign Relations of the Confederate States of America,* 2nd rev. ed. (Chicago: Univ. of Chicago Press, 1959), 261.

3. Theodore S. Woolsey, "Blockade," in John J. Lalor, ed., *Cyclopaedia of Political Science...,* I, 293; John Bassett Moore, *A Digest of International Law...,* VII (Washington, D.C.: Government Printing Office, 1906), 748; George Grafton Wilson, ed., *Elements of International Law* by Henry Wheaton (Oxford: Clarendon Press, 1936), 560–80.

4. Theodore Dwight Woolsey, *Introduction to the Study of International Law* (Boston: James Munroe and Co., 1860), 416–17; Halleck quoted in Moore, *Digest of International Law,* I, 837.

5. Levi C. Turner to Andrew Sterett Ridgeley, July 20, 1863, Turner-Baker Papers (microfilm) Cases #878, #4066, #1050.

6. Moore, *Digest of International Law,* I, 838; *O.R.,* II, 2, p. 307.

7. Peter J. Parish, *The American Civil War* (New York: Holmes and Meier, 1975), 406; Francis Wharton gave the earliest and most scathing denunciation of "continuous voyage" in his *Digest of the International Law of the United States,* III (Washington, D.C.: Government Printing Office, 1886), 394–406. The doctrine of "continuous voyage" aimed at punishing attempts to evade the blockade by transferring cargo to Mexico and carrying it by land into the Confederacy.

8. Turner-Baker Papers (microfilm) Case #1027.

9. E.R.S. Canby, MS. opinion, Dec. 10, 1863, Turner-Baker Papers (microfilm) Case #2486.

10. Dana to Welles, Dec. 26, 1863, in *Official Records of the Union and Confederate Navies in the War of the Rebellion,* I, 9 (Washington: Government Printing Office, 1899), 279 (hereafter cited as *O.R., Navy*).

11. William H. Seward to Edwin M. Stanton, Jan. 11, 1864, Turner-Baker Papers (microfilm) Case #2490.

12. Printed order pasted on back of cover of "Prisoners Captured on Blockade Runners, 1861–1865," MS. ledger, U.S. Navy Dept., RG 45, NA (hereafter cited PCBR); John Wilkinson, *The Narrative of a Blockade-Runner* (orig. pub. 1877; New York: Time-Life Books, 1981), 155.

13. Opinion quoted in Levi C. Turner to James A. Hardie, June 4, 1864, *O.R.,* II, 7, pp. 194–95.

14. Turner-Baker Papers (microfilm) Case #2496 and PCBR, p. 124; Turner to Hardie, 4 June 1864, *O.R.,* II, 7, pp. 194–95.

15. Clement Eaton, *History of the Southern Confederacy* (New York: Free Press,

1965), 146; William Watson, *Adventures of a Blockade Runner* (London: T.F. Unwin, 1892), 96–97.

16. Turner to Hardie, June 6, 1864, Turner-Baker Papers (microfilm) Case #2208.

17. Printed order pasted in "Prisoners Captured on Blockade Runners, 1861–1865"; Gideon Welles to Edwin M. Stanton, Mar. 18, 1865, *O.R.,* II, 8, p. 406.

18. *Diary of Gideon Welles,* I, 86.

19. Henry W. Halleck, *Elements of International Law and Laws of War* (orig. pub. 1866; Philadelphia: J.B. Lippincott, 1872), 3.

20. *Diary of Gideon Welles,* I, 82.

21. Ibid., I, 398, 451.

22. Ibid., 453; *Coll. Works of Lincoln,* VI, 380n.

23. Turner-Baker Papers (microfilm) Case #3969.

24. Mark E. Neely, Jr., "The Perils of Running the Blockade: The Influence of International Law in an Era of Total War," *Civil War History,* XXXII (June 1986), 109.

25. *Report of the Secretary of the Navy* (Washington, D.C.: Government Printing Office, 1865), 452; *Report of the Secretary of the Navy* (Washington, D.C.: Government Printing Office, 1863), X; J. Russell Soley, *The Blockade and the Cruisers* (New York: Charles Scribner's Sons, 1883), 87.

26. Case of Charles Armstrong, Turner-Baker Papers (microfilm) Case #3186.

27. S.P. Lee to Gideon Welles, Nov. 12, 1863, quoted in *O.R., Navy,* I, 9, p. 291.

28. J.B. Breck to S.P. Lee, Nov. 12, 1863, in *O.R., Navy,* I, 9, p. 295.

29. Turner-Baker Papers (microfilm) Case #2412.

30. Turner-Baker Papers (microfilm) Case #3048.

31. Turner-Baker Papers (microfilm) Cases #2486, #3048; see also PCBR, p. 99.

32. Turner-Baker Papers (microfilm) Cases #2486, #3048; see also PCBR, p. 100.

33. Turner-Baker Papers (microfilm) Case #3207. See also, for example, the case of Rudolphus D. Crittenden, Case #2479.

34. S.P. Chase to S.P. Lee, Nov. 2, 1863, and S.P. Lee to Gideon Welles, Nov. 5, 1863, in *O.R., Navy,* I, 9, p. 261.

35. See, for example, *O.R., Navy,* I, 9, pp. 347, 406, 607–8.

36. William G. Thompson to Edwin M. Stanton, Jan. 1, 1864, Turner-Baker Papers (microfilm) Cases #2496, #2529.

37. Turner-Baker Papers (microfilm) Case #3610.

38. Turner-Baker Papers (microfilm) Case #1301.

39. Don Higginbotham, *The War of American Independence: Military Attitudes, Policies, and Practices, 1763–1789* (New York: Macmillan, 1971), passim.

40. "Next Month," *History Today,* XXXIX (Jan. 1989), 64.

41. Woolsey, *Introduction to the Study of International Law,* pp. 251–52.

42. LeRoy P. Graf and Ralph W. Haskins, eds., *The Papers of Andrew Johnson,* V (Knoxville: Univ. of Tennessee Press, 1979), 410–11.

43. Ibid., V, 445–46; *Coll. Works of Lincoln,* V, 264.

44. Robert Ould to William H. Ludlow, Dec. 11, 1862, in *O.R.,* II, 5, pp. 71–72; Record of Prisoners of State, Turner-Baker Papers (unmicrofilmed), National Archives.

45. Turner-Baker Papers (microfilm) Case #3198.

46. Richard Shelly Hartigan, *Lieber's Code and the Law of War* (Chicago: Precedent, 1983), 56.

47. Ibid., p. 50.

48. Ibid., p. 58.

49. Turner-Baker Papers (microfilm) Case #3198.

50. S.J. Quinn, *History of the City of Fredericksburg, Virginia* (Richmond: Hermitage Press, 1908), 99–106.

51. *O.R.,* I, 36, pt. 2, pp. 829, 934–35; pt. 3, pp. 26–27.

52. George Mitchell to Mr. Seymour, May 23, 1864, with endorsement by Joseph Holt, in Turner-Baker Papers (microfilm) Case #2663.

53. Record of Prisoners of State, Turner-Baker Papers (unmicrofilmed).

54. *Coll. Works of Lincoln,* VII, 358.

55. Turner-Baker Papers (microfilm) Case #3686; James Hamilton to Edwin M. Stanton, Sept. 2, 1864, and Hoffman to Schoepf, Sep. 20, 1864, in *O.R.,* II, 7, pp. 849–50.

56. John Y. Simon, ed., *Papers of Ulysses S. Grant,* XIII, 429–30, 430n–31n.

8. *The Irrelevance of the* Milligan *Decision*

1. Francis Lieber to Henry W. Halleck, June 13, 1864, quoted in Thomas Sergeant Perry, ed., *The Life and Letters of Francis Lieber* (Boston: James R. Osgood and Co., 1882), 347–48; *Coll. Works of Lincoln,* VI, 265–66.

2. *Coll. Works of Lincoln,* V, 437.

3. See Joseph George, Jr., "The North Affair: A Lincoln Administration Military Trial, 1864," *Civil War History,* XXXIII (Sep. 1987), 201–18; David Miller DeWitt, *The Judicial Murder of Mary E. Surratt* (Baltimore: John Murphy & Co., 1895); Frank L. Klement, "The Indianapolis Treason Trials and *Ex Parte Milligan,*" in Michael Belknap, ed., *American Political Trials* (Westport: Greenwood Press, 1981), 101–27; Gilbert R. Tredway, *Democratic Opposition to the Lincoln Administration in Indiana* (Indianapolis: Indiana Historical Bureau, 1973), pp. 224–48; Frank L. Klement, *The Limits of Dissent: Clement L. Vallandigham & the Civil War.*

4. Registers of the Records of the Proceedings of the United States Army General Courts-Martial, 1809–1890, Records of the Office of the Judge Advocate General [Army], RG 153, National Archives (microfilm). After counting all, I sampled 226 cases.

5. Charles Fairman, "The Law of Martial Rule and the National Emergency," *Harvard Law Review,* LV (June 1942), 1253–1302.

6. W. Winthrop, ed., *Digest of Opinions of the Judge Advocate General of the Army....,* 3rd ed. (Washington: Government Printing Office, 1868), 222. Winthrop became deputy judge advocate general in 1884 and later taught law at West Point. James Grant Wilson and John Fiske, eds., *Appleton's Cyclopaedia of American Biography,* VI (New York: D. Appleton, 1889), 577.

7. Case MM618, Records of the Proceedings of the United States Army General Courts-Martial, 1809–1890, RG 153 (hereafter, RCM).

8. Case MM601, RCM; Winthrop, *Digest of Opinions of the Judge Advocate General of the Army,* p. 366.

9. Cases MM451, MM452, RCM.

10. Cases MM603, MM625, RCM.

11. Case NN3156, RCM; Lincoln's endorsement was signed Jan. 11, 1865.

12. Case NN1822, RCM; *Coll. Works of Lincoln,* VIII, 546. Lincoln's endorsement was signed Oct. 13, 1864.

13. Case NN1820, RCM.

14. Drawer 537, Reference Room, Manuscripts Division, Library of Congress.

15. Case NN1823, RCM; Holt's recommendation to Lincoln was dated May 30, 1864. See also the case of Fountain Brown, who sold people into slavery in Arkansas after

the issuance of the Emancipation Proclamation; *Coll. Works of Lincoln*, VII, 357n, and *O.R.*, II, 7, pp. 159–62. Brown's conviction by a military commission was upheld. My thanks to Howard Westwood for calling the Brown case to my attention.

16. For examples of a teamster in 1855 see Case HH572, RCM; and of a carriage maker in the Ordnance Department in 1856 see case HH683, RCM.

17. Winthrop, *Digest of Opinions of the Judge Advocate General of the Army*, p. 20. Conversely, General Order No. 100, in Article 49, defined "all those who are attached to the Army for its efficiency and promote directly the object of the war," if captured, as prisoners of war, "and as such exposed to the inconveniences as well as entitled to the privileges of a prisoner of war." Lieber extended it even further in Article 50: "Moreover, citizens who accompany an army for whatever purpose, such as sutlers, editors, or reporters of journals, or contractors, if captured, may be made prisoners of war and be detained as such." Richard Shelly Hartigan, *Lieber's Code and the Law of War*, p. 55.

18. Winthrop, *Digest of Opinions of the Judge Advocate General*, p. 118n.

19. All trials dated up to and including Apr. 1865 were counted as occurring during the Civil War. Those occurring in May 1865 or afterward were counted as postwar.

20. Case MM451, RCM. See Fellman, *Inside War*, pp. 97–112.

21. Case of Henry E. Dixon, numbered MM454, RCM. Oddly enough, the specification leaves a blank for the day, month, and year the oath was taken.

22. Case LL646, RCM.

23. Case numbered LL646, RCM.

24. William E. Parrish, *A History of Missouri*, Volume III: *1860–1875*, pp. 52–53.

25. Case MM617, RCM.

26. Cases of Elijah Burris (MM454), Jordan O'Bryan (MM454), Saunders M. Utterbach (LL646), and James H. Woods (MM451), RCM.

27. Cases of William H. Walthal (MM451), John Bevens (MM454), and Francis M. Armstrong (MM617), RCM.

28. Cases of David Tomlinson (LL647), James M. Forester (MM451), Aaron Dean (LL646), James Sutton (MM623), James C. Moore (LL647), James E. Hicks (MM451), Joshua Tucker (LL647), John Colbert (LL647), Henry Garner (MM451), James H. Freeman (MM451), and Samuel Bryant (MM451), RCM.

29. Cases of John H. Jones (NN2660) and A. G. Hurst (MM455), RCM.

30. Case numbered LL647, RCM.

31. Cases NN2660 and LL647, RCM.

32. LL646, RCM.

33. Cases of John Bauvais (NN2070), R. Bauvais (NN2070), Horace E. Dimick (NN2070), John Murphy (NN2660), Isaac Dale (MM3212), Benjamin F. Williams (NN2660), William P. Ferris (MM454), and William J. Kribben (LL647), RCM.

34. See, for example, cases numbered MM601, RCM.

35. Henry W. Halleck to James H. Carleton, Feb. 4, 1864, in *O.R.*, I, 34, pt. 2, pp. 245–46; Kirby Benedict to James H. Carleton, July 8, 1864, in *O.R.*, I, 41, pt. 2, pp. 112–14. Bates's letter, dated by the source as Sept. 16, 1863, is quoted in the *Illinois State Register*, Feb. 10, 1864.

36. Arnold M. Shankman, *The Pennsylvania Antiwar Movement, 1861–1865* (Rutherford: Fairleigh Dickinson Univ. Press, 1980), 154–57; see also cases numbered NN3348, RCM.

37. See cases numbered NN1478, RCM.

38. *Coll. Works of Lincoln*, VI, 303.

39. Charles Warren, *The Supreme Court in United States History,* III (Boston: Little, Brown, 1923), 149, 154.

40. Allan Nevins, "The Case of the Copperhead Conspirator," in John A. Garraty, ed., *Quarrels That Have Shaped the Constitution,* p. 118.

41. Frank L. Klement, "The Indianapolis Treason Trials and *Ex Parte Milligan,"* in Michael Belknap, ed., *American Political Trials,* pp. 119, 120.

42. Patrick J. Furlong et al., *We the People: Indiana and the United States Constitution* (Indianapolis: Indiana Historical Society, 1987), 2.

43. Quoted in Charles Warren, *The Supreme Court in United States History,* III, 154, 161.

44. *Ex parte Milligan,* 4 Wallace 2 (1866).

45. James E. Sefton, *The United States Army and Reconstruction, 1865–1877* (orig. pub. 1967; Westport: Greenwood Press, 1980), 35. General Grant created a new class of defendants by proclaiming that Southerners found in arms after the surrender would be regarded as guerrillas.

46. Case OO 1323, RCM. See also cases numbered OO 1319, RCM.

47. Sefton, *U.S. Army and Reconstruction,* pp. 35–36; Case OO 1323, RCM.

48. Case OO 1025, RCM.

49. Case OO 1321, RCM.

50. See, for example, the cases of Sidney Summers, Jonas Reidley, T.D. Schlachter, John W. Dey, C. Sinkard, and Anne Kennedy, all tried in Morganton, North Carolina, in 1867 (cases numbered OO 2419, RCM).

51. Sefton, *U.S. Army and Reconstruction,* pp. 36–41.

52. Case OO 1320, RCM.

53. Case OO 1320, RCM.

54. Cases of Frank Fitzgerald and Richard Lawton, numbered OO 1320, RCM.

55. Sefton, *U.S. Army and Reconstruction,* pp. 77–79. Albert Castel in *The Presidency of Andrew Johnson* (Lawrence: Regents Press of Kansas, 1979), 73, states that Johnson thus ordered an end to military arrests and trials, but Sefton's more careful judgment is accurate.

56. See finding by Joseph Holt in the case of William C. Allen, numbered OO 1323, RCM.

57. Sefton, *U.S. Army and Reconstruction,* pp. 109–11, 225.

58. John J. Lalor, ed., *Cyclopaedia of Political Science...,* II, 837–38. Clampitt's language followed almost word-for-word the text of Winthrop's *Digest of Opinions of the Judge Advocate General of the Army,* pp. 229–30.

59. Lalor, *Cyclopaedia,* II, 837–38.

60. John W. Clampitt, "The Trial of Mrs. Surratt," *North American Review,* CXXXI (Sept. 1880), 234; Lalor, *Cyclopaedia,* II, 839.

61. Burgess, *Political Science and Comparative Constitutional Law,* I (Boston: Ginn, 1890), 250. After the turn of the century, Burgess made a similarly hard-nosed assessment of the *Milligan* decision: "There is no question that the practice of the Administration and the opinion of the Court were at variance, and there is little doubt that in spite of the opinion of the Court the practices of the Administration would be repeated under like circumstances. The practices of the Administration are, therefore, to be considered as the precedents of the Constitution in civil war rather than the opinion of the Court. They are justified by the necessity under which the Government must act in executing its powers during a state of civil war." Burgess, *The Civil War and the Constitution, 1859–1865,* II (New York: Charles Scribners' Sons, 1901), 218–

19. The precision of use of the term "civil war" rather than "war" in Burgess's statement here, of course, made his prediction entirely irrelevant to the future course of American history.

62. Charles Warren, in fact, drafted the Espionage Act. See Leonard W. Levy, "Charles Warren," in Levy et al., *Encylopedia of the American Constitution*, IV (New York: Macmillan, 1986), 1029. For precise figures on arrests see Harry N. Scheiber, *The Wilson Administration and Civil Liberties, 1917–1921* (Ithaca: Cornell Univ. Press, 1960), 19, 63.

63. Scheiber, *Wilson and Civil Liberties*, p. 45; Paul L. Murphy, *World War I and the Origin of Civil Liberties in the United States* (New York: W.W. Norton, 1979), 222–25.

64. James G. Randall, *Constitutional Problems under Lincoln*, p. 513; J.G. Randall and David Donald, *Civil War and Reconstruction*, 2nd ed. (Boston: D.C. Heath, 1961), 304. Samuel Klaus's *The Milligan Case* (New York: Alfred A. Knopf, 1929), also proved to be the only volume issued in the proposed series on famous trials.

65. William A. Dunning, "Disloyalty in Two Wars," *American Historical Review*, XXIV (Oct. 1918), 630, 628. In keeping with the dispassionate tone of writing in political science in this early period, Dunning made it difficult for his readers to infer criticism of either Lincoln or Wilson. Though Dunning concluded that the reign of law was safe only because the threat to American security was less than that presented in the Civil War, he also pointed out that "President Wilson's authority, actually exercised, surpassed in variety and scope the wildest dreams of 1861–1865." Such emphasis would be continued by constitutional commentator Edward S. Corwin later in the century.

66. Klaus, ed., *The Milligan Case*, unnumbered pages at beginning and pp. 3, 59–60.

67. See Paul L, Murphy, *World War I and the Origin of Civil Liberties in the United States*, pp. 217, 225n; Scheiber, *Wilson and Civil Liberties*, pp. 49–50.

68. "Excerpts from Summary of Report on Internments in U.S. in World War II," New York *Times*, Feb. 25, 1983, p. 10; "Internment of Japanese Eyed in '36," Chicago *Tribune*, Feb. 11, 1983, p. 4.

69. Peter Irons, *Justice at War* (New York: Oxford Univ. Press, 1983), 144, 146, 148, 150–51.

70. Charles Fairman, *The Oliver Wendell Holmes Devise History of the Supreme Court of the United States*, Volume VI: *Reconstruction and Reunion, 1864–88*; part one (New York: Macmillan, 1971), 212.

71. Richard Loss, "Edward S. Corwin," in Leonard W. Levy, Kenneth L. Karst, and Dennis J. Mahoney, eds., *Encyclopedia of the American Constitution*, II (New York: Macmillan, 1986), 508–9; Paul L. Murphy, *World War I and the Origin of Civil Liberties in the United States*, p. 110, 110n; Stephen Vaughn, *Holding Fast the Inner Lines: Democracy, Nationalism, and the Committee on Public Information* (Chapel Hill: Univ. of North Carolina Press, 1980), 54–55, 229–30; Edward S. Corwin, *The President: Office and Powers: History and Analysis of Practice and Opinion* (New York: New York Univ. Press, 1940), 165–66.

9. The Democratic Opposition

1. *New York Freeman's Journal and Catholic Register*, May 11, 1861.
2. Springfield *Illinois State Register*, Sept. 2, 1861.
3. See, for example, Raphael Semmes, "Civil War Song Sheets," *Maryland Historical*

Magazine, XXXVII (Sept. 1943), 205–29.

4. Published by E.S. Riley, Printer, in Frederick, Maryland. Quotations from p. 24.

5. On Johnson, see [Edward Ingersoll], *Personal Liberty and Martial Law: A Review of Some Pamphlets of the Day* (Philadelphia: no pub., 1862), p. 10n. Carroll's pamphlets included *The War Powers of the General Government* (Washington: Henry Polkinhorn, 1861).

6. [S.S. Nicholas], *Martial Law* (n.p., [1861]). The pamphlet was out early enough for Robert L. Breck to review it in an article written in 1861. See Breck, *The Habeas Corpus and Martial Law* (Cincinnati: R.H. Collins, 1862), 5. Nicholas's work was republished in Philadelphia by John Campbell, Booksellers, in 1862.

7. Gettysburg *Compiler,* June 10, 1861; Aug. 26, 1861; Sept. 16 1861; Fort Wayne, Indiana, *Weekly Sentinel,* Sept. 21, 1861; Chicago *Times* quoted in *Illinois State Register,* Dec. 28, 1861; J.S. Havens, *The Usurpations of the Federal Government: Speech of... of Suffolk, on the Resolutions of Judge Dean to Raise a Select Committee to Investigate the Subject of Arbitrary Arrests and the Suspension of the Writ of Habeas Corpus* (n.p., 1863), 5.

8. Robert L. Breck, *The Habeas Corpus and Martial Law,* p. 32.

9. Joel Parker, *Habeas Corpus and Martial Law: A Review of the Opinion of Chief Justice Taney, in the Case of John Merryman,* 2nd ed. (Philadelphia: John Campbell, 1862), 24, 32, 35. On Parker, see Dale Baum, *The Civil War Party System: The Case of Massachusetts, 1848–1876* (Chapel Hill: Univ. of North Carolina Press, 1984), 63–64 and *Proceedings of the Convention of the People of Massachusetts, Holden at Faneuil Hall, Boston, October 7th 1862* (Boston: J.C. Peters, 1862), 20–29. Parker had not been as eager to defend habeas corpus before the war, when he attacked the North's personal-liberty laws and urged their repeal to appease the South. See Thomas D. Morris, *Free Men All,* pp. 209–10.

10. *Illinois State Register,* Oct. 29, 1861.

11. *Cong. Globe,* 37 Cong., 2 sess., pt. I, pp. 90–98 (Dec. 16, 1861).

12. Horace Binney, *The Privilege of the Writ of Habeas Corpus under the Constitution* (Philadelphia: C. Sherman & Son, 1862). For the pamphlet's date, see Frank Freidel, *Francis Lieber: Nineteenth-Century Liberal* (Baton Rouge: Louisiana State Univ. Press, 1947), 311–12 and 311n.

13. John T. Montgomery, *The Writ of Habeas Corpus and Mr. Binney* ([Philadelphia: 1862]). See also a second edition published by Philadelphia bookseller John Campbell in Apr. and Edward Ingersoll's tribute to Montgomery in the beginning of Ingersoll's own pamphlet, *Personal Liberty and Martial Law: A Review of Some Pamphlets of the Day* (Philadelphia: no pub., 1862). Nicholas B. Wainwright, ed., *A Philadelphia Perspective: The Diary of Sidney George Fisher Covering the Years 1834–1871* (Philadelphia: Historical Society of Pennsylvania, 1967), 495.

14. S.S. Nicholas, *Habeas Corpus: A Response to Mr. Binney* (Louisville: Bradley & Gilbert, 1862), 15. Nicholas accurately characterized J.C. Bullitt's pamphlet, *A Review of Mr. Binney's Pamphlet on "The Privilege of the Writ of Habeas Corpus under the Constitution"* (Philadelphia: John Campbell, 1862). Another pamphlet with narrow focus on the question, "Who had the power?" was James F. Johnson, *The Suspending Power and the Writ of Habeas Corpus* (Philadelphia: John Campbell, 1862). Johnson's identity remains obsure.

15. S.S. Nicholas, *Martial Law* (Philadelphia: John Campbell, 1862). On Nicholas,

see E. Merton Coulter. *The Civil War and Readjustment in Kentucky*, pp. 32–33, 53.

16. G.M. Wharton, *Remarks on Mr. Binney's Treatise on the Writ of Habeas Corpus*. On Wharton, see William Dusinberre, *Civil War Issues in Philadelphia, 1856–1865,* pp. 31n, 109, 121, 158, and Wainwright, ed., *A Philadelphia Perspective,* p. 496.

17. *Illinois State Register,* Mar. 24, 1862.

18. *Cong. Globe,* 37 cong., 2 sess., pt. II, pp. 1370–1371 (Mar. 27, 1862): *Illinois State Register,* Apr. 9, 1862.

19. *Illinois State Register,* Aug. 28, 1862.

20. New York *Tribune,* Sept. 3, 1862, quoted in *Illinois State Register,* Sept. 6, 1862.

21. *Illinois State Register,* Sept. 26, 1862; Nov. 6, 1862.

22. George Clarke Sellery, *Lincoln's Suspension of Habeas Corpus as Viewed by Congress,* p. 247 and 247n.

23. Stewart Mitchell, *Horatio Seymour of New York* (Cambridge: Harvard Univ. Press, 1938), 265–71; on Pruyn, see Allen Johnson, ed., *Dictionary of American Biography,* VIII (New York: Charles Scribner's Sons, 1927), 253–54.

24. Indianapolis *Daily State Sentinel,* Jan. 13, 1863.

25. Mitchell, *Horatio Seymour of New York,* p. 270; *Illinois State Register,* Jan. 29, 1863; J.S. Havens, *The Usurpations of the Federal Government,* p. 7.

26. *Illinois State Register,* Jan. 16, 1863.

27. Allan G. Bogue, *The Earnest Men: Republicans of the Civil War Senate* (Ithaca: Cornell Univ. Press, 1981), 266; *Cong. Globe,* 37 Cong., 3 sess., pt. I, p. 30.

28. Horace White, *The Life of Lyman Trumbull* (Boston: Houghton Mifflin, 1913), 199.

29. *Cong. Globe,* 37 Cong., 3 sess., pt. I, p. 30.

30. Francis Key Howard, *Fourteen Months in American Bastiles* (Baltimore: Kelly, Hedian & Piet, 1863), 4–5, 18–19, 29. Howard was an editor of the *Daily Exchange* and admitted later:

... when Mr. LINCOLN, on his accession to office, appointed some of the most extreme partisans to high office at home, and selected others to represent the country abroad, and gave ample evidence of his incapacity to understand the questions at issue, and of his determination neither to conciliate the Southern people, nor to deal with what he called the "rebellion" according to the mode provided by the Constitution and laws, then a large portion of the people of Maryland expressed their sympathy for the South, and their conviction of the justice of its cause. They then asserted that the conquest of the South was an impossibility, that the Union was in point of fact dissolved, and they insisted that in such case the people of the State had the right to decide their own destiny for themselves. These views I also entertained and expressed.

31. Sidney Cromwell, *Political Opinions in 1776 and 1863: A Letter to a Victim of Arbitrary Arrests and "American Bastiles"* (New York: Anson D.F. Randolph, 1863), 9n, 10–12, 14–15.

32. Philadelphia *Inquirer,* Sept. 17, 1863.

33. *Illinois State Register,* May 14, 1863.

34. Frank L. Klement, *The Limits of Dissent: Clement L. Vallandigham and the Civil War,* pp. 180–81, 181n.

35. John V. L. Pruyn et al., *Reply to President Lincoln's Letter of 12th June 1863,* Papers from the Society for the Diffusion of Political Knowledge, no. 10 (New York: 1863). The pamphlet is reproduced in Frank Freidel, ed., *Union Pamphlets of the Civil*

War, 1861–1865, II (Cambridge: Harvard Univ. Press, 1967), 752–65. Page references are to that reprint.

36. Pruyn, et al., *Reply*, pp. 756, 758–59; *"After Some Time Be Past." Speech of Hon. C.L. Vallandigham, of Ohio, on Executive Usurpation...July 10, 1861* (n.p.; n.d.), 6–7.

37. Pruyn et al., *Reply,* p. 756.

38. Ibid., p. 759; Randall, *Constitutional Problems under Lincoln,* pp. 186, 214. The *Reply* appeared also bound in the *Hand-Books of the Democracy for 1863 & 1864,* a compilation of pamphlets. The German translation was *Erwiederung des Schreibens des Präsidenten Lincoln,* also published in 1863.

39. Suspending the writ may not have eliminated the possibility of suing for wrongful arrest, but the Habeas Corpus Act of March 3, 1863, with the text of which the Albany committee proved itself closely familiar, did in certain cases—a fact not alluded to in the *Reply* to the president. Years later, historian James G. Randall saw as the main feature of that law the point ignored in the *Reply*. He wrote a chapter on the bill in his famous book and entitled it "The Indemnity Act of 1863," not "The Habeas Corpus Act of 1863," as it is more commonly known.

40. *Coll. Works of Lincoln,* VI, 302.

41. Fort Wayne *Weekly Sentinel,* Oct. 3, 1863.

42. Harvey Palmer, *The Writ of Habeas Corpus, the Governor's Message, and Policy of the War: Speech of... March 13th 1863* (n.p.: N.Y. State Union Central Committee, [1863]), 5.

43. *Coll. Works of Lincoln,* VI, 268–69; on Douglas's elaborate role in defending the bill in 1844, see Robert W. Johannsen, *Stephen A. Douglas* (New York: Oxford Univ. Press, 1973).

44. *Coll. Works of Lincoln,* I, 226.

45. Daniel Gardner, *A Treatise on the Law of the American Rebellion, and Our True Policy, Domestic and Foreign* (New York: John W. Amerman, Printer, 1862), 6, 8, 9. Gardner's pamphlet was provoked by the issuance of writs of habeas corpus in the Benedict case by Judge Hall of the Northern District of New York and in the *Merryman* case as well as by the pamphlet of Benjamin R. Curtis denouncing martial law.

46. Charles Ingersoll, *An Undelivered Speech on Executive Arrests* (Philadelphia: no pub., 1862), 89, 91. William Dusinberre, *Civil War Issues in Philadelphia, 1856–1865,* pp. 142–43; Stephen A. Douglas, *Speech of... on Refunding the Fine Assessed upon Gen. Jackson by Judge Hall in 1815* (Washington: Towers, [1844]), 1–2.

47. Nicholas B. Wainwright, ed., *A Philadelphia Perspective,* pp. 460–61.

48. D.C. Cloud and John H. Power, *The Powers of the President of the United States in Times of War* (Muscatine: Daily Journal, 1863), 19.

49. *Opinion of Judge N.K. Hall... on Habeas Corpus in the Case of Rev. Judson Benedict....* (Buffalo: Joseph Warren & Co., Printers, Courier Office, 1863), 22. Two of three justices of the Supreme Court of New York had refused to issue a writ of habeas corpus before Benedict's lawyer turned to Judge Hall. See Marshall, *American Bastile,* p. 186. Hall's opinion, buttressed with other documents, was widely circulated as a pamphlet—exhausting two editions of 10,000 and 8,000 copies each.

50. *Mr. Lincoln's Arbitrary Arrests: The Acts which the Baltimore Platform Approves* ([New York]: no pub., [1864]), 5, 6, 7, 11, 12.

51. Ibid., pp. 2, 23.

52. A.D. Duff, *Footprints of Despotism... Speech of Hon..., May 28th, A.D. 1864*

(Benton: Standard Office Print., 1864), 3, 6, 10–11. I am indebted to Thomas Schwartz, Lincoln Curator of the Illinois State Historical Library, for sending me a photocopy of this rare habeas-corpus pamphlet.

53. Robert S. Harper, *Lincoln and the Press* (New York: McGraw-Hill, 1951), 193; *Illinois State Register,* July 21, 1864; Mitchell, *Horatio Seymour of New York,* pp. 370–71.

54. Fort Wayne *Daily Gazette,* Oct. 6, 1864.

55. See, for example, the *Illinois State Register,* June 22, 1864.

56. *Illinois State Register,* Oct. 21, 1864.

57. William Starr Myers, *General George Brinton McClellan: A Study in Personality* (New York: D. Appleton, Century, 1934), 307–8. What he said about military arrests in the letter contradicted his own earlier actions in Maryland.

58. *Illinois State Register,* Sept. 7, 1864; Myers, *General George Brinton McClellan,* p. 429.

59. Joel H. Silbey, *A Respectable Minority: The Democratic Party in the Civil War Era, 1860–1868* (New York: W.W. Norton. 1977), 70; Harold M. Hyman, *A More Perfect Union: The Impact of the Civil War and Reconstruction on the Constitution,* p. 77.

60. Thomas D. Morris, *Free Men All,* pp. 8–10, 125.

10. Lincoln and the Constitution

1. *Coll. Works of Lincoln,* I, 480–81, 485, 486, 488.

2. Ibid., I, 5–8, 48, 62–63, 170–71.

3. Ibid., I, 171–72; II, 255.

4. Ibid., I, 312.

5. Ibid., III, 358; I, 347–48, 476.

6. Ibid., I, 476; IV, 66.

7. Ibid., I, 451–52.

8. Daniel Walker Howe, *The Political Culture of the American Whigs* (Chicago: Univ. of Chicago Press, 1979), 23; Don E. Fehrenbacher, *The Dred Scott Case: Its Significance in American Law and Politics* (New York: Oxford Univ. Press, 1978), 465.

9. *Coll. Works of Lincoln,* II, 274; IV, 11.

10. Robert A. Ferguson, *Law and Letters in American Culture* (Cambridge: Harvard Univ. Press, 1984), 305. *Coll. Works of Lincoln,* I, 112, 115. For opinions of the Lyceum address see Albert J. Beveridge, *Abraham Lincoln, 1809–1858,* I (Boston: Houghton Mifflin, 1928), 227; James G. Randall, *Lincoln the President: Springfield to Gettysburg,* I, 20–23; Benjamin P. Thomas, *Abraham Lincoln: A Biography* (New York: Alfred A. Knopf, 1952), 71–73; and (quoted here) Reinhard H. Luthin, *The Real Abraham Lincoln: A Complete One Volume History of His Life and Times* (Englewood Cliffs: Prentice Hall, 1960), 49–50.

11. George M. Fredrickson, "The Search for Order and Community," in Cullom Davis et al, *The Public and Private Lincoln: Contemporary Perspectives* (Carbondale: Southern Illinois Univ. Press, 1979), 91–93, 96–97.

12. *Coll. Works of Lincoln,* II, 245. Like most of Lincoln's speeches in the pre-presidential period, this one is known only from newspaper reports; the possibility of inaccuracy is, unfortunately, high.

13. Ibid., II, 249, 267.

14. Ibid., II, 274, 501.

15. Ibid., II, 276, 500–1; III, 327, 543–44.

16. Ibid., II, 231.

17. Ibid., IV, 531–32. Perhaps the first historian to state the question as it is stated above was the Englishman, K.C. Wheare, in *Abraham Lincoln and the United States* (New York: Macmillan, 1949), 158.

18. Leonard L. Richards, *The Life and Times of Congressman John Quincy Adams* (New York: Oxford Univ. Press, 1986), 123, 164; Charles Francis Adams, Jr., "John Quincy Adams and Martial Law," *Massachusetts Historical Society, Proceedings,* 2 ser., XV (Jan. 1902), 437–78.

19. S.S. Nicholas, *Martial Law,* pp. 1–2.

20. David Donald was the first historian to sense the importance of Adams's precedent for the Emancipation Proclamation; see Donald, *Lincoln Reconsidered: Essays on the Civil War Era,* 2nd ed. (New York: Vintage Books, 1961), 204–5.

21. William Whiting, *The War Powers of the President, and the Legislative Powers of Congress in Relation to Rebellion, Treason and Slavery* (Boston: John L. Shorey, 1862).

22. *Coll. Works of Lincoln,* V, 421.

Epilogue

1. Phillip S. Paludan, "Toward a Lincoln Conversation," *Reviews in American History,* XVI (Mar. 1988), 40–41.

2. *Coll. Works of Lincoln,* VI, 264–65.

3. Ibid., VI, 267.

4. Ibid., VI, 269.

5. James G. Randall, *Constitutional Problems under Lincoln,* pp. viii–ix.

6. John A. Marshall, *American Bastile;* Allan Nevins, James I. Robertson, Jr., and Bell I. Wiley, eds., *Civil War Books: A Critical Bibliography,* I (Baton Rouge: Louisiana State Univ. Press, 1967), 197. For so influential a book, its author remains surprisingly obscure. The leading student of Copperhead literature, Frank L. Klement, identifies Marshall as "a Marylander who was arrested arbitrarily in 1861." I found record of a J.A. Marshall arrested in January 1862, incarcerated in Old Capitol prison, and paroled in March. The authorship is uncertain, in truth. Marshall maintained in his introduction that he had been selected as historian of the Association of State Prisoners. In the only other reference to such an organization uncovered to date, Dennis A. Mahony founded the "State Prisoners Association" in February 1863. Some of the text of *American Bastile* is very close to Mahony's in *The Prisoner of State*— Appendix E, for example, or the long chapter on "The Old Capitol Prison: Its History and Incidents." And Mahony consistently spelled "bastille" with only one "l." Mahony was probably responsible for a substantial part of the text of *American Bastile.* Frank L. Klement, *The Copperheads in the Middle West,* p. 254; *O.R.,* II, 2, pp. 271–78; D.A. Mahony, "A Proclamation Convening Prisoners of State, Victims of Despotism, in General Convention," New York, Feb. 10, 1863, broadside in Turner-Baker Papers (microfilm) Case #413.

7. Marshall, *American Bastile,* 2nd ed., p. 843.

8. William Hanchett, *The Lincoln Murder Conspiracies* (Urbana: Univ. of Illinois Press, 1983), 93–94, 104–15. There were problems for Democrats with the Surratt case as well. Andrew Johnson suspended the writ of habeas corpus when a judge issued one to find out why she was being held by military authority, and Johnson was not a Republican. For an article emphasizing Johnson's action "in a time of peace, when the national life was not endangered by armed invasion or rebellion" as "an example to

be lamented and execrated by all men who love liberty," see John W. Clampitt, "Habeas Corpus," in John J. Lalor, ed., *Cyclopaedia of Political Science,* II, 431.

9. Sydney George Fisher, "The Suspension of Habeas Corpus during the War of the Rebellion," *Political Science Quarterly,* III (Sept. 1888), 454–88.

10. Thorpe quoted in Harold M. Hyman, *A More Perfect Union: The Impact of the Civil War and Reconstruction on the Constitution,* p. xvii. See also Herman Belz, "The Realist Critique of Constitutionalism in the Era of Reform," *American Journal of Legal History,* XV (1971), 288–94.

11. Herman Belz first noticed the Progressive admiration for Lincoln's executive energy in *Lincoln and the Constitution: The Dictatorship Question Reconsidered* (Fort Wayne: Louis A. Warren Lincoln Library and Museum, 1984), 3. He ignored the nationalism so evident in works like Nathaniel W. Stephenson, *Lincoln and the Progress of Nationality in the North* (Washington: Government Printing Office, 1923), 361, 362. Stephenson's work was reprinted as a pamphlet from the Annual Report of the American Historical Association for 1919.

12. Belz, *Lincoln and the Constitution,* pp. 2–3; William A. Dunning, "Disloyalty in Two Wars," *American Historical Review,* XXIV (Oct. 1918), 625, 630. On the difficulties in characterizing Dunning's ideas, see Phillip R. Muller, "Look Back without Anger: A Reappraisal of William A. Dunning," *Journal of American History,* LXI (Sept. 1974), 325–38.

13. James G. Randall, "Captured and Abandoned Property during the Civil War," *American Historical Review,* XIX (Oct. 1913), 65; James G. Randall to Col. C.W. Weeks, June 3, 1918; "Capturing Politics for God" and "An Outworn World Idea," notes for religious talks, Randall Papers, Archives, Univ. of Illinois, Champaign.

14. James G. Randall to Prof. Samuel B. Harding, Jan. 6, 1918, Randall Papers. Harding was a member of the Committee on Public Information. For a rather favorable appraisal of its work, see Stephen Vaughn, *Holding Fast the Inner Lines: Democracy, Nationalism, and the Committee on Public Information* (Chapel Hill: Univ. of North Carolina Press, 1980).

15. James G. Randall, "A Constitutional View of the Civil War," Randall Papers.

16. James G. Randall to R.O. Gattel, Aug. 24, 1918; James G. Randall to C.W. Weeks, June 3, 1918, Randall Papers.

17. Edwin P. Tanner to James G. Randall, Sept. 1, 1924, and Randall to Tanner, Sept. 20, 1924 (carbon), Randall Papers.

18. See Michael Kammen, *A Machine That Would Go of Itself: The Constitution in American Culture* (New York: Alfred A. Knopf, 1986).

19. Stephenson, *Lincoln and the Progress of Nationality in the North,* p. 355; Randall, *Constitutional Problems under Lincoln,* pp. 82–84, 184. See also Arthur C. Cole, "Lincoln and the American Tradition of Civil Liberty," *Journal of the Illinois State Historical Society,* XIX (Oct. 1926–Jan. 1927), 108–9.

20. Andrew C. McLaughlin, *Constitutional History of the United States* (New York: D. Appleton-Century, 1936), 623.

21. See Thomas J. Pressly, *Americans Interpret Their Civil War,* paperback ed. (New York: Free Press, 1966), 291–328; Frank L. Klement, *The Copperheads in the Middle West* and *Dark Lanterns: Secret Political Societies, Conspiracies, and Treason Trials in the Civil War* (Baton Rouge: Louisiana State Univ. Press, 1984) summarize the work of his long and influential career.

22. Donald L. Smith, *Zechariah Chafee, Jr.: Defender of Liberty and Law* (Cambridge: Harvard Univ. Press, 1986), 13–14.

23. McLaughlin, *Constitutional History of the United States,* p. 639.

24. McLaughlin, "Lincoln, the Constitution, and Democracy," *Abraham Lincoln Association Papers* (Springfield: Abraham Lincoln Association, 1937), 26. See also Herman Belz, "The Realist Critique of Constitutionalism in the Era of Reform," *American Journal of Legal History,* XV (1971), 288–94.

25. Harold M. Hyman, *A More Perfect Union: The Impact of the Civil War and Reconstruction on the Constitution,* p. xvi; Hyman and Wiecek, *Equal Justice under Law: Constitutional Development, 1835–1875* (New York: Harper & Row, 1982).

26. Dean Sprague, *Freedom under Lincoln;* Kenneth A. Bernard, "Lincoln and Civil Liberties," *Abraham Lincoln Quarterly,* VI (Sept. 1951), 375–99.

27. Edmund Wilson, *Letters on Literature and Politics, 1912–1972,* ed. by Elena Wilson (New York: Farrar, Straus and Giroux, 1977), 621; Edmund Wilson, *Patriotic Gore: Studies in the Literature of the American Civil War* (New York: Oxford, 1962) xvi, 99–130.

28. Dwight G. Anderson, *Abraham Lincoln: The Quest for Immortality* (New York: Alfred A. Knopf, 1982), pp. 5, 61, 193.

29. Gore Vidal, *Lincoln: A Novel* (New York: Random House, 1984), 459.

30. William Safire, *Freedom* (Garden City: Doubleday, 1987), 66, 496.

31. David Donald, *Lincoln's Herndon* (New York: Alfred A. Knopf, 1948), 213.

Index of Prisoners of State*

*The names in this index represent less than 2 percent of the cases examined for this book.

Index